Modelling Photovoltaic
Systems using PSpice®

Modelling Photovoltaic Systems using PSpice®

Luis Castañer and **Santiago Silvestre**
Universidad Politecnica de Cataluña, Barcelona, Spain

JOHN WILEY & SONS, LTD

Other Wiley Editorial Offices

John Wiley & Sons Inc., 111 River Street, Hoboken, NJ 07030, USA

Jossey-Bass, 989 Market Street, San Francisco, CA 94103-1741, USA

Wiley-VCH Verlag GmbH, Boschstr. 12, D-69469 Weinheim, Germany

John Wiley & Sons Australia Ltd, 33 Park Road, Milton, Queensland 4064, Australia

John Wiley & Sons (Asia) Pte Ltd, 2 Clementi Loop #02-01, Jin Xing Distripark, Singapore 129809

John Wiley & Sons Canada Ltd, 22 Worcester Road, Etobicoke, Ontario, Canada M9W 1L1

Library of Congress Cataloging-in-Publication Data

Castañer, Luis.
    Modelling photovoltaic systems using PSpice / Luis Castañer, Santiago Silvestre.
        p.   cm.
    Includes bibliographical references and index.
    ISBN 0-470-84527-9 (alk. paper) – ISBN 0-470-84528-7 (pbk. : alk. paper)
        1. Photovoltaic power systems—Mathematical models.   2. Photovoltatic power
    systems—Computer simulation.   3. PSpice.   I. Silvestre, Santiago.   II. Title.

    TK1087 .C37 2002
    621.31′244—dc21   200202741

British Library Cataloguing in Publication Data

A catalogue record for this book is available from the British Library

ISBN 0-470-845279 (HB)   0-470-84528-7 (PB)

To
Our Wives and Children

Contents

8 Grid-connected PV Systems   215

9 Small Photovoltaics   245

Foreword

Photovoltaics is rapidly becoming a mature industry as the performance of photovoltaic system components and the leverage of large-scale industrial production steadily decrease costs. Measures taken by governments and other agencies to subsidize the costs of installations and regulations in several countries concerning the purchase of electricity produced by grid-connected systems are promoting public awareness and the widespread use of environmentally friendly solar electricity.

The rapid growth of the industry over the past few years has expanded the interest and the need for education and training worldwide at all levels of photovoltaics, but especially at the system level where more people are involved. Photovoltaic system engineering and system design is often treated only superficially in existing books on photovoltaics. This book appears to be the first including computer modelling of photovoltaic systems with a detailed and quantitative analysis. PSpice is a good choice of simulator for this modelling because of its worldwide use in electric and electronic circuit simulation and its widespread availability. It can be used synergistically to learn about photovoltaic systems or to solve technical problems of design, sizing or analysis.

In universities and technical schools, students are computer literate and increasingly familiar with learning in a computer-assisted environment. The interest that photovoltaics awakes in them can be further boosted by the availability of a friendly tool which gets into the details and solves complicated and nonlinear equations easily. Teaching can then focus on the concepts, criteria and the results. Educators in electrical engineering or power electronics can also use photovoltaic examples from this book.

The book will also be useful in assisting professional engineers gain specialization in photovoltaics and allows them to use the techniques and models described in their activity.

The book covers photovoltaic system engineering starting from Chapter 1 where solar radiation availability and standard spectra are described along with some of the basics of PSpice. Chapter 2 follows with simplified PSpice modelling of the short-circuit current of solar cells where the basic concepts of quantum efficiency and spectral response are introduced. The electrical current–voltage characteristics of a solar cell are described in Chapter 3 where the effects of series and shunt resistances, irradiance, temperature and space radiation are described and PSpice models developed. A new 'behavioural' solar cell model is introduced allowing the simulation of responses under arbitrary time series of irradiances and ambient temperatures. Chapter 4 scales the solar cell models to arrays, terrestrial modules and generators. Specific issues are modelled and illustrated, such as the use of bypass diodes, the partial shadowing of solar cell arrays and the safe operation area is introduced. The solar cell behavioural model is extended to a generator of arbitrary size. Chapter 5 describes PSpice models for water pumping systems and for batteries, allowing the simulations of the performance of a photovoltaic array battery in association with the transient response of a water pumping system. Chapter 6 describes models for charge regulators, maximum power point trackers and inverters.

Data from measurements over several days are used to compare simulations to measurements. Inverters are modelled both topologically, for standard inverter circuits, and behaviourally, for long-term simulation. Chapter 7 is devoted to standalone systems, and sizing procedures are described. Details of daily energy balance and long-term system response are provided using hourly radiation time series for one year. 'Loss of load' probability is then introduced and calculated for realistic examples. Finally, the sizing procedure is refined by the use of long-term simulation. Chapter 8 is devoted to grid-connected photovoltaic systems making use of the inverter models described in Chapter 6, as well as new ones. The particular case of AC photovoltaics modules is also described and used to size these systems. Finally, Chapter 9 considers a number of cases of 'small photovoltaics' including systems working under artificial light with random values of the effective irradiance, such as for a pocket calculator. Monte Carlo analysis is used in this chapter to produce daily radiation time series and then to simulate performance over the worst month.

I welcome this addition to the photovoltaic literature and trust that many will benefit from the practical approach to photovoltaics that it provides.

Martin A. Green
Scientia Professor
Centre for Photovoltaic Engineering
University of New South Wales, Sydney
April, 2002

Preface

Photovoltaic engineering is a multidisciplinary speciality deeply rooted in semiconductor physics for solar cell theory and technology, and heavily relying on electrical and electronic engineering for system design and analysis.

The conception, design and analysis of photovoltaic systems are important tasks often requiring the help of computers to perform fast and accurate computations or simulations. Today's engineers and professionals working in the field and also students of different technical disciplines know how to use computers and are familiar with running specialized software. Computer-aided technical work is of great help in photovoltaics because most of the system components are described by nonlinear equations, and the node circuit equations that have to be solved to find the values of the currents and voltages, most often do not have analytical solutions. Moreover, the characteristics of solar cells and PV generators strongly depend on the intensity of the solar radiation and on the ambient temperature. As these are variable magnitudes with time, the system design stage will be more accurate if an estimation of the performance of the system in a long-term scenario with realistic time series of radiation and temperature is carried out.

The main goal of this book is to help understand PV systems operation gathering concepts, design criteria and conclusions, which are either defined or illustrated using computer software, namely PSpice.

The material contained in the book has been taught for more than 10 years as an undergraduate semester course in the UPC (Universidad Politecnica de Cataluña) in Barcelona, Spain and the contents refined by numerous interactions with the students. PSpice was introduced as a tool in the course back in 1992 to model a basic solar cell and since then more elaborated models, not only for solar cells but also for PV generators, battery, converters, inverters, have been developed with the help of MSc and PhD students. The impression we have as instructors is that the students rapidly jump into the tool and are ready to use and apply the models and procedures described in the book by themselves. Interaction with the students is helped by the universal availability of PSpice or more advanced versions, which allow the assignments to be tailored to the development of the course and at the same time providing continuous feedback from the students on the difficulties they find. We think that a key characteristic of the teaching experience is that quantitative results are readily available and data values of PV modules and batteries from web pages may be fed into problems and exercises thereby translating a sensation of proximity to the real world.

PSpice is the most popular standard for analog and mixed-signal simulation. Engineers rely on PSpice for accurate and robust analysis of their designs. Universities and semiconductor manufacturers work with PSpice and also provide PSpice models for new devices. PSpice is a powerful and robust simulation tool and also works with Orcad Capture®, Concept® HDL, or PSpice schematics in an integrated environment where engineers create designs, set up and run simulations, and analyse their simulation results. More details and information about PSpice can be found at http://www.pspice.com/.

At the same web site a free PSpice, PSpice 9.1 student version, can be downloaded. A request for a free Orcad Lite Edition CD is also available for PSpice evaluation from http://www.pspice.com/download/default.asp.

PSpice manuals and other technical documents can also be obtained at the above web site in PDF format. Although a small introduction about the use of PSpice is included in Chapter 1 of this book, we strongly encourage readers to consult these manuals for more detailed information. An excellent list of books dedicated to PSpice users can also be found at http://www.pspice.com/publications/books.asp.

All the models presented in this book, developed for PSpice simulation of solar cells and PV systems behaviour, have been specially made to run with version 9 of PSpice. PSpice offers a very good schematics environment, Orcad Capture for circuit designs that allow PSpice simulation, despite this fact, all PSpice models in this book are presented as text files, which can be used as input files. We think that this selection offers a more comprehensive approach to the models, helps to understand how these models are implemented and allows a quick adaptation of these models to different PV system architectures and design environments by making the necessary file modifications. A second reason for the selection of text files is that they are transportable to other existing PSpice versions with little effort.

All models presented here for solar cells and the rest of the components of a PV system can be found at www.esf.upc.es/esf/, where users can download all the files for simulation of the examples and results presented in this book. A set of files corresponding to stimulus, libraries etc. necessary to reproduce some of the simulations shown in this book can also be found and downloaded at the above web site. The login, esf and password, esf, are required to access this web site.

Acknowledgements

Several persons have contributed to the effort to build models in PSpice code for photovoltaics in our research group GDS at the UPC (Polytechnical University of Cataluña), Spain, and the authors of the book would like to specially acknowledge the work by Raimond Aloy supervised by Daniel Carles. They did wonderful, systematic and comprehensive work, principally in battery modelling. This work was further continued by Andreu Moreno and Javier Julve who developed revised models, and also behavioural macromodels, and were able to compare long-term simulations with the monitoring results in our experimental University PV installation. This helped to create confidence in the validity of the models and on the utility of the PSpice environment and stimulated the authors to write this book. We would like to acknowledge the interest demonstrated by the students who have been taking the elective course on photovoltaics at the UPC since 1992. Their criticisms, comments and enthusiasm have helped our effort.

The authors also received support from the Comision Interministerial de Ciencia y Tecnología under research contract TIC97-0949 on the PV system modelling area, boosting the commitment of the authors to gather all the material to transform it in a readable book, adding more fundamental and basic chapters.

Luis Castañer and **Santiago Silvestre**

1

Introduction to Photovoltaic Systems and PSpice

Summary

This chapter reviews some of the basic magnitudes of solar radiation and some of the basics of PSpice. A brief description of a photovoltaic system is followed by definitions of spectral irradiance, irradiance and solar radiation. Basic commands and syntax of the sentences most commonly used in this book are shortly summarized and used to write PSpice files for the AM1.5G and AM0 sun spectra, which are used to plot the values of the spectral irradiance as a function of the wavelength and compare them with a black body radiation. Solar radiation availability at the earth's surface is next addressed, and plots are shown for the monthly and yearly radiation received in inclined surfaces. Important rules, useful for system design, are described.

1.1 The Photovoltaic System

A photovoltaic (PV) system generates electricity by the direct conversion of the sun's energy into electricity. This simple principle involves sophisticated technology that is used to build efficient devices, namely solar cells, which are the key components of a PV system and require semiconductor processing techniques in order to be manufactured at low cost and high efficiency. The understanding of how solar cells produce electricity from detailed device equations is beyond the scope of this book, but the proper understanding of the electrical output characteristics of solar cells is a basic foundation on which this book is built.

A photovoltaic system is a modular system because it is built out of several pieces or elements, which have to be scaled up to build larger systems or scaled down to build smaller systems. Photovoltaic systems are found in the Megawatt range and in the milliwatt range producing electricity for very different uses and applications: from a wristwatch to a communication satellite or a PV terrestrial plant, grid connected. The operational principles though remain the same, and only the conversion problems have specific constraints. Much is gained if the reader takes early notice of this fact.

The elements and components of a PV system are the photovoltaic devices themselves, or solar cells, packaged and connected in a suitable form and the electronic equipment required to interface the system to the other system components, namely:

- a storage element in standalone systems;
- the grid in grid-connected systems;
- AC or DC loads, by suitable DC/DC or DC/AC converters.

Specific constraints must be taken into account for the design and sizing of these systems and specific models have to be developed to simulate the electrical behaviour.

1.2   Important Definitions: Irradiance and Solar Radiation

The radiation of the sun reaching the earth, distributed over a range of wavelengths from 300 nm to 4 micron approximately, is partly reflected by the atmosphere and partly transmitted to the earth's surface. Photovoltaic applications used for space, such as satellites or spacecrafts, have a sun radiation availability different from that of PV applications at the earth's surface. The radiation outside the atmosphere is distributed along the different wavelengths in a similar fashion to the radiation of a 'black body' following Planck's law, whereas at the surface of the earth the atmosphere selectively absorbs the radiation at certain wavelengths. It is common practice to distinguish two different sun 'spectral distributions':

(a) AM0 spectrum outside of the atmosphere.

(b) AM 1.5 G spectrum at sea level at certain standard conditions defined below.

Several important magnitudes can be defined: spectral irradiance, irradiance and radiation as follows:

(a) **Spectral irradiance** I_λ – the power received by a unit surface area in a wavelength differential $d\lambda$, the units are W/m^{2}μm.

(b) **Irradiance** – the integral of the spectral irradiance extended to all wavelengths of interest. The units are W/m^2.

(c) **Radiation** – the time integral of the irradiance extended over a given period of time, therefore radiation units are units of energy. It is common to find radiation data in J/m^2-day, if a day integration period of time is used, or most often the energy is given in kWh/m^2-day, kWh/m^2-month or kWh/m^2-year depending on the time slot used for the integration of the irradiance.

Figure 1.1 shows the relationship between these three important magnitudes.

Example 1.1

Imagine that we receive a light in a surface of 0.25 m^2 having an spectral irradiance which can be simplified to the rectangular shape shown in Figure 1.2, having a constant value of

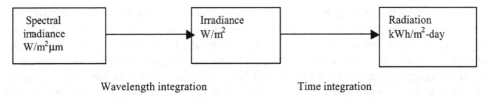

Wavelength integration   Time integration

Figure 1.1   Relationship between spectral irradiance, irradiance and radiation

Spectral irradiance

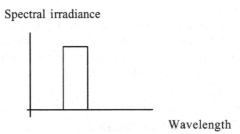

Wavelength

Figure 1.2   Spectrum for Example 1.1

1000 W/m²µm from 0.6 µm to 0.65 µm and zero in all other wavelengths. Calculate the value of the irradiance received at the surface and of the radiation received by the same surface after 1 day.

Solution

The irradiance is calculated by integration of the spectral irradiance over the wavelength range (0.6 to 0.65 µm)

$$\text{Irradiance} = \int_{0.60\mu}^{0.65\mu} 1000 \, d\lambda = 0.05 \times 1000 \frac{W}{m^2} = 50 \frac{W}{m^2}$$

As the irradiance is defined by unit of area, the result is independent of the amount of area considered. The radiation received at the 0.25 m² area, comes now after integration of the irradiance over the period of time of the exercise, that is one day:

$$\text{radiation} = \text{Area} \int_{0}^{24h} \text{Irradiance} \cdot dt = 0.25 \, m^2 \, 24 \, h \times 50 \frac{W}{m^2} = 300 \, \text{Wh-day}$$

As can be seen from Example 1.1, the calculation of the time integral involved in the calculation of the irradiance is very straightforward when the spectral irradiance is constant, and also the calculation of the radiation received at the surface reduces to a simple product when the irradiance is constant during the period of time considered.

It is obvious that this is not the case in photovoltaics. This is because the spectral irradiance is greater in the shorter wavelengths than in the longer, and of course, the irradiance received at a given surface depends on the time of the day, day of the year, the site location at the earth's surface (longitude and latitude) and on the weather conditions. If the calculation is performed for an application outside the atmosphere, the irradiance depends on the mission, the orientation of the area towards the sun and other geometric, geographic and astronomical parameters.

It becomes clear that the calculation of accurate and reliable irradiance and irradiation data has been the subject of much research and there are many detailed computation methods. The photovoltaic system engineer requires access to this information in order to know the availability of sun radiation to properly size the PV system. In order to make things easier, standard spectra of the sun are available for space and terrestrial applications. They are named AM0 and AM1.5 G respectively and consist of the spectral irradiance at a given set of values of the wavelength as shown in Annex 1.

1.3   Learning Some PSpice Basics

The best way to learn about PSpice is to practise performing a PSpice simulation of a simple circuit. We have selected a circuit containing a resistor, a capacitor and a diode in order to show how to:

- describe the components.

- connect them.

- write PSpice sentences.

- perform a circuit analysis.

First, nodes have to be assigned from the schematics. If we want to simulate the electrical response of the circuit shown in Figure 1.3 following an excitation by a pulse voltage source we have to follow the steps:

1. Node assignation
According to Figure 1.3 we assign

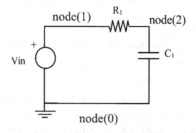

Figure 1.3   Circuit used in file learning.cir

(0) GROUND

(1) INPUT

(2) OUTPUT

In Spice NODE (0) is always the reference node.

2. Circuit components syntax

Resistor syntax

> rxx node_a node_b value

Capacitor syntax

> cxx node_a node_b value

According to the syntax and the nodes assignation we must write:

> r1 1 2 1 K; resistor between node (1) and node (2) value 1 KOhm

> c1 2 0 1 n; capacitor between node (2) and node (0) value 1nF

Comments can be added to the netlist either by starting a new line with a * or by adding comments after a semicolon (;).

Sources syntax

A voltage source is needed and the syntax for a pulsed voltage source is as follows.

Pulse voltage source

vxx node+ node− pulse (initial_value pulse_value delay risetime falltime pulse_length period)

where node+ and node− are the positive and negative legs of the source, and all other parameters are self-explanatory. In the case of the circuit in Figure 1.3, it follows:

> vin 1 0 pulse (0 5 0 1u 1u 10u 20u)

meaning that a voltage source is connected between nodes (1) and (0) having an initial value of 0 V, a pulse value of 5 V, a rise and fall time of 1 μs, a pulse length of 10 μs and a period of 20 μs.

3. Analysis

Several analysis types are available in PSpice and we begin with the transient analysis, which is specified by a so-called 'dot command' because each line has to start with a dot.

Transient analysis syntax (dot command)

.tran tstep tstop tstart tmax

where:

first character in the line *must* be a dot

tstep: printing increment

tstop: final simulation time

tstart: (optional) start of printing time

tmax: (optional) maximum step size of the internal time step

In the circuit in Figure 1.3 this is written as:

.tran 0.1u 40u

setting a printing increment of 0.1 μs and a final simulation time of 40 μs.

4. Output (more dot commands)

Once the circuit has been specified the utility named 'probe' is a post processor, which makes available the data values resulting from the simulation for plotting and printing. This is run by a dot command:

.probe

Usually the user wants to see the results in graphic form and then wants some of the node voltages or device currents to be plotted. This can be perfomed directly at the probe window using the built-in menus or specifying a dot command as follows:

.plot tran variable_1 variable_2

In the case of the example shown in Figure 1.3, we are interested in comparing the input and output waveforms and then:

.plot tran v(1) v(2)

The file has to be terminated by a final dot command:

.end

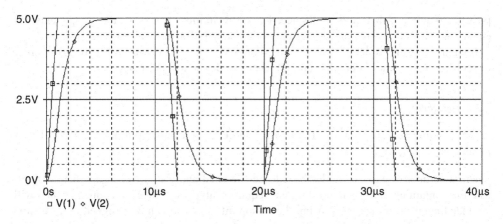

Figure 1.4   Input and output waveforms of simulation of circuit learning.cir

The file considered as a start-up example runs a simulation of the circuit shown in Figure 1.3, which is finally written as follows, using the direct application of the rules and syntax described above.

```
*learning.cir

r1 1 2 1K; resistor between node (1) and node (2) value 1 KOhm

c1 2 0 1n; capacitor between node (2) and node (0) value 1 nF

vin 1 0 pulse (0 5 0 1u 1u 10u 20u); voltage source between node (1) and node (0)

.tran 0 40u

.probe

.plot tran v(1) v(2)

.end
```

The result is shown in Figure 1.4 where both input and output signals have been plotted as a function of time. The transient analysis generates, as a result of the simulation graphs, where the variables are plotted against time.

1.4 Using PSpice Subcircuits to Simplify Portability

The above example tells us about the importance of node assignation and, of course, care must be taken to avoid duplicities in complex circuits unless we want an electrical connection. In order to facilitate the portability of small circuits from one circuit to another, or to replicate the same portion of a circuit in several different parts of a larger circuit without having to renumber all the nodes every time the circuit is added to or changed, it is possible to define 'subcircuits' in PSpice. These subcircuits encapsulate the components and electrical connections by considering the node numbers for internal use only.

Imagine we want to define a subcircuit composed of the RC circuit in Figure 1.3 in order to replicate it in a more complex circuit. Then we define a subcircuit as:

Subcircuit syntax

.subckt name external_node_1 external_node_2 params: parameter_1 = value_1 parameter_2 = value_2

where a number of external nodes have to be specified and also a list of parameter values. The file containing the subcircuit does not require analysis sentences or sources, which will be added later on to the circuit using the subcircuit. This file can be assigned an extension .lib indicating that it is a subcircuit of a library for later use.

For the RC circuit we need two external nodes: input and output, and two parameters: the resistor value and the capacitor value to be able to change the values at the final circuit stage.

Warning: inside a subcircuit the node (0) is forbidden.

We will name the nodes for external connection – (11) for the input, (12) for the output and (10) for the reference.

```
* rc.lib
.subckt rc 12 11 10 params: r = 1 c = 1

r1 11 12 {r}
c1 12 10 {c}
.ends rc
```

Now, every time an RC circuit is to be included in a larger circuit, such as the one depicted in Figure 1.5 where two RC circuits of different component values are used, the RC circuit described in the subcircuit is used twice by means of a sentence, where a new component with first letter 'x' – a description given by the subcircuit name – is introduced as follows:

Syntax for a part of a circuit described by a subcircuit file

x_name node_1 node_2 node_i subcircuit_name params: param_1 = value_1

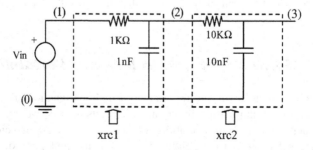

Figure 1.5 Circuit using the same RC subcircuit twice

Applying this syntax to the circuit in Figure 1.5 for the RC number 1 and number 2 it follows:

$$xrc1\ 2\ 1\ 0\ rc\ params: r = 1\,k\ c = 1\,n$$

$$xrc2\ 3\ 2\ 0\ rc\ params: r = 10\,k\ c = 10\,n$$

indicating that the subcircuits named xrc1 and xrc2, with the contents of the file rc.lib and the parameter values shown, are called and placed between the nodes 2, 1 and 0 for xrc1 and between 3 2 0 for xrc2. Finally the netlist has to include the file describing the model for the subcircuit and this is done by another dot command:

$$.include\ rc.lib$$

So the total file will now be:

```
* learning_subckt.cir
xrc1 2 1 0 rc params: r = 1k c = 1n
xrc2 3 2 0 rc params: r = 10k c = 10n
.include rc.lib
vin 1 0 pulse (0 5 0 1u 1u 10u 20u); voltage source between node (1) and node (0)

.tran 0.1u 40u
.probe
.plot tran v(1) v(2) v(3)
.end
```

1.5 PSpice Piecewise Linear (PWL) Sources and Controlled Voltage Sources

In photovoltaic applications the inputs to the system are generally the values of the irradiance and temperature, which cannot be described by a pulse kind of source as the one used above. However, an easy description of arbitrarily shaped sources is available in PSpice under the denomination of piecewise linear (PWL) source.

Syntax for piecewise linear voltage source

$$Vxx\ node+\ node-\ pwl\ time\_1\ value\_1\ time\_2\ value\_2\ ...$$

This is very convenient for the description of many variables in photovoltaics and the first example is shown in the next section.

A PSpice device which is very useful for any application and for photovoltaics in particular is the E-device, which is a voltage-controlled voltage source having a syntax as follows.

Syntax for E-device

$$e\_name\ node+\ node-\ control\_node+\ control\_node-\ gain$$

As can be seen this is a voltage source connected to the circuit between nodes node+ and node−, with a value given by the product of the gain by the voltage applied between control_node+ and control_node−.

A simplification of this device consists of assigning a value which can be mathematically expressed as follows:

E_name node+ node− value = {expression}

These definitions are used in Sections 1.6 and 1.7 below in order to plot the spectral irradiance of the sun.

1.6   Standard AM1.5 G Spectrum of the Sun

The name given to these standard sun spectra comes from Air Mass (AM) and from a number which is 0 for the outer-space spectrum and 1.5 for the sea-level spectrum. In general we will define a spectrum AM*x* with *x* given by:

$$x = \frac{1}{\cos \vartheta_z} \tag{1.1}$$

where ϑ_z is the zenith angle of the sun. When the sun is located at the zenith of the receiving area $x = 1$, meaning that a spectrum AM1 would be spectrum received at sea level on a clear day with the sun at its zenith. It is generally accepted that a more realistic terrestrial spectrum for general use and reference is provided by a zenith angle of 48.19° (which is equivalent to $x = 1.5$). The spectrum received at a surface tilted 37° and facing the sun is named a 'global–tilt' spectrum and these data values, usually taken from the reference [1.1] are commonly used in PV engineering.

An easy way to incorporate the standard spectrum into PSpice circuits and files is to write a subcircuit which contains all the data points in the form of a PWL source. This is achieved by using the diagram and equivalent circuit in Figure 1.6 which implements the PSpice file. The complete file is shown in Annex 1 but the first few lines are shown below:

```
* am15g.lib
.subckt am15 g 11 10
v_am15 g 11 10 pwl    0.295u 0
+  0.305u 9.2
+  0.315u 103.9
+  0.325u 237.9
+  0.335u 376
+  0.345u 423
+  all data points follow here (see Annex 1 for the complete netlist)

.ends am15 g
```

It is important to notice here that the time, which is the default axis for PSpice transient analysis, has been replaced by the value of the wavelength in microns.

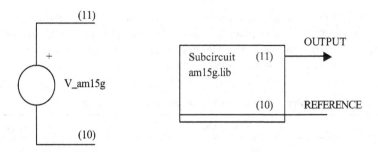

Figure 1.6 PSpice subcircuit for the spectral irradiance AM1.5 G

In order to plot a graph of the spectral irradiance we write a .cir file as follows:

```
*am15 g.cir
xspectr_irrad 11 0 am15 g
.include am15 g.lib
e_spectr_irrad_norm 12 0 value = {1000/962.5*v(11)}
.tran 0.1u 4u
.probe
.plot tran v(12)
.end
```

which calls the 'am15 g.lib' subcircuit and runs a transient simulation where the time scale of the x-axis has been replaced by the wavelength scale in microns.

A plot of the values of the AM1.5 G spectral irradiance in W/m^{2}μm is shown in Figure 1.7.

Care must be taken throughout this book in noting the axis units returned by the PSpice simulation because, as is shown in the plot of the spectral irradiance in Figure 1.7, the y-axis returns values in volts which have to be interpreted as the values of the spectral irradiance in

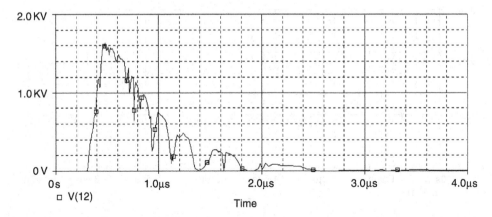

Figure 1.7 PSpice plot of AM1.5 G sun spectrum normalized to 1 kW/m^2 total irradiance. *Warning*, x-axis is the wavelength in μm and the y-axis is the spectral irradiance in W/m^{2}μm

$W/m^2\mu m$. So 1 V in the y-axis of the graph means 1 $W/m^2\mu m$. The same happens to the x-axis: 1 μs in the graph means in practice 1 μm of wavelength. The difference between the internal PSpice variables and the real meaning is an important convention used in this book. In the example above, this is summarized in Table 1.1.

Table 1.1 Internal PSpice units and real meaning

	Internal PSpice variable	Real meaning
Horizontal x-axis	Time (μs)	Wavelength (μm)
Vertical y-axis	Volts (V)	Spectral irradiance ($W/m^2\mu m$)

Throughout this book warnings on the real meaning of the axis in all graphs are included in figure captions to avoid misinterpretations and mistakes.

1.7 Standard AM0 Spectrum and Comparison to Black Body Radiation

The irradiance corresponding to the sun spectrum outside of the atmosphere, named AM0, with a total irradiance of is 1353 W/m^2 is usually taken from the reported values in reference [1.2]. The PSpice subcircuit corresponding to this file is entirely similar to the am15 g.lib subcircuit and is shown in Annex 1 and plotted in Figure 1.8 (files 'am0.lib' and 'am0.cir').

The total irradiance received by a square metre of a surface normal to the sun rays outside of the atmosphere at a distance equal to an astronomical unit (1AU $= 1.496 \times 1011$ m) is called the solar constant S and hence its value is the integral of the spectral irradiance of the AM0, in our case 1353 W/m^2.

The sun radiation can also be approximated by the radiation of a black body at 5900 K. Planck's law gives the value of the spectral emisivity E_λ, defined as the spectral power

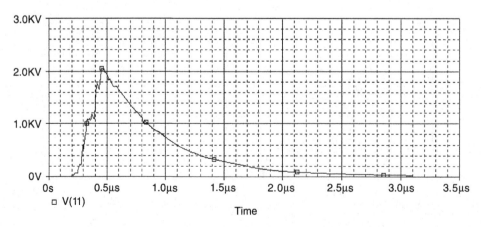

Figure 1.8 PSpice plot of AM0 spectrum of the sun. *Warning*, x-axis is the wavelength in μm and the y-axis is the spectral irradiance in $W/m^2\mu m$

radiated by unit of area and unit of wavelength, as

$$E_\lambda = \frac{2\pi h C_o^2}{\lambda^5 (e^{\frac{h C_o}{\lambda k T}} - 1)} \left[\frac{W}{m^3}\right]$$ (1.2)

where h is the Planck's constant ($h = 6.63 \times 10^{-34}$ J s) and

$$2\pi h C_o^2 = 3.74 \times 10^{-16} W m^2$$

$$\frac{h C_o}{k} = 0.0143 \, mK$$ (1.3)

are the first and second radiation Planck's constants. The total energy radiated by a unit area of a black body for all values of wavelengths is given by

$$\int_0^\infty E_\lambda \, d\lambda = \sigma T^4 = 5.66 \times 10^{-8} T^4 \left(\frac{W}{m^2}\right)$$ (1.4)

with the temperature T in K°.

Assuming that the black body radiates isotropically, the spectral irradiance received from the sun at an astronomical unit of distance (1 AU) will be given by

$$I_\lambda = E_\lambda \frac{S}{\int_0^\infty E_\lambda \, d\lambda}$$ (1.5)

where S is the solar constant. Finally, from equation (1.2), I_λ can be written as:

$$I_\lambda = \frac{8.925 \times 10^{-12}}{\lambda^5 [e^{0.0143/\lambda T} - 1] T^4} \left(\frac{W}{m^2 \mu m}\right)$$ (1.6)

In order to be able to plot the spectral irradiance of a black body at a given temperature, we need to add some potentialities to the subcircuit definition made in the above sections. In fact what we want is to be able to plot the spectral irradiance for *any* value of the temperature and, moreover we need to provide the value of the wavelength. To do so, we first write a subcircuit containing the wavelength values, for example in microns as shown in Annex 1, 'wavelength.lib'.

This will be a subcircuit as

.subckt wavelength 11 10

having two pins: (11) is the value of the wavelength in metres and (10) is the reference node. Next we have to include equation (1.6) which is easily done in PSpice by assigning a value to an E-device as follows:

```
*black_body.lib

.subckt black_body 12 11 10 params:t=5900

e_black_body 11 10 value={8.925e-12/(((v(12)*1e-6)**5)*(t**4)
+*(exp(0.0143/(v(12)*1e-6*t))-1))}

.ends black_body
```

The factor 1×10^{-6} converts the data of the wavelength from micron to metre.

Once we have the subcircuit files we can proceed with a black_body.cir file as follows:

```
*black_body.cir

.include black_body.lib

.include wavelength.lib

x_black_body 12 11 0 black_body

x_wavelength 12 0 wavelength

.tran 0.1u 4u

.probe

.plot tran v(11)

.end
```

where the wavelength is written in metres and the temperature T in K and V(11) is the spectral irradiance. This can also be plotted using a PSpice file. Figure 1.9 compares the black body spectral irradiance with the AM0 and AM1.5 G spectra.

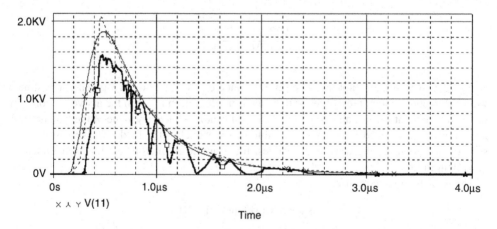

Figure 1.9 Black body spectral irradiance at 5900 K (middle) compared with AM0 spectral irradiance (upper) and AM1.5 (lower). *Warning*, x-axis is the wavelength in μm and the y-axis is the spectral irradiance in W/m²μm

1.8   Energy Input to the PV System: Solar Radiation Availability

The photovoltaic engineer is concerned mainly with the radiation received from the sun at a particular location at a given inclination angle and orientation and for long periods of time. This solar radiation availability is the energy resource of the PV system and has to be known as accurately as possible. It also depends on the weather conditions among other things such as the geographic position of the system. It is obvious that the solar radiation availability is subject to uncertainty and most of the available information provides data processed using measurements of a number of years in specific locations and complex algorithms. The information is widely available for many sites worldwide and, where there is no data available for a particular location where the PV system has to be installed, the databases usually contain a location of similar radiation data which can be used (see references 1.3, 1.4, and 1.5 for example).

The solar radiation available at a given location is a strong function of the orientation and inclination angles. Orientation is usually measured relative to the south in northern latitudes and to the north in southern latitudes, and the name of the angle is the 'azimuth' angle. Inclination is measured relative to the horizontal. As an example of the radiation data available at an average location, Figure 1.10 shows the radiation data for San Diego (CA), USA, which has a latitude angle of 33.05° N. These data have been obtained using Meteonorm 4.0 software [1.5]. As can be seen the yearly profile of the monthly radiation values strongly depends on the inclination for an azimuth zero, that is for a surface facing south. It can be seen that a horizontal surface receives the largest radiation value in summer. This means that if a system has to work only in summer time the inclination should be chosen to be as horizontal as possible. Looking at the curve corresponding to 90° of

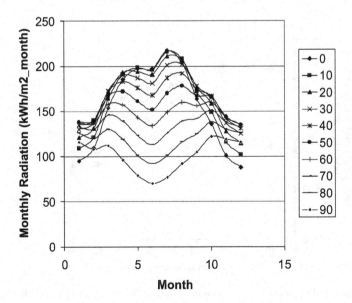

Figure 1.10 Monthly radiation data for San Diego. Data adapted from results obtained using Meteonorm 4.0 [1.5] as a function of the month of the year and of the inclination angle

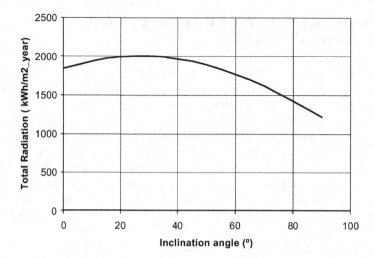

Figure 1.11 Total radiation at an inclined surface in San Diego, for a surface facing south, as a function of the inclination angle

inclination at the vertical surface (bottom graph) it can be seen that this surface receives the smallest radiation of all angles in most of the year except in winter time.

Most PV applications are designed in such a way that the surface receives the greatest radiation value integrated over the whole year. This is seen in Figure 1.11 where the yearly radiation values received in San Diego are plotted as a function of the inclination angle for a surface facing south. As can be seen, there is a maximum value for an inclination angle of approximately 30°, which is very close to the latitude angle of the site.

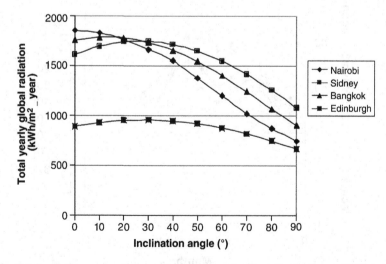

Figure 1.12 Total radiation received at an inclined surface for four sites in the world (Nairobi, 1.2° S, Sidney 33.45° S, Bangkok 13.5° N and Edinburgh 55.47° N. North hemisphere sites facing south and south hemisphere facing north

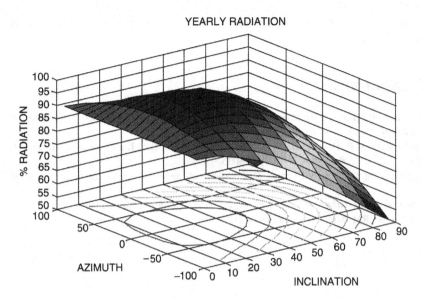

YEARLY RADIATION

Figure 1.13 Plot of the yearly radiation in Agoncillo, Logroño, Spain for arbitrary values of inclination and azimuth angles. Expressed in % of the maximum. The bottom projection are iso-ratio curves from 95% to 50%. Data values elaborated from Meteonorm 4.0 results

It can be concluded, as a general rule, that an inclination angle close to the value of the latitude maximizes the total radiation received in one year. This can also be seen in Figure 1.12 where the total yearly radiation received at inclined surfaces at four different sites in the world are shown, namely, Nairobi, 1.2° S, Sidney 33.45° S, Bangkok 13.5° N and Edinburgh 55.47° N, north hemisphere sites facing south and south hemisphere facing north. As can be seen the latitude rule is experimentally seen to hold approximately in different latitudes. Moreover, as the plots in Figures 1.11 and 1.12 do not have a very narrow maximum, but a rather wide one, not to follow the latitude rule exactly does not penalize the system to a large extent. To quantify this result and to involve the azimuth angle in the discussion, we have plotted the values of the total yearly radiation received for arbitrary inclination and orientation angles, as shown in Figure 1.13 for a different location, this time a location in southern Europe at 40° latitude (Logroño, Spain).

The results shown have been derived for a static flat surface, this means a PV system located in a surface without concentration, and that does not move throughout the year. Concentration and sun tracking systems require different solar radiation estimations than the one directly available at the sources in References 1.3 to 1.6 and depend very much on the concentration geometry selected and on the one or two axis tracking strategy.

1.9 Problems

1.1   Draw the spectral irradiance of a black body of 4500 K by writing a PSpice file.

1.2   From the data values in Figure 1.11, estimate the percentage of energy available at a inclined surface of 90° related to the maximum available in that location.

1.3   From the data of the AM1.5 G spectrum calculate the energy contained from $\lambda = 0$ to $\lambda = 1.1$ µm.

1.10 References

[1.1] Hulstrom, R., Bird, R. and Riordan, C., 'Spectral solar irradiance data sets for selected terrestrial conditions' in *Solar Cells*, vol. **15**, pp. 365–91, 1985.

[1.2] Thekaekara, M.P., Drummond, A.J., Murcray, D.G., Gast, P.R., Laue E.G. and Wilson, R.C., *Solar Electromagnetic Radiation* NASA SP 8005, 1971.

[1.3] Ministerio de Industria y Energía, *Radiación Solar sobre superficies inclinadas* Madrid, Spain, 1981.

[1.4] Censolar, *Mean Values of Solar Irradiation on Horizontal Surface*, 1993.

[1.5] METEONORM, http://www.meteotest.ch.

2

Spectral Response and Short-Circuit Current

Summary

This chapter describes the basic operation of a solar cell and uses a simplified analytical model which can be implemented in PSpice. PSpice models for the short circuit current, quantum efficiency and the spectral response are shown and used in several examples. An analytical model for the dark current of a solar cell is also described and used to compute an internal PSpice diode model parameter: the reverse saturation current which along with the model for the short-circuit current is used to generate an ideal $I(V)$ curve. PSpice DC sweep analysis is described and used for this purpose.

2.1 Introduction

This chapter explains how a solar cell works, and how a simple PSpice model can be written to compute the output current of a solar cell from the spectral irradiance values of a given sun spectrum. We do not intend to provide detailed material on solar cell physics and technology; many other books are already available and some of them are listed in the references [2.1], [2.2], [2.3], [2.4] and [2.5]. It is, however, important for the reader interested in photovoltaic systems to understand how a solar cell works and the models describing the photovoltaic process, from photons impinging the solar cell surface to the electrical current produced in the external circuit.

Solar cells are made out of a semiconductor material where the following main phenomena occur, when exposed to light: photon reflection, photon absorption, generation of free carrier charge in the semiconductor bulk, migration of the charge and finally charge separation by means of an electric field. The main semiconductor properties condition how effectively this process is conducted in a given solar cell design. Among the most important are:

(a) Absorption coefficient, which depends on the value of the bandgap of the semiconductor and the nature, direct or indirect of the bandgap.

(b) Reflectance of the semiconductor surface, which depends on the surface finishing: shape and antireflection coating.

(c) Drift-diffusion parameters controlling the migration of charge towards the collecting junction, these are carrier lifetimes, and mobilities for electron and holes.

(d) Surface recombination velocities at the surfaces of the solar cell where minority carriers recombine.

2.1.1   Absorption coefficient α(λ)

The absorption coefficient is dependent on the semiconductor material used and its values are widely available. As an example, Figure 2.1 shows a plot of the values of the absorption coefficient used by PC1D for silicon and GaAs [2.6]. Values for amorphous silicon are also plotted.

As can be seen the absorption coefficient can take values over several orders of magnitude, from one wavelength to another. Moreover, the silicon coefficient takes values greater than zero in a wider range of wavelengths than GaAs or amorphous silicon. The different shapes are related to the nature and value of the bandgap of the semiconductor. This fact has an enormous importance in solar cell design because as photons are absorbed according to Lambert's law:

$$\phi(x) = \phi(0)\, e^{-\alpha x} \tag{2.1}$$

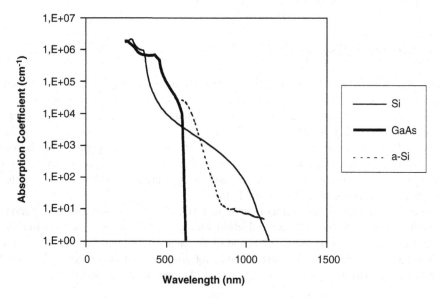

Figure 2.1 Absorption coefficient for silicon, GaAs and amorphous silicon as a function of the wavelength. Data values taken from PC1D [2.6]

if the value of α is high, the photons are absorbed within a short distance from the surface, whereas if the value of α is small, the photons can travel longer distances inside the material. In the extreme case where the value of α is zero, the photons can completely traverse the material, which is then said to be transparent to that particular wavelength. From Figure 2.1 it can be seen that, for example, silicon is transparent for wavelengths in the infrared beyond 1.1 micron approximately. Taking into account the different shapes and values of the absorption coefficient, the optical path length required inside a particular material to absorb the majority of the photons comprised in the spectrum of the sun can be calculated, concluding that a few microns are necessary for GaAs material and, in general, for direct gap materials, whereas a few hundreds microns are necessary for silicon. It has to be said that modern silicon solar cell designs include optical confinement inside the solar cell so as to provide long photon path lengths in silicon wafers thinned down to a hundred micron typically.

2.1.2 Reflectance R(λ)

The reflectance of a solar cell surface depends on the surface texture and on the adaptation of the refraction coefficients of the silicon to the air by means of antireflection coatings. It is well known that the optimum value of the refraction index needed to minimize the reflectance at a given wavelength has to be the geometric average of the refraction coefficients of the two adjacent layers. In the case of a solar cell encapsulated and covered by glass, an index of refraction of 2.3 minimizes the value of the reflectance at 0.6 μm of wavelength. Figure 2.2 shows the result of the reflectance of bare silicon and that of a silicon solar cell surface described in the file Pvcell.prm in the PC1D simulator (surface textured 3 μm deep and single AR coating of 2.3 index of refraction and covered by 1 mm glass).

As can be seen great improvements are achieved and more photons are absorbed by the solar cell bulk and thus contribute to the generation of electricity if a proper antireflection design is used.

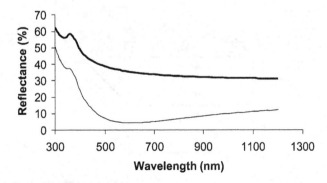

Figure 2.2 Reflectance of bare silicon surface (thick line) and silicon covered by an antireflection coating (thin line), data values taken from PC1D [2.6]

2.2 Analytical Solar Cell Model

The calculation of the photo-response of a solar cell to a given light spectrum requires the solution of a set of five differential equations, including continuity and current equations for both minority and majority carriers and Poisson's equation. The most popular software tool used to solve these equations is PC1D [2.6] supported by the University of New South Wales, and the response of various semiconductor solar cells, with user-defined geometries and parameters can be easily simulated. The solution is numerical and provides detailed information on all device magnitudes such as carrier concentrations, electric field, current densities, etc. The use of this software is highly recommended not only for solar cell designers but also for engineers working in the photovoltaic field. For the purpose of this book and to illustrate basic concepts of the solar cell behaviour, we will be using an analytical model for the currents generated by a illuminated solar cell, because a simple PSpice circuit can be written for this case, and by doing so, the main definitions of three important solar cell magnitudes and their relationships can be illustrated. These important magnitudes are:

(a) spectral short circuit current density;

(b) quantum efficiency;

(c) spectral response.

A solar cell can be schematically described by the geometry shown in Figure 2.3 where two solar cell regions are identified as emitter and base; generally the light impinges the solar cell by the emitter surface which is only partially covered by a metal electrical grid contact. This allows the collection of the photo-generated current as most of the surface has a low reflection coefficient in the areas not covered by the metal grid.

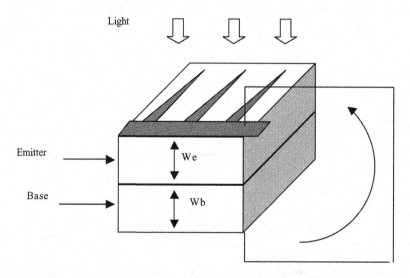

Figure 2.3   Schematic view of an externally short circuited solar cell

As can be seen, in Figure 2.3, when the solar cell is illuminated, a non-zero photocurrent is generated in the external electric short circuit with the sign indicated, provided that the emitter is an n-type semiconductor region and the base is a p-type layer. The sign is the opposite if the solar cell regions n-type and p-type are reversed.

The simplified model which we will be using, assumes a solar cell of uniform doping concentrations in both the emitter and the base regions.

2.2.1   Short-circuit spectral current density

Our model gives the value of the photocurrent collected by a 1 cm^2 surface solar cell, and circulating by an external short circuit, when exposed to a monochromatic light. Both the emitter and base regions contribute to the current and the analytical expression for both are given as follows (see Annex 2 for a summary of the solar cell basic analytical model).

Emitter short circuit spectral current density

$$J_{scE}(\lambda) = \frac{q\alpha\phi_0(1-R)L_p}{(\alpha L_p)^2 - 1}\left[-\alpha L_p e^{-\alpha W_e} + \frac{S_e \frac{L_p}{D_p} + \alpha L_p - e^{-\alpha W_e}\left(S_e \frac{L_p}{D_p} Ch\frac{W_e}{L_p} + Sh\frac{W_e}{L_p}\right)}{Ch\frac{W_e}{L_p} + S_e \frac{L_p}{D_p} Sh\frac{W_e}{L_p}}\right]$$

$$(2.2)$$

Base short circuit spectral current density

$$J_{scB}(\lambda) = \frac{q\alpha\phi'_0(1-R)L_n}{(\alpha L_n)^2 - 1}\left[-\alpha L_n - \frac{S_b \frac{L_n}{D_n}\left(Ch\frac{W_b}{L_n} - e^{-\alpha W_b}\right) + Sh\frac{W_b}{L_n} + \alpha L_n e^{-\alpha W_b}}{Ch\frac{W_b}{L_n} + S_b \frac{L_n}{D_n} Sh\frac{W_b}{L_n}}\right] \quad (2.3)$$

where the main parameters involved are defined in Table 2.1.

Table 2.1   Main parameters involved in the analytical model

Symbol	Name	Units
α	Absorption coefficient	cm^{-1}
ϕ_0	Photon spectral flux at the emitter surface	Photon/cm^{2}μm s
ϕ'_0	Photon spectral flux at the base-emitter interface	Photon/cm^{2}μm s
L_n	Electron diffusion length in the base layer	cm
L_p	Hole diffusion length in the emitter layer	cm
D_n	Electron diffusion constant in the base layer	cm^2/s
D_p	Hole diffusion constant in the emitter layer	cm^2/s
S_e	Emitter surface recombination velocity	cm/s
S_b	Base surface recombination velocity	cm/s
R	Reflection coefficient	—

The sign of the two components is the same and they are positive currents going out of the device by the base layer as shown in Figure 2.3.

As can be seen the three magnitudes involved in equations (2.2) and (2.3) are a function of the wavelength: absorption coefficient α, see Figure 2.1, reflectance $R(\lambda)$, see Figure 2.2 and the spectral irradiance I_λ, see Chapter 1, Figure 1.9. The spectral irradiance is not explicitly involved in equations (2.2) and (2.3) but it is implicitly through the magnitude of the spectral photon flux, described in Section 2.2.2, below.

The units of the spectral short-circuit current density are A/cm^{2}μm, because it is a current density by unit area and unit of wavelength.

2.2.2   Spectral photon flux

The spectral photon flux ϕ_0 received at the front surface of the emitter of a solar cell is easily related to the spectral irradiance and to the wavelength by taking into account that the spectral irradiance is the power per unit area and unit of wavelength. Substituting the energy of one photon by hc/λ, and arranging for units, it becomes:

$$\phi_0 = 10^{16} \frac{I_\lambda \lambda}{19.8} \left[\frac{\text{photon}}{\text{cm}^2 \mu\text{m} \cdot \text{s}} \right] \tag{2.4}$$

with I_λ written in W/m^{2}μm and λ in μm.

Equation (2.4) is very useful because it relates directly the photon spectral flux per unit area and unit of time with the spectral irradiance in the most conventional units found in textbooks for the spectral irradiance and wavelength. Inserting equation (2.4) into equation (2.2) the spectral short circuit current density originating from the emitter region of the solar cell is easily calculated.

The base component of the spectral short circuit current density depends on ϕ'_0 instead of ϕ_0 because the value of the photon flux at the emitter–base junction or interface has to take into account the absorption that has already taken place in the emitter layer. ϕ'_0 relates to ϕ_0 as follows,

$$\phi'_0 = \phi_0 e^{-\alpha W_e} = e^{-\alpha W_e} 10^{16} \frac{I_\lambda \lambda}{19.8} \left[\frac{\text{photon}}{\text{cm}^2 \mu\text{m} \cdot \text{s}} \right] \tag{2.5}$$

where the units are the same as in equation (2.4) with the wavelength in microns.

2.2.3   Total short-circuit spectral current density and units

Once the base and emitter components of the spectral short-circuit current density have been calculated, the total value of the spectral short-circuit current density at a given wavelength is calculated by adding the two components to give:

$$J_{sc\lambda} = J_{scE\lambda} + J_{scB\lambda} \tag{2.6}$$

with the units of A/cm^{2}μm. The photocurrent collected at the space charge region of the solar cell has been neglected in equation (2.6).

It is important to remember that the *spectral* short-circuit current density is a different magnitude than the total short circuit current density generated by a solar cell when illuminated by an spectral light source and not a monochromatic light. The relation between these two magnitudes is a wavelength integral as described in Section 2.3 below.

2.3 PSpice Model for the Short-Circuit Spectral Current Density

The simplest PSpice model for the short-circuit spectral current density can be easily written using PWL sources to include the files of the three magnitudes depending on the wavelength: spectral irradiance, absorption coefficient and reflectance. In the examples shown below we have assumed a constant value of the reflectance equal to 10% at all wavelengths.

2.3.1 Absorption coefficient subcircuit

The absorption coefficient for silicon is described by a subcircuit file, 'silicon_abs.lib' in Annex 2, having the same structure as the spectral irradiance file 'am15g.lib' and two access nodes from the outside: the value of the absorption coefficient at the internal node (11) and the reference node (10). The block diagram is shown in Figure 2.4.

As can be seen a PWL source is assigned between internal nodes (11) and (10) having all the list of the couples of values wavelength-absorption coefficient in cm^{-1}.

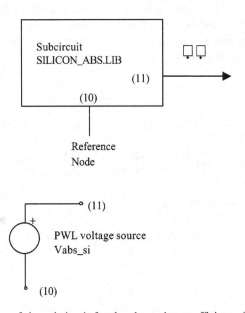

Figure 2.4 Block diagram of the subcircuit for the absorption coefficient of silicon and the internal schematic representation

2.3.2 Short-circuit current subcircuit model

The PSpice short-circuit model is written in the file 'jsc.lib', shown in Annex 2, where the implementation of equations (2.2) and (2.3) using equations (2.4) and (2.5) is made using voltage controlled voltage sources (e-devices). This is shown below.

```
egeom0 230 200 value={1.6e-19*v(202)*v(203)*(1000/962.5)*v(201)*(1e16/19.8)*
+lp*(1-V(204))/(v(202)*lp+1)}
egeom1 231 200 value={cosh(we/lp)+se*(lp/dp)*sinh(we/lp)}
egeom2 232 200 value={se*(lp/dp)*cosh(we/lp)+sinh(we/lp)}
egeom3 233 200 value={(se*(lp/dp)+v(202)*lp-exp(-v(202)*we)*v(232))}
ejsce 205 200 value={v(230)/(v(202)*lp-1)*(-v(202)*lp*
+exp(-v(202)*we)+v(233)/v(231))};short circuit
```

Note that equation (2.2) has been split in four parts and it is very simple to recognize the parts by comparison. The e-source named 'ejsce' returns the value of the emitter short circuit spectral current density, this means the value of $J_{scE\lambda}$ for every wavelength for which values of the spectral irradiance and of the absorption coefficient are provided in the corresponding PWL files. For convergence reasons the term ($\alpha^2 Lp^2 - 1$) has been split into ($\alpha Lp + 1$) ($\alpha Lp - 1$) being the first term included in 'egeom3' and the second in 'ejsce' sources.

A similar approach has been adopted for the base. It is worth noting that the value of the photon flux has been scaled up to a $1000\ W/m^2$ AM1.5 G spectrum as the file describing the spectrum has a total integral, that means a total irradiance, of $962.5\ W/m^2$. This is the reason why the factor (1000/962.5) is included in the e-source egeom0 for the emitter and egeom33 for the base. The complete netlist can be found in Annex 2 under the heading of 'jsc.lib' and the details of the access nodes of the subcircuit are shown in Figure 2.5. The meaning of the nodes QE and SR is described below.

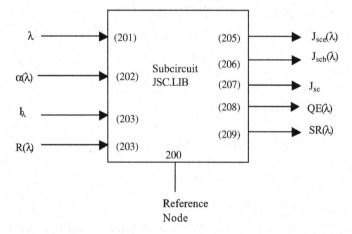

Figure 2.5 Node description of the subcircuit jsc.lib

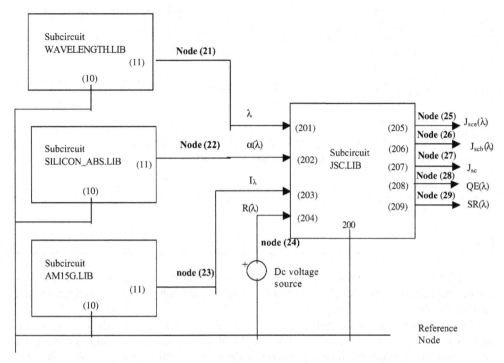

Figure 2.6 Structure of the file JSC_SILICON.CIR. The nodes of the circuit are shown in bold letters

The complete PSpice file to calculate the spectral emitter and base current densities is a '.cir' file where all the required subcircuits are included and the transient analysis is performed. The block diagram is shown in Figure 2.6 where a general organization of a circuit PSpice file containing subcircuits can be recognized. In particular it should be noticed that the internal node numbers of a subcircuit do not conflict with the same internal node numbers of another subcircuit and that the whole circuit has a new set of node numbers, shown in bold numbers Figure 2.6. This is a very convenient way to write different PSpice circuit files, for instance to compute the spectral current densities for different sun spectra or different absorption coefficient values, because only the new subcircuit has to substitute the old. This will be illustrated below.

Example 2.1

Consider a silicon solar cell with the dimensions given below. Write a PSpice file to compute the emitter and base spectral short-circuit current densities.

DATA: emitter thickness $W_e = 0.3$ mm, Base thickness $W_b = 300\,\mu m$, $L_p = 0.43\,\mu m$, $S_e = 2 \times 10^5$ cm/s, $D_p = 3.4\,cm^2/s$, $L_n = 162\,\mu m$, $S_b = 1000\,cm/s$, $D_n = 36.33\,cm^2/s$.

Solution

The PSpice file is given by

```
* JSC_SILICON.CIR
.include silicon_abs.lib
.include am15g.lib
.include wavelength.lib
.include jsc.lib
****** circuit
xwavelength 21 0 wavelength
xabs 22 0 silicon_abs
xsun 23 0 am15g
xjsc 0 21 22 23 24 25 26 27 28 29 jsc params: we=0.3e-4 lp=0.43e-4 dp=3.4
+ se=20000 wb=300e-4 ln=162e-4 dn=36.63 sb=1000
vr 24 0 dc 0.1
****** analysis
.tran 0.1u 1.2u 0.3u 0.01u
.option stepgmin
.probe
.end
```

As can be seen the absorption coefficient, the AM1.5 G spectrum and the wavelength files are incorporated into the file by '.include' statements and the subcircuits are connected to the nodes described in Figure 2.6.

The simulation transient analysis is carried out by the statement '.tran' which simulates up to 1.2 μs. As can be seen from the definition of the files PWL, the unit of time μs is assigned to the unit of wavelength μm which becomes the internal PSpice time variable. For this reason the transient analysis in this PSpice netlist becomes in fact a wavelength sweep from 0 to 1.2 μm. The value of the reflection coefficient is included by means of a DC voltage source having a value of 0.1, meaning a reflection coefficient of 10%, constant for all wavelengths.

The result can be seen in Figure 2.7.

As can be seen the absorption bands of the atmosphere present in the AM1.5 spectrum are translated to the current response and are clearly seen in the corresponding wavelengths in Figure 2.7. The base component is quantitatively the main component contributing to the total current in almost all wavelengths except in the shorter wavelengths, where the emitter layer contribution dominates.

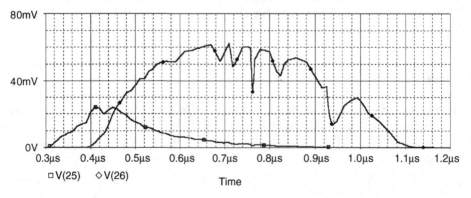

Figure 2.7 Spectral short circuit current densities for emitter (squares) and base (diamonds). Warning: The units of the y-axis are mA/cm^{2}μ, and that of the x-axis are microns representing the wavelength

2.4 Short-Circuit Current

Section 2.3 has shown how the spectral short-circuit current density generated by a mono-chromatic light is a function of the wavelength. As all wavelengths of the sun spectrum shine on the solar cell surface, the total short-circuit current generated by the solar cell is the wavelength integral of the short-circuit spectral density current, as follows:

$$J_{sc} = \int_0^\infty J_{sc\lambda} \, d\lambda = \int_0^\infty (J_{scE\lambda} + J_{scB\lambda}) \, d\lambda \qquad (2.7)$$

The units of the short-circuit current density are then A/cm^2. One important result can be made now. As we have seen that the emitter and base spectral current densities are linearly related to the photon flux ϕ_0 at a given wavelength, if we multiply by a constant the photon flux at all wavelengths, that means we multiply by a constant the irradiance, but we do not change the spectral distribution of the spectrum, then the total short-circuit current will also be multiplied by the same constant. This leads us to the important result that the short-circuit current density of a solar cell is proportional to the value of the irradiance. Despite the simplifications underlying the analytical model we have used, this result is valid for a wide range of solar cell designs and irradiance values, provided the temperature of the solar cell is the same and that the cell does not receive high irradiance values as could be the case in a concentrating PV system, where the low injection approximation does not hold.

In order to compute the wavelength integration in equation (2.7) PSpice has a function named sdt() which performs time integration. As in our case we have replaced time by wavelength, a time integral means a wavelength integral (the result has to be multiplied by 10^6 to correct for the units). This is illustrated in Example 2.2.

Example 2.2

Considering the same silicon solar cell of Example 2.1, calculate the short-circuit current density.

Solution

According to the solar cell geometry and parameter values defined in Example 2.1, the integral is performed by the sdt() PSpice function, adding a line to the code:

```
ejsc 207 200 value={1e6*sdt(v(205)+v(206))}
```

where v(207) returns the value of the total short-circuit current for the solar cell of 31.811 mA/cm^2. Figure 2.8 shows the evolution of the value of the integral from 0.3 μm and 1.2 μm.

Of course, as we are interested in the total short-circuit current density collected integrating over all wavelengths of the spectrum, the value of interest is the value at 1.2 μm.

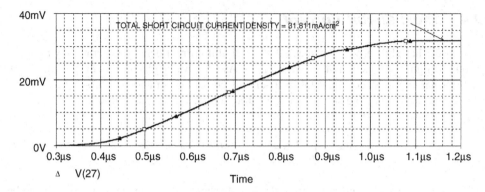

Figure 2.8   Wavelength integral of spectral short circuit current density in Figure 2.4. Warning: x-axis is the wavelength in microns and the y-axis is the $0-\lambda$ integral of the spectral short circuit current density (in mA/cm^2 units). Overall spectrum short circuit current density is given by the value at 1.2 µm

Warning

The constant 1×10^6 multiplying the integral value in the e-device ejsc, comes from the fact that PSpice performs a 'time' integration whereas we are interested in a wavelength integration and we are working with wavelength values given in microns.

2.5   Quantum Efficiency (QE)

Quantum efficiency is an important solar cell magnitude which is defined as the number of electrons produced in the external circuit by the solar cell for every photon in the incident spectrum. Two different quantum efficiencies can be defined: internal and external. In the internal quantum efficiency the incident spectrum considered is only the non-reflected part whereas in the definition of the external quantum efficiency the total spectral irradiance is considered.

$$IQE = \frac{J_{sc\lambda}}{q\phi_0(1-R)} \qquad (2.7)$$

$$EQE = \frac{J_{sc\lambda}}{q\phi_0} \qquad (2.8)$$

Quantum efficiencies have two components: emitter and base, according to the two components of the short-circuit spectral current density. Quantum efficiency is a magnitude with no units and is generally given in %.

Example 2.3

For the same solar cell described in Example 2.1, calculate the total internal quantum efficiency IQE.

Solution

The spice controlled e-sources 'eqee' and 'equeb' return the values of the internal quantum efficiency:

```
eqee 234 200 value={v(205)*19.8/(q*v(203)*(1000/962.5)*(1-V(204))*v(201)*1e16)}
eqeb 244 200 value={v(206)*19.8/(q*v(203)*(1000/962.5)*(1-v(204))*v(201)*1e16)}
```

As can be seen v(205) and v(206) return the values of the spectral short-circuit current density for the AM1.5 spectrum normalized to 1000 W/m², hence the value of the denominator in equations (2.7) and (2.8) has also been normalized to 1000 W/m² by using the normalization factor (1000/962.5) in the e-sources.

Figure 2.9 is a plot of the total quantum efficiency, that is the plot of v(28) giving the quantum efficiency in % against the wavelength.

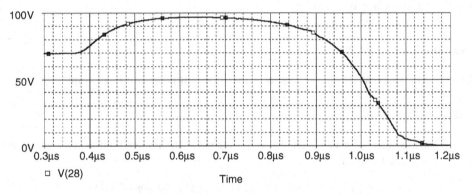

□ V(28)

Time

Figure 2.9 Plot of the total internal quantum efficiency for the solar cell described in Example 2.1. Warning: *x*-axis is % and *y*-axis is the wavelength in microns

As can be seen in Example 2.3, the internal quantum efficiency has a maximum of around 95% at 0.7 μm of wavelength and fades away from the maximum value at the two ends of the spectrum. Of course the solar cell geometry and design has much influence in the QE shape and values, and a solar cell is more ideal as the IQE becomes as large as possible at all wavelengths, the maximum value of the IQE being 100%.

These PSpice simulation results are close to the ones returned by numerical analysis and models of PC1D. The origin of the differences comes from the analytical nature of the equations used, which do not take into account many important solar cell operation features, only numerically treatable features, and hence the PSpice results have to be considered as first-order results.

2.6 Spectral Response (SR)

The spectral response of a solar cell is defined as the ratio between the short circuit spectral current density and the spectral irradiance. Taking into account equation (2.3):

$$ISR = \frac{J_{sc\lambda}}{I_\lambda(1-R)} \tag{2.9}$$

$$ESR = \frac{J_{sc\lambda}}{I_\lambda} \tag{2.10}$$

where the units are A/W. The relationship between the spectral response and quantum efficiency is easily found by taking into account equations (2.2), (2.3), (2.6) and (2.9), (2.10).

$$ISR = 0.808 \cdot (IQE) \cdot \lambda \tag{2.11}$$

and

$$ESR = 0.808 \cdot (EQE) \cdot \lambda \tag{2.12}$$

Equations (2.11) and (2.12) give the values of the spectral response in A/W provided the wavelength is written in microns.

Example 2.4

Calculate the internal spectral response of the solar cell described in Example 2.1.

Solution

Using equation (2.11) for the internal spectral response, the PSpice statement to be added to the netlist is the following:

$$\text{esr } 209\ 200\ \text{value} = \{(v(234) + v(244))^* 0.808^* v(201)\}$$

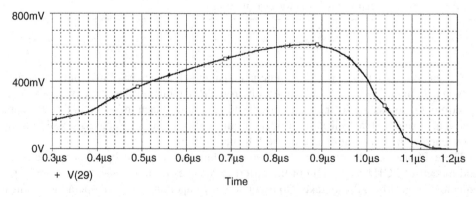

Figure 2.10 Spectral response. Warning: *x*-axis is the wavelength in microns and the *y*-axis is the SR in mA/W

that is v(209) returns the value of the spectral response. The results are shown in Figure 2.10.

As can be seen the greater sensitivity in terms of ampere per watt is located around 0.9 μm wavelength.

2.7 Dark Current Density

Photovoltaic devices are about generating power out of light and, of course, the short circuit condition we have used for illustrative purposes in the previous sections, is not an operational condition, because in a short circuit the power delivered to the load is zero even if a non-zero short circuit current flows. Chapter 3 will describe the whole range of operating points of a solar cell and the full current–voltage output characteristics, thus enabling us to calculate the power actually delivered to a load by an illuminated solar cell. This current–voltage curve is similar to the current–voltage characteristic of a photodiode and the analytical equation describing it requires the model of the dark current–voltage curve, that is under no light.

As can be presumed, this dark current–voltage curve is entirely similar to that of a conventional diode. The resulting current–voltage curve has two components: emitter and base, similar to the short-circuit current density calculated in previous sections. The interested reader can find a summary of the derivation in Annex 10. Considering that the two solar cell regions: base and emitter, are formed by uniformly doped semiconductor regions, the two components can be written as:

$$J_{darkE} = q \frac{n_i^2 D_p}{N_{Deff} L_p} \left[\frac{S_e \frac{L_p}{D_p} Ch \frac{W_e}{L_p} + Sh \frac{W_e}{L_p}}{S_e \frac{L_p}{D_p} Sh \frac{W_e}{L_p} + Ch \frac{W_e}{L_p}} \right] \left[e^{\frac{V}{V_T}} - 1 \right] \tag{2.13}$$

and

$$J_{darkB} = q \frac{n_i^2 D_n}{N_{Aeff} L_n} \left[\frac{S_b \frac{L_n}{D_n} Ch \frac{W_b}{L_n} + Sh \frac{W_b}{L_n}}{S_b \frac{L_n}{D_n} Sh \frac{W_b}{L_n} + Ch \frac{W_b}{L_n}} \right] \left[e^{\frac{V}{V_T}} - 1 \right] \tag{2.14}$$

Now the total current–voltage curve is the sum of the two – emitter and base – components and is written as:

$$J_{dark} = J_{darkE} + J_{darkB} = J_0 \left[e^{\frac{V}{V_T}} - 1 \right] \tag{2.15}$$

where J_0 is known as the saturation current density. Figure 2.11 shows the diode symbol representing the dark characteristics of the solar cell in equation (2.15). The signs for the dark current and for the voltage are also shown.

The saturation density current is then given by the sum of the pre-exponential terms of emitter and base dark components.

As can be seen, the dark saturation current obviously does not depend on any light parameters, but depends on geometrical parameters or transport semiconductor parameters,

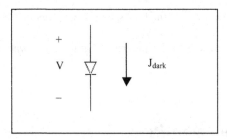

Figure 2.11   Convention of signs for the dark characteristics

and on a very important parameter the intrinsic carrier concentration n_i, which is a parameter that depends on the conduction and valence band density of states and on the energy bandgap of the semiconductor. This creates large differences in the values of n_i for semiconductors of practical interest in photovoltaics. If we notice that the dark saturation current density depends on the intrinsic concentration squared, the differences in the values of J_0 can be of several orders of magnitude difference among different semiconductors. It must also be said that heavy doping effects modify the value of the effective intrinsic carrier concentration because of the bandgap narrowing effect. This is easily modelled by replacing the value of the doping concentrations by 'effective' doping concentration values which are described in Annex 10.

2.8   Effects of Solar Cell Material

We have seen so far that several important magnitudes in the photovoltaic conversion depend on the semiconductor material properties: absorption coefficient, reflection coefficient, intrinsic carrier concentration, mobilities and lifetimes. Perhaps the value of the intrinsic carrier concentration is where very significant differences are found, which produce important differences in the photovoltaic response of solar cells.

For the sake of comparison Table 2.2 summarizes the main parameter values of two solar cells, one is our silicon baseline solar cell already used in Examples 2.1 to 2.3 and the second is a GaAs solar cell with parameter values taken from reference [2.2].

Table 2.2 also shows the values for J_{sc} and J_0 calculated using the analytical model in equations (2.2) and (2.3) for the short circuit current densities and by equations (2.13) and (2.14) for the dark current. Bandgap narrowing and mobility values for silicon have been taken from the model described in Annex 10.

As can be seen, the most important difference comes from the values of the dark saturation density current J_0 which is roughly eight orders of magnitude smaller in the GaAs solar cell than in the silicon solar cell. The origin of such an enormous difference comes mainly from the difference of almost eight orders of magnitude of the values of n_i^2, namely 1×10^{20} for silicon, 8.6×10^{11} for GaAs.

The differences in the values of the short circuit current densities J_{sc} are much smaller. Even if more sophisticated simulation models are used, these differences never reach one order of magnitude. This result allows to conclude that the semiconductor used in building a solar cell has a very important effect on the electrical characteristics in addition to geometrical and technological implications.

Table 2.2 Comparison between silicon and GaAs solar cells

Parameter	Silicon baseline solar cell*	GaAs solar cell**
n_i (cm^{-3})	1×10^{10}	9.27×10^5
W_e (µm)	0.3	0.2
L_p (µm)	0.43	0.432
S_e (cm/s)	2×10^5	5×10^3
D_p (cm^2/V)	3.4	7.67
W_b (µm)	300	3.8
L_n (µm)	162	1.51
S_b (cm/s)	1×10^3	5×10^3
D_n (cm^2/V)	36.63	46.8
J_0 (A/cm^2)	1×10^{-12}	9.5×10^{-20}
J_{sc} (A/cm^2)	31.188×10^{-3}	25.53×10^{-3}

* geometry from PC1D5 file Pvcell.prm, and calculation using equations in Annex 2
** From L.D. Partain, *Solar Cells and their applications* Wiley 1995 page 16 [2.2]

2.9. Superposition

Once the dark and illuminated characteristics have been calculated separately, a simple way, although it may be inaccurate in some cases, to write the total current–voltage characteristic is to assume that the superposition principle holds, that is to say that the total characteristic is the sum of the dark (equation (2.15)) and illuminated characteristics (equation (2.7)):

$$J = J_{sc} - J_{dark} \qquad (2.16)$$

It should be noted that the dark and illuminated components have opposite signs as it becomes clear from the derivation shown in Annex 10 and from the sign convention described in Figures 2.3 and 2.9.

Substituting in equation (2.16) the dark current by equation (2.15),

$$J = J_{sc} - J_0 \left(e^{\frac{V}{V_T}} - 1 \right) \qquad (2.17)$$

It should be noted that the sign convention described so far is a photovoltaic convention where the short circuit current originated in the photovoltaic conversion is considered positive. This has a more practical origin than a technical one.

2.10. DC Sweep Plots and I(v) Solar Cell Characteristics

If a PSpice file is to be written to plot equation (2.17) we can simply connect a DC constant current source and a diode. It is now pertinent to remember the syntax of these two PSpice elements.

Syntax for a DC curent source

$$\text{iname node+ node− DC value}$$

Where the first letter always has to be an 'i' and by convention the current is flowing inwards from the external circuit into the node+ terminal.

Syntax for a diode

$$\text{Dname anode cathode model\_name}$$

where anode is the positive diode terminal, cathode is the negative diode terminal and a model_name has to be called. By default the diode model is named d and has several internal model parameters, one of which is the saturation current, named is (this is the saturation current and not the saturation density current). The way the model is called for is by means of a Dot command:

$$\text{.model model\_name spice\_model}$$

Moreover, we do not now want any transient analysis but rather a DC sweep analysis because we would like to see the values of the solar cell current for any value of the voltage across the cell. This is implemented in PSpice code by a DC analysis.

Syntax for a DC analysis

The syntax is:

$$\text{.dc voltage\_source start\_value end\_value step}$$

where voltage_source is the name of the voltage source we want to sweep from a start_value to an end_value by given step size increments.

Example 2.5

Write a PSpice netlist for the solar cell equation (2.17) having a $1\,\text{cm}^2$ area and the data shown in Table 2.2 for J_{SC} and J_0 for the silicon solar cell.

Solution

```
******* solar cell characteristics
*cel_1_si.cir
isc 0 1 31.188e-3
dcell 1 0 diode
.model diode d(is=1e-12)
vbias 1 0 dc 0
.dc vbias 0 0.7 0.01
.probe
.end
```

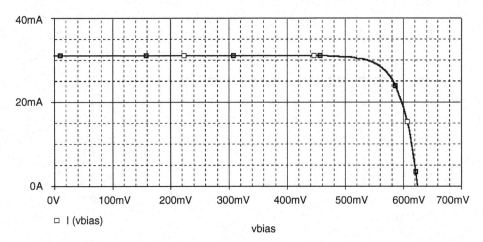

Figure 2.12 Si solar cell $I(V)$ characteristics for the data in Table 2.2 and model in equation (2.17)

The result of the $I(V)$ characteristic is plotted in Figure 2.12. As can be seen the exponential term in equation (2.17) has been modelled by a diode having a saturation current of 1×10^{-12} A/cm^2 and the short circuit current has been modelled by a constant current source 'isc' of 31.188 mA.

Example 2.6

Simulate the $I(V)$ characteristic of a GaAs solar cell using the device equation (2.17) and the data values in Table 2.2.

Solution

The spice model for the GaAs solar cells are formally the same as the silicon solar cell where the data values of the parameters J_0 and J_{SC} are to be changed and taken from Table 2.1 ($J_{SC} = 25.53 \times 10^{-3}$ A/cm^2 and $J_0 = 9.5 \times 10^{-20}$ A/cm^2).

```
******cell_1_gaas
******* solar cell characteristics
isc 0 1 25.53e-3
dcell 1 0 diode
.model diode d(is=9.5e-20)
vbias 1 0 dc 0
.dc vbias 0 1.2 0.01
.probe
.end
```

The result of the simulation is shown in Figure 2.13.

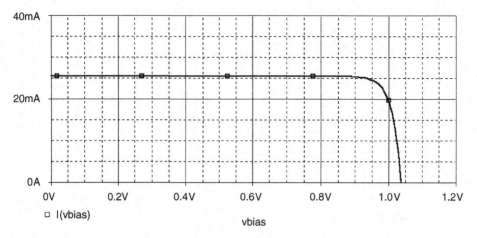

Figure 2.13   GaAs solar cell I(V) characteristics for the data in Table 2.2 and model in equation (2.17)

As can be seen, the comparison of the results shown in Figures 2.12 and 2.13 illustrate how different solar cell materials and geometries can produce different levels of current and voltage. In particular, it can be noticed that the voltage difference is the most important in this case: the maximum voltage value of the GaAs solar cell is almost double that of a silicon solar cell, and this can be attributed in a major part to the large difference in the values of J_0, which come from the large difference in the values of n_i^2. This allows us to conclude that fundamental semiconductor parameter values are at the origin of the differences observed in the electrical characteristics.

2.11. Failing to Fit to the Ideal Circuit Model: Series and Shunt Resistances and Recombination Terms

The analytical models described so far, and due to the many simplifying assumptions made, are not able to explain or model important device effects such as for instance 2D effects in the current flow, detailed surface recombination models, light confinement, back surface reflectance, back surface field, etc.

In addition to that, the process to fabricate real solar cells and modules introduce not only many design compromises, but also technology compromises, in order to produce cost-effective industrial products. Fabrication reproducibility and reliability are also statistical magnitudes conditioning the 'average characteristics' of a production series. It is commonly accepted that a simplifying model taking into account resistive losses in the solar cell circuit model is indispensable to simulate correctly the operation of PV systems. The origin of the resistive losses basically comes from two causes:

1. the series resistive losses due to the semiconductor resistivity, 2D conduction mechanisms in the emitter layer, contact resistance; and

2. the shunt resistance which takes into account all parallel losses across the semiconductor junction due to partial junction short circuits.

Sometimes, simply considering the series and shunt resistances as circuit constants is not enough to model accurately the illuminated $I(V)$ characteristics of a solar cell. This may be originated by different phenomena:

(a) Diode non-ideality. This means that a real measurement of the dark $I(V)$ curve does not exactly fit to the dark part of equation (2.17). This requires the use of a 'diode factor' n, which is different from the unity, in the denominator of the exponent.

(b) Recombination current in the space charge region of the junction which behaves as an exponential term but with a diode factor greater than unity and generally considered as equal to two.

(c) In amorphous silicon solar cells which are fabricated using a p-i-n structure, the recombination current at the intrinsic layer is not an exponential function but a non-linear function of the voltage and of the current.

The effects of the series and shunt resistances and of the non-idealities are considered in Chapter 3.

2.12 Problems

2.1 Compare the quantum efficiency of a solar cell with an arbitrarily long base layer (assume 2 mm long) with that of the same solar cell having a base layer thickness of 170 µm and a base surface recombination velocity of 10000 cm/s. The other parameter values are the same as in Example 2.1.

2.2 With the same parameter values for the base layer as in Example 2.1, adjust the emitter thickness to have an emitter quantum efficiency at 500 nm of 50%.

2.13 References

[2.1] Green, M.A., *Silicon solar cells*, Centre for Photovoltaic Devices and Systems, University of New South Wales, Sydney, 1995.
[2.2] Partain, L.D., *Solar Cells and their Applications*, Wiley, 1995.
[2.3] Green, M.A., *Solar Cells*, University of New South Wales, Sydney, 1992.
[2.4] Fahrenbruch, A.L. and Bube, R.T., *Fundamentals of Solar Cells*, Academic Press, 1983.
[2.5] Van Overtraeten, R. and Mertens, R., *Physics, Technology and Use of Photovoltaics*, Adam Hilger, 1986.
[2.6] PC1D Photovoltaics Special Research Centre at the University of New South Wales, Sydney, Australia.

3

Electrical Characteristics of the Solar Cell

Summary

The basic equations of a solar cell are described in this chapter. The dark and illuminated $I(V)$ characteristics are analytically described and PSpice models are introduced firstly for the simplest model, composed of a diode and a current source. The fundamental electrical parameters of the solar cell are defined: short circuit current (I_{sc}), open circuit voltage (V_{oc}), maximum power (P_{max}) and fill factor (FF). This simple model is then generalized to take into account series and shunt resistive losses and recombination losses. Temperature effects are then introduced and the effects of space radiation are also studied with a modification of the PSpice model. A behavioural model is introduced which allows the solar cell simulation for arbitrary time profiles of irradiance and temperature.

3.1 Ideal Equivalent Circuit

As shown in Chapter 2, a solar cell can in a first-order model, be described by the superposition of the responses of the device to two excitations: voltage and light. We start by reproducing here the simplified equation governing the current of the solar cell, that is equation (2.16)

$$J = J_{sc} - J_0(e^{\frac{V}{V_T}} - 1) \tag{3.1}$$

which gives the current density of a solar cell submitted to a given irradiance and voltage. The value of the current generated by the solar cell is given by

$$I = I_{sc} - I_0(e^{\frac{V}{V_T}} - 1) \tag{3.2}$$

where I_{sc} and I_0 relate to their respective current densities J_{sc} and J_0 as follows:

$$I_{sc} = AJ_{sc} \qquad (3.3)$$
$$I_0 = AJ_0 \qquad (3.4)$$

where A is the total area of the device. The metal covered area has been neglected.

As can be seen, both the short circuit current and the dark current scale linearly with the solar cell area, and this is an important result which facilitates the scaling-up or down of PV systems according to the requirements of the application.

This is the simplest, yet the most used model of a solar cell in photovoltaics and its applications, and can be easily modelled in PSpice code by a current source of value I_{sc} and a diode.

Although I_0 is a strong function of the temperature, we will consider first that I_0 can be given a constant value. Temperature effects are addressed later in this chapter.

3.2   PSpice Model of the Ideal Solar Cell

As has already been described, one way to handle a PSpice circuit is to define subcircuits for the main blocks. This is also the case of a solar cell, where a subcircuit facilitates the task of connecting several solar cells in series or in parallel as will be shown later.

The PSpice model of the subcircuit of an ideal solar cell is shown in Figure 3.1(a) which is the circuit representation of equation (3.2). The case in photovoltaics is that a solar cell receives a given irradiance value and that the short circuit current is proportional to the irradiance. In order to implement that in PSpice the value of the short circuit current, is assigned to a G-device which is a voltage-controlled current source, having a similar syntax to the e-devices:

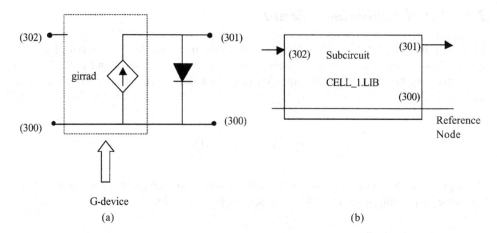

Figure 3.1   (a) Cell_1.lib subcircuit a solar cell and (b) block diagram

Syntax for G-device

$$g\_name\ node+\ node-\ control\_node+\ control\_node-\ gain$$

As can be seen, this is a current source connected to the circuit between nodes 'node+' and 'node−', with a value given by the product of the gain by the voltage applied between control_node+ and control_node−.

A simplification of this device consists of assigning a value which can be a mathematical expression as follows:

$$g\_name\ node+\ node-\ value = \{expression\}$$

In our case the G-device used is named 'girrad' and is given by:

$$girrad = \frac{J_{sc}A}{1000}G \tag{3.5}$$

where G is the value of the irradiance in W/m^2. Equation (3.5) considers that the value of J_{sc} is given at standard (AM1.5G, 1000 W/m^2 $T_{cell} = 25\,°C$) conditions, which are the conditions under which measurements are usually made. Solar cell manufacturer's catalogues provide these standard values for the short circuit current. Equation (3.5) returns the value of the short circuit current at any irradiance value G, provided the proportionality between irradiance and short circuit current holds. This is usually the case provided low injection conditions are satisfied.

The subcircuit netlist follows:

```
.subckt cell_1 300 301 302 params: area=1, j0=1, jsc=1

girrad 300 301 value={(jsc/1000)*v(302)*area}
d1 301 300 diode
.model diode d(is={j0*area})
.ends cell_1
```

As can be seen, dummy values, namely unity, are assigned to the parameters at the subcircuit definition; the real values are specified later when the subcircuit is included in a circuit. The diode model includes the definition of the parameter 'is' of a PSpice diode as the result of the product of the saturation current density J_0 multiplied by the area A. The block diagram is shown in Figure 3.1(b). As can be seen there is a reference node (300), an input node (302) to input a voltage numerically equal to the irradiance value, and an output node (301) which connects the solar cell to the circuit.

The solar cell subcircuit is connected in a measurement circuit in order to obtain the $I(V)$ characteristic. This is accomplished by the circuit shown in Figure 3.2, where the solar cell area corresponds to a 5″ diameter device. A short circuit density current of 34.3 mA/cm^2 has

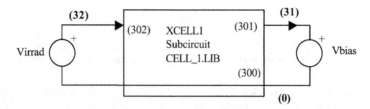

Figure 3.2   Measurement circuit of the $I(V)$ characteristics of a solar cell

been assumed and $J_0 = 1 \times 10^{-11}$ A/cm^2 is chosen as an example leading to a PSpice file as follows:

```
*cell_1.cir
.include cell_1.lib
xcell1 0 31 32 cell_1 params:area=126.6  j0=1e-11  jsc=0.0343
vbias 31 0 dc 0
virrad 32 0 dc 1000

.plot dc i(vbias)
.probe
.dc vbias -0.1 0.6 0.01
.end
```

As can be seen the circuit includes a DC voltage source 'vbias' which is swept from -0.5 V to $+0.6$ V. The result of the simulation of the above netlist is shown in Figure 3.3. The intersection of the graph with the y-axis provides the value of the short circuit current of the cell, namely 4.342 A, which is, of course, the result of $0.0343 \times 126.6 = 4.342$ A. Some other important points can be derived from this $I(V)$ plot as described in the next section.

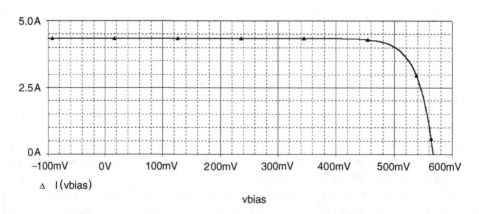

Figure 3.3 $I(V)$ characteristics of the solar cell model in Figure 3.2

3.3 Open Circuit Voltage

Besides the short circuit current, a second important point in the solar cell characteristics can be defined at the crossing of the $I(V)$ curve with the voltage axis. This is called the open circuit point and the value of the voltage is called the open circuit voltage, V_{oc}.

Applying the open circuit condition, $I = 0$, to the $I(V)$ equation (3.2) as follows:

$$I = 0 = I_{sc} - I_0 \left(e^{\frac{V_{oc}}{V_T}} - 1 \right) \tag{3.6}$$

the open circuit voltage is given by:

$$V_{oc} = V_T \ln \left(1 + \frac{I_{sc}}{I_0} \right) \tag{3.7}$$

From equation (3.7), it can be seen than the value of the open circuit voltage depends, logarithmically on the I_{sc}/I_0 ratio. This means that under constant temperature the value of the open circuit voltage scales logarithmically with the short circuit current which, in turn scales linearly with the irradiance resulting in a logarithmic dependence of the open circuit voltage with the irradiance. This is also an important result indicating that the effect of the irradiance is much larger in the short circuit current than in the open circuit value.

Substituting equations (3.3) and (3.4) in equation (3.7), results in:

$$V_{oc} = V_T \ln \left(1 + \frac{J_{sc}}{J_0} \right) \tag{3.8}$$

The result shown in equation (3.8) indicates that the open circuit voltage is *independent* of the cell area, which is an important result because, regardless of the value of the cell area, the open circuit voltage is always the same under the same illumination and temperature conditions.

Example 3.1

Consider a circular solar cell of $6''$ diameter. Assuming that $J_{sc} = 34.3$ A/cm^2 and $J_0 = 1 \times 10^{-11}$ A/cm^2, plot the $I(V)$ characteristic and calculate the open circuit voltage for several irradiance values, namely $G = 200$, 400, 600, 800 and 1000 W/m^2.

The netlist in this case has to include a instruction to solve the circuit for every value of the irradiance. This is performed by the instructions below:

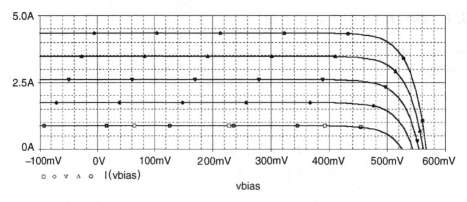

Figure 3.4 $I(V)$ plots of a solar cell under several irradiance values: 200, 400, 600, 800 and 1000 W/m^2

```
*IRRADIANCE.CIR
.include cell_1.lib
xcell1 0 31 32 cell_1 params:area=126.6  j0=1e-11  jsc=0.0343
vbias 31 0 dc 0
.param IR=1
virrad 32 0 dc {IR}
.step param IR list 200 400 600 800 1000
.plot dc i(vbias)
.probe
.dc vbias -0.1 0.6 0.01
.end
```

It can be seen that a new parameter is defined, IR, which is assigned to the value of the voltage source virrad. The statement '.step param' is a dot command which makes PSpice repeat the simulation for all values of the list.

The plots in Figure 3.4 are obtained and, using the cursor utility in Probe, the values of V_{oc} and I_{sc} are measured. The results are shown in Table 3.1.

Table 3.1 Short circuit current and open circuit voltages for several irradiance values

Irradiance G(W/m^2)	Short circuit current I_{sc}(A)	Open circuit voltage I_{oc}(V)
1000	4.34	0.567
800	3.47	0.561
600	2.60	0.554
400	1.73	0.543
200	0.86	0.525

3.4 Maximum Power Point

The output power of a solar cell is the product of the output current delivered to the electric load and the voltage across the cell. It is generally considered that a positive sign indicates power being delivered to the load and a negative sign indicates power being consumed by the solar cell. Taking into account the sign definitions in Figure 3.1, the power at any point of the characteristic is given by:

$$P = V \times I = V\left[I_L - I_0\left(e^{\frac{V}{V_T}} - 1\right)\right] \tag{3.9}$$

Of course, the value of the power at the short circuit point is zero, because the voltage is zero, and also the power is zero at the open circuit point where the current is zero. There is a positive power generated by the solar cell between these two points. It also happens that there is a maximum of the power generated by a solar cell somewhere in between. This happens at a point called the maximum power point (MPP) with the coordinates $V = V_m$ and $I = I_m$. A relationship between V_m and I_m can be derived, taking into account that at the maximum power point the derivative of the power is zero:

$$\frac{dP}{dV} = 0 = I_L - I_0\left(e^{\frac{V_m}{V_T}} - 1\right) - \frac{V_m}{V_T}I_0 e^{\frac{V_m}{V_T}} \tag{3.10}$$

at the MPP,

$$I_m = I_L - I_0\left(e^{\frac{V_m}{V_T}} - 1\right) \tag{3.11}$$

it follows that,

$$V_m = V_{oc} - V_T \ln\left(1 + \frac{V_m}{V_T}\right) \tag{3.12}$$

which is a transcendent equation. Solving equation (3.13) V_m can be calculated provided V_{oc} is known.

Using PSpice the coordinates of the MMP can be easily found plotting the $I \times V$ product as a function of the applied voltage. We do not need to write a new PSpice '.cir' file but just draw the plot. This is shown in Figure 3.5 for the solution of Example 3.1.

Table 3.2 shows the values obtained using equation (3.12) compared to the values obtained using PSpice. As can be seen the accuracy obtained by PSpice is related to the step in voltage we have used, namely 0.01 V. If more precision is required, a shorter simulation step should be used.

Other models can be found in the literature [3.11] to calculate the voltage V_m, which are summarized below.

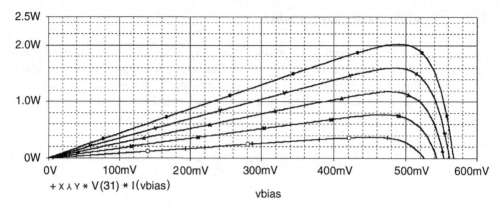

Figure 3.5 Plot of the product of the current by the voltage across the solar cell for several values of the irradiance, namely: 200, 400, 600, 800 and 1000 W/m²

Alternative model number 1:

$$V_m = V_{oc} - 3V_T \tag{3.13}$$

Alternative model number 2:

$$\frac{V_m}{V_{oc}} = 1 - \left(\frac{1 + \ln \beta}{2 + \ln \beta}\right) \frac{\ln(1 + \ln \beta)}{\ln \beta} \tag{3.14}$$

with

$$\beta = \frac{I_{sc}}{I_0} \tag{3.15}$$

Table 3.2 PSpice results for several irradiance values

Irradiance (W/m²)	V_m (PSpice) (V)	V_m from equation (3.14) (V)	I_m (PSpice) (A)	P_{max} (PSpice) (W)
1000	0.495	0.4895	4.07	2.01
800	0.485	0.4825	3.28	1.59
600	0.477	0.476	2.47	1.18
400	0.471	0.466	1.63	0.769
200	0.45	0.4485	0.820	0.37

3.5 Fill Factor (FF) and Power Conversion Efficiency (η)

A parameter called fill factor (FF) is defined as the ratio between the maximum power P_{max} and the $I_{sc}V_{oc}$ product:

$$FF = \frac{V_m I_m}{V_{oc} I_{sc}} \tag{3.16}$$

The fill factor has no units, indicating how far the product $I_{sc}V_{oc}$ is from the power delivered by the solar cell.

The FF can also be approximated by an empirical relationship as follows [3.2],

$$FF_0 = \frac{v_{oc} - \ln(v_{oc} + 0.72)}{1 + v_{oc}} \tag{3.17}$$

Sub-index 0 indicates that this is the value of the FF for the ideal solar cell without resistive effects to distinguish it from the FF of a solar cell with arbitrary values of the losses resistances, which will be addressed in the sections below. The parameter v_{oc} is the normalized value of the open circuit voltage to the thermal potential V_T, as

$$v_{oc} = \frac{V_{oc}}{V_T} \tag{3.18}$$

Equation (3.17) gives reasonable accuracy for v_{oc} values greater than 10.

Example 3.2

From the PSpice simulations in Example 3.1 obtain the values of the FF for several values of the irradiance.

Taking the data for I_{sc} and V_{oc} from Table 3.1 and the data for V_m and I_m from Table 3.2 the values for FF are easily calculated and are shown in Table 3.3.

As can be seen, the FF is reasonably constant for a wide range of values of the irradiance and close to 0.8.

Table 3.3 Fill factor values for Example 3.1

Irradiance (W/m^2)	Fill factor
1000	0.816
800	0.816
600	0.819
400	0.818
200	0.819

Table 3.4 Fill factor values for Example 3.1

Irradiance (W/m²)	Power conversion efficiency (%)
1000	15.8
800	15.68
600	15.53
400	15.17
200	14.6

The power conversion efficiency η is defined as the ratio between the solar cell output power and the solar power impinging the solar cell surface, P_{in}. This input power equals the irradiance multiplied by the cell area:

$$\eta = \frac{V_m I_m}{P_{in}} = FF \frac{V_{oc} I_{sc}}{P_{in}} = FF \frac{V_{oc} I_{sc}}{G \times Area} = FF \frac{V_{oc} J_{sc}}{G} \tag{3.19}$$

As can be seen the power conversion efficiency of a solar cell is proportional to the value of the three main photovoltaic parameters: short circuit current density, open circuit voltage and fill factor, for a given irradiance G.

Most of the time the efficiency is given in %.

In Example 3.2 above, the values of the efficiency calculated from the results in Tables 3.2 and 3.3 are given in Table 3.4.

As can be seen the power conversion efficiency is higher at higher irradiances. This can be analytically formulated by considering that the value of the irradiance is scaled by a 'scale factor S' to the standard conditions: 1 Sun, AM1.5 1000 W/m², as follows:

$$G = SG_1 \tag{3.20}$$

with $G_1 = 1000$ W/m². The efficiency is now given by:

$$\eta = FF \frac{V_{oc} I_{sc}}{SG_1 \times Area} \tag{3.21}$$

If proportionality between irradiance and short circuit current can be assumed:

$$
\begin{aligned}
I_{sc} &= SI_{sc1} \\
V_{oc} &= V_T \ln\left(1 + \frac{SI_{sc1}}{I_0}\right) \approx V_T \ln\left(\frac{SI_{sc1}}{I_0}\right) = V_T \ln\left(\frac{I_{sc1}}{I_0}\right) + V_T \ln S
\end{aligned}
\tag{3.22}
$$

and finally

$$\eta = FF \frac{SI_{sc1}}{SG_1 \times Area}\left[V_T \ln\left(\frac{I_{sc1}}{I_0}\right) + V_T \ln S\right] = FF \frac{I_{sc1} V_{oc1}}{G_1 \times Area}\left[1 + \frac{V_T \ln S}{V_{oc1}}\right] \tag{3.23}$$

Applying the efficiency definition in equation (3.21) to the one-sun conditions

$$\eta_1 = FF_1 \frac{V_{oc1} I_{sc1}}{G_1 \times Area} \tag{3.24}$$

it follows

$$\eta = \eta_1 \left[1 + \frac{V_T \ln S}{V_{oc1}} \right] \tag{3.25}$$

provided that $FF = FF_1$, thereby indicating that the efficiency depends logarithmically on the value of the scale factor S of the irradiance. The assumptions made to derive equation (3.25), namely constant temperature is maintained, that the short circuit current is proportional to the irradiance and that the fill factor is independent of the irradiance value, have to be fulfilled. This is usually the case for irradiance values smaller than one sun.

3.6 Generalized Model of a Solar Cell

The equivalent circuit and the PSpice model of the solar cell described so far takes into account an ideal behaviour of a solar cell based on an ideal diode and an ideal current source. Sometimes this level-one model is insufficient to accurately represent the maximum power delivered by the solar cell. There are several effects which have not been taken into account and that may affect the solar cell response.

(a) Series resistance

One of the main limitations of the model comes from the series resistive losses which are present in practical solar cells. In fact, the current generated in the solar cell volume travels to the contacts through resistive semiconductor material, both, in the base region, not heavily doped in general, and in the emitter region, which although heavily doped, is narrow. Besides these two components, the resistance of the metal grid, contacts and current collecting bus also contribute to the total series resistive losses. It is common practice to assume that these series losses can be represented by a lumped resistor, R_s, called the series resistance of the solar cell.

(b) Shunt resistance

Solar cell technology in industry is the result of mass production of devices generally made out of large area wafers, or of large area thin film material. A number of shunt resistive losses are identified, such as localized shorts at the emitter layer or perimeter shunts along cell borders are among the most common. This is represented generally by a lumped resistor, R_{sh}, in parallel with the intrinsic device.

(c)   Recombination

Recombination at the space charge region of solar cells explains non−ohmic current paths in parallel with the intrinsic solar cell. This is relevant at low voltage bias and can be represented in an equivalent circuit by a second diode term with a saturation density current J_{02}, which is different from the saturation density current of the ideal solar cell diode, and a given ideality diode factor different to 1, it is most often assumed to equal 2. This can be added to the solar cell subcircuit by simply adding a second diode 'diode2' with $n = 2$ in the description of the diode model, as follows:

```
.model diode2 d(is={j02*area}, n=2)
```

(d)   Non-ideality of the diffusion diode

In practice few devices exhibit a totally ideal $I(V)$ characteristic with ideality coefficient equal to unity. For this reason it is common practice to also add a parameter 'n' to account for hese non−idealities. In the same way as described above for the recombination diode, the main diode model can be modified to take into account this effect as follows:

```
.model diode d(is={j0*area},n=1.1)
```

where $n = 1.1$ is an example.

In summary, a new relationship between current and voltage can be written taking into account these effects as follows:

$$I = I_L - I_0\left(e^{\frac{V+IR_s}{nV_T}} - 1\right) - I_{02}\left(e^{\frac{V+IR_s}{2V_T}} - 1\right) - \frac{V + IR_s}{R_{sh}} \tag{3.26}$$

This equation comes from the equivalent circuit shown in Figure 3.6. It must be emphasized that the meaning of the short circuit current has to change in this new circuit due to the fact that the short circuit conditions are applied to the external solar cell terminals, namely nodes (303) and (300) in Figure 3.6, whereas equations (2.2), (2.3) and (2.6) in Chapter 2 were derived for a short circuit applied to the internal solar cell nodes, namely (301) and (300) in Figure 3.6. This is the reason why the current generator is now named I_L, or photogenerated current, to differentiate it from the new short circuit current of the solar cell. Finally, the diode D1 implements the first diode term in equation (3.26), and the diode D2 implements the second diode. It will be shown in the examples below how the values of the several parameters involved in the model influence the solar cell response, in particular the differences between photogenerated current and short circuit current.

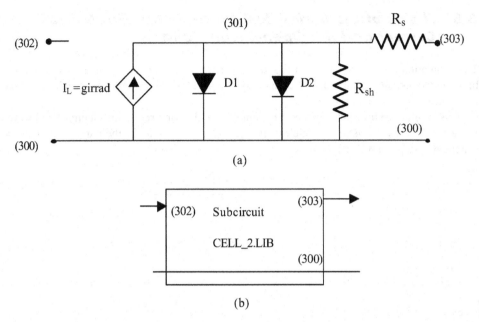

Figure 3.6   Subcircuit cell_2.lib including two diodes and series and shunt resistors

3.7 Generalized PSpice Model of a Solar Cell

The circuit in Figure 3.6, which includes the new subcircuit and the series and shunt resistors, can be described in PSpice code by the following netlist:

```
*CELL_2.LIB*
.subckt cell_2 300 303 302 params:area=1, j0=1, jsc=1, j02=1, rs=1, rsh=1

girrad 300 301 value={(jsc/1000)*v(302)*area}
d1 301 300 diode
.model diode d(is={j0*area})
d2 11 10 diode2
.model diode2 d(is={j02*area}, n=2)
.ends cell_2
```

where new model parameters are j02, rs and rsh which are given appropriate values. In order to plot the $I(V)$ curve and analyse the effects of the values of the new parameters a .cir file has to be written calling the new subcircuit. This is shown in the next section.

3.8   Effects of the Series Resistance on the Short-Circuit Current and the Open-Circuit Voltage

The simulation results are shown in Figure 3.7 for several values of the series resistance at a high and constant shunt resistance, namely $1 \times 10^5 \, \Omega$, and at equivalent values of the irradiance and the temperature.

This is achieved by changing the statement calling the subcircuit, introducing {RS} as the value for the series resistance, which is then given several values at the parametric analysis statement '.step param', as follows,

```
* cell_2.cir
.include cell_2.lib
xcell2 0 31 32 cell_2 params:area=126.6  j0=1e-11  j02=1E-9
+ jsc=0.0343 rs={RS} rsh=100000
.param RS=1
vbias 31 0 dc 0
virrad 32 0 dc 1000
.plot dc i(vbias)
.dc vbias −0.1 0.6 0.01
.step param RS list 0.0001 0.001 0.01 0.1 1
.probe
.end
```

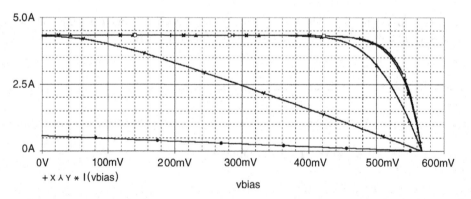

Figure 3.7   Series resistance effects. $I(V)$ characteristics for series resistance values from $1 \, \Omega$ (bottom graph), $0.1 \, \Omega, 0.01 \, \Omega$, $0.001 \, \Omega$, $0.0001 \, \Omega$ (top graph)

As can be seen in Figure 3.7, large differences are observed in the $I(V)$ characteristics as the value of the series resistance increases, in particular the values of the short-circuit current and of the fill factor can be severely reduced. Some of these changes can be explained using equation (3.26) and are summarized and explained next.

Short-circuit current

In contrast with the ideal result, at short circuit the value of the short-circuit current is not equal to the photocurrent I_L. By replacing $V = 0$ in equation (3.26) it follows that:

$$I_{sc} = I_L - I_0\left(e^{\frac{I_{sc}R_s}{nV_T}} - 1\right) - I_{02}\left(e^{\frac{I_{sc}R_s}{2V_T}} - 1\right) - \frac{I_{sc}R_s}{R_{sh}} \qquad (3.27)$$

In practice, the difference between I_{SC} and I_L in a real solar cell is small because the series resistance is kept low by a proper design of the metal grid and doping levels, and because the parallel resistance is large in devices passing the post-fabrication tests. This is clearly seen in the several plots in Figure 3.7 where the value of the short-circuit current remains sensibly constant, provided that values for the series resistance equal to or smaller than 0.01 Ω are used. Taking into account that the solar cell has an area, in this example, of 126.6 cm^2, this means that very small values of the series resistive losses can be tolerated if the short-circuit current is to be maintained.

Open-circuit voltage

The open-circuit voltage can be derived from equation (3.26) setting $I = 0$.

$$0 = I_L - I_0\left(e^{\frac{V_{oc}}{nV_T}} - 1\right) - I_{02}\left(e^{\frac{V_{oc}}{2V_T}} - 1\right) - \frac{V_{oc}}{R_{sh}} \qquad (3.28)$$

It is clear that the open-circuit voltage given by equation (3.28) is independent of the series resistance value and this is true regardless of the value of the shunt resistance, and the values of the parameters of the recombination diode. If we neglect the third and fourth terms in equation (3.28), the open-circuit voltage is given by

$$V_{oc} = nV_T \ln\left(1 + \frac{I_{sc}}{I_0}\right) \qquad (3.29)$$

which is sensibly the same result found using the ideal equivalent circuit in equation (3.7) with the exception of the non-ideality factor of the main diode, n.

In Figure 3.7 the graphs shown confirm that the open-circuit voltage is independent of the series resistance value, as all curves cross at the same point on the voltage axis.

3.9 Effect of the Series Resistance on the Fill Factor

From the results shown in Figure 3.7 it becomes clear that one of the most affected electrical parameters of the solar cell by the series resistance is the fill factor.

If we restrict the analysis to the effect of the series resistance only, the solar cell can be modelled by the diffusion diode and the series resistance, simplifying equation (3.26), which then becomes,

$$I = I_L - I_0 \left(e^{\frac{V+IR_s}{nV_T}} - 1 \right)$$

(3.30)

in open circuit conditions:

$$0 = I_L - I_0 \left(e^{\frac{V_{oc}}{nV_T}} - 1 \right)$$

(3.31)

from equation (3.31), I_0 can be written as

$$I_0 = \frac{I_L}{\left(e^{\frac{V_{oc}}{nV_T}} - 1 \right)} \approx I_L e^{-\frac{V_{oc}}{nV_T}}$$

(3.32)

and substituting in equation (3.30)

$$I = I_L - I_L e^{-\frac{V_{oc}}{nV_T}} \left(e^{\frac{V+IR_s}{nV_T}} - 1 \right) \approx I_L \left(1 - e^{\frac{V+IR_s-V_{oc}}{nV_T}} \right)$$

(3.33)

at the maximum power point

$$I_m \approx I_L \left(1 - e^{\frac{V_m+IR_s-V_{oc}}{nV_T}} \right)$$

(3.34)

Multiplying equation (3.34) by the voltage V and making the derivative of the product equal to zero, the maximum power point coordinates are related by

$$I_m + (I_m - I_L) \left(\frac{V_m - I_m R_s}{nV_T} \right) = 0$$

(3.35)

if V_{oc} and I_L are known, equations (3.35) and (3.34) give the values of I_m and V_m.

A simplified formulation of the same problem assumes that the maximum power delivered by the solar cell can be calculated from

$$P'_m = P_m - I_m^2 R_s$$

(3.36)

where P'_m is the maximum power when R_s is not zero. This equation implies that the maximum power point shifts approximately at the same current value I_m.

Multiplying and dividing the second term of the right-hand side in equation (3.36) by V_m, it follows that

$$P'_m = P_m\left(1 - \frac{I_m}{V_m}R_s\right) \tag{3.37}$$

If we further assume that

$$\frac{I_m}{V_m} \approx \frac{I_{sc}}{V_{oc}} \tag{3.38}$$

then equation (3.38) becomes

$$P'_m = P_m\left(1 - \frac{I_{sc}}{V_{oc}}R_s\right) = P_m(1 - r_s) \tag{3.39}$$

where

$$r_s = \frac{R_s}{V_{oc}/I_{sc}} \tag{3.40}$$

is the normalized value of the series resistance. The normalization factor is the ratio of the open-circuit voltage to the short-circuit current.

The result in equation (3.40) is easily translated to the fill factor:

$$FF = \frac{P'_m}{V_{oc}I_{sc}} = \frac{P_m(1 - r_s)}{V_{oc}I_{sc}} = FF_0(1 - r_s) \tag{3.41}$$

relating the value of the FF for a non-zero value of the series resistance FF with the value of the FF when the series resistance is zero, FF_0. This result is valid provided that both the value of the short-circuit current and of the open-circuit voltage are independent of the value of the series resistance.

Example 3.3

Considering the solar cell data and PSpice model given above in this section, simulate a solar cell with the following parameter values: area $= 126.6$, $J_0 = 10^{-11}$, $J_{02} = 0$, $J_{sc} = 0.0343$, $R_{sh} = 100$, for the values of the series resistance, 10^{-4}, 10^{-3}, 2×10^{-3}, 5×10^{-3}, 10^{-2}, 2×10^{-2}, 5×10^{-2}, 10^{-1} Ω. Calculate the values of the FF and compare with the results given by equation (3.41).

We first start by writing the file shown in Annex 3 with name 'example3_3'. After running PSpice the values of the maximum power point are calculated and given in Table 3.5. A plot

Table 3.5 Results of Example 3.3

Series resistance R_s	Maximum power P_m	FF	Normalized series resistance r_s	FF_o $(1 - r_s)$
0.0001	2.02	0.82	7.69×10^{-4}	0.819
0.001	2.004	0.814	7.69×10^{-3}	0.813
0.002	1.98	0.8	0.0153	0.807
0.005	1.93	0.78	0.038	0.788
0.01	1.85	0.75	0.076	0.757
0.02	1.689	0.686	0.15	0.697

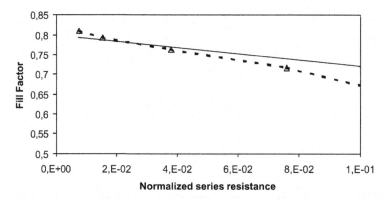

Figure 3.8 Values of the fill factor plotted against the normalized value of the series resistance in Example 3.3. Solid line is calculated from values from equation (3.41) and dashed line is PSpice values

of the FF value calculated after the PSpice simulation against the normalized value of the series resistance is shown in Figure 3.8 and compared with the value given by equation (3.41). As can be seen for small values of the series resistance, the results agree quite well, and spread out as the series resistance increases.

3.10 Effects of the Shunt Resistance

The shunt resistance also degrades the performance of the solar cell. To isolate this effect from the others, the series resistance and the second diode can be eliminated from equation (3.26) by setting for R_s a very small value $R_s = 1 \times 10^{-6}$ Ω and $J_{02} = 0$. The PSpice circuit can now be solved for several values of the shunt resistance as shown in Annex 3, file 'shunt.cir'.

The results are shown in Figure 3.9. It is clear that unless the parallel resistance takes very small values, the open circuit voltage is only very slightly modified. On the other hand at

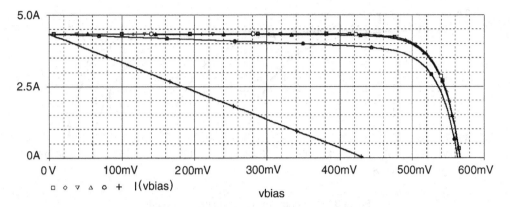

Figure 3.9   Effects of the shunt resistance on the $I(V)$ characteristics

short circuit conditions, all plots cross at the same point. This is not unusual because from equation (3.26), with $R_s = 0$ and $J_{02} = 0$, it results at $V = 0$:

$$I_{sc} = I_L \tag{3.42}$$

which is independent of R_{sh}. Additionally if we look at the $I(V)$ characteristic under reverse bias, $V \ll 0$

$$I = I_L - \frac{V}{R_{sh}} \tag{3.43}$$

indicating that the $I(V)$ characteristic should be a straight line with slope equal to $(-1/R_{sh})$. Small values of the shunt resistance also heavily degrade the fill factor.

3.11   Effects of the Recombination Diode

The open-circuit voltage is also degraded when the recombination diode becomes important. This can be seen in Figure 3.10 where in order to isolate the recombination diode effect, a high value of the parallel resistance and a low value of the series resistance have been selected.

Several values for the parameter J_{02} are plotted, namely 1×10^{-8} A/cm^2, 1×10^{-7} A/cm^2, 1×10^{-6} A/cm^2, 1×10^{-5} A/cm^2, 1×10^{-4} A/cm^2. The corresponding netlist is shown in Annex 3 called 'diode_rec.cir'.

The result indicates that when the recombination diode dominates, the characteristic is also heavily degraded, both in the open-circuit voltage and in the FF. The short-circuit current remains constant.

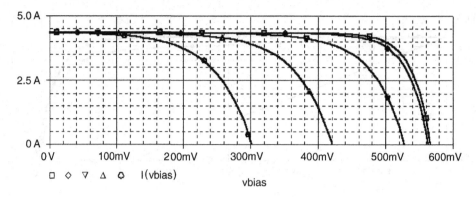

Figure 3.10   Effects of the recombination diode

3.12   Temperature Effects

Operating temperature has a strong effect on the electrical response of solar cells. Taking into account that in, terrestrial applications, solar cells can easily warm up to 60–65 °C and that in space or satellite applications temperatures can be even higher, it follows that a proper modelling of the temperature coefficients of the main electrical parameters is mandatory.

Temperature effects in a solar cell can be included in the PSpice model by using the built-in parameters of the diode model included in the equivalent circuit. Namely the saturation density current of a diode has a strong dependence on temperature and it is usually given by:

$$J_0 = BT^{XTI} e^{-\frac{E_g}{kT}} \tag{3.44}$$

where B is a constant independent of the temperature and XTI is a PSpice parameter also independent of the temperature. Equation (3.44) is valid for any arbitrary value of temperature. It can also be written for a nominal or reference temperature T_{nom}, as follows

$$J_0(T_{nom}) = BT_{nom}^{XTI} e^{-\frac{E_g(T_{nom})}{kT_{nom}}} \tag{3.45}$$

Dividing equation (3.44) by equation (3.45):

$$\ln\left[\frac{J_0}{J_0(T_{nom})}\right] = XTI \ln\left(\frac{T}{T_{nom}}\right) + \frac{E_g(T_{nom})}{kT_{nom}} - \frac{E_g}{kT} \tag{3.46}$$

The value for the energy bandgap of the semiconductor E_g at any temperature T is given by:

$$E_g = E_{g0} - \frac{GAP1 \cdot T^2}{GAP2 + T} \tag{3.47}$$

where E_{g0} is the value of the band gap extrapolated to $T=0\,\mathrm{K}$ for the semiconductor considered.

Example 3.4

Plot the $I(V)$ characteristics of a silicon solar cell 5″ diameter, $J_{sc} = 0.343$ A/cm², $J_0 = 1 \times 10^{-11}$ assumed ideal for various temperatures, namely 27 °C, 35 °C, 40 °C, 45 °C, 50 °C, 55 °C and 60 °C. This analysis assumes that the short-circuit current is independent of temperature.

The netlist now has to be modified to include temperature analysis. This is highlighted in the netlist that follows:

```
*temp.cir
.include cell_3.lib
xcell3 0 31 32 cell_3 params:area=126.6 j0=1e-11 j02=0
+ jsc=0.0343 rs=1e-6 rsh=1000
vbias 31 0 dc 0
virrad 32 0 dc 1000
.plot dc i(vbias)
.dc vbias 0 0.6 0.01
.temp 27 35 40 45 50 55 60

.probe
.end
```

We have simplified the solar cell model setting $J_{02} = 0$, the series resistance has been set to a very small value and the shunt resistance to a large value so as not to obscure the temperature effects, which we highlight in this section. As can also be seen the subcircuit model has been changed by a 'cell_3.lib' file shown in Annex 3 to include, in the diode model, the silicon value for the parameter $E_{g0} = 1.17$ eV for silicon.

The results of the simulation are shown in Figure 3.11.

Using the cursor on the probe plots, the values of the open-circuit voltage for the various temperatures can be easily obtained. They are shown in Table 3.6.

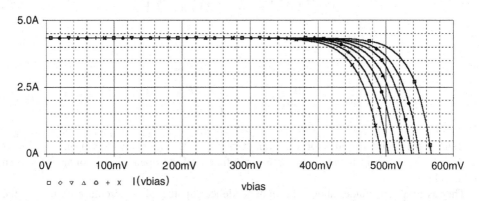

Figure 3.11 Effects of temperature

Table 3.6 Values of the open-circuit voltage and temperature coefficient for several temperatures

Temperature (°C)	27	35	40	45	50	55	60
V_{oc}(mV)	567.5	550	538.4	526.9	515.3	503.8	492.9
$\Delta V_{oc}/\Delta T$ (mV/°C)		−2.18	−2.3	−2.3	−2.31	−2.31	−2.18

As can be seen the temperature coefficient of the open-circuit voltage, calculated in the second row in Table 3.6 is close to −2.3 mV/°C for a wide range of operating temperatures. This result is consistent with a theoretical deduction that can be made from the temperature derivative in equation (3.8) of the open-circuit voltage:

$$\frac{dV_{oc}}{dT} = \frac{V_T}{T}\ln\frac{J_{sc}}{J_0} + V_T \frac{d}{dT}\left(\ln\frac{J_{sc}}{J_0}\right) \tag{3.48}$$

Assuming that the short-circuit current is independent of the temperature, at least compared with the other contributions in equation (3.48), it follows:

$$\frac{dV_{oc}}{dT} = \frac{V_T}{T}\ln\frac{J_{sc}}{J_0} - V_T \frac{d}{dT}(\ln J_0) \tag{3.49}$$

Taking into account equation (3.44)

$$J_0 = CT^\gamma e^{-\frac{E_{g0}}{kT}} \tag{3.50}$$

where C and γ are constants independent of the temperature. Replacing equation (3.50) in equation (3.49) and taking into account

$$V_{oc} = V_T \ln\left(1 + \frac{J_{sc}}{J_0}\right) \approx V_T \ln\left(\frac{J_{sc}}{J_0}\right) \tag{3.51}$$

it follows,

$$\frac{dV_{oc}}{dT} = \frac{V_{oc}}{T} - \frac{\gamma V_T}{T} - \frac{E_{g0}}{qT} \tag{3.52}$$

Using typical silicon values for the parameters in equation (3.52), values around −2.3 mV/°C for $\gamma = 3$ are found for the temperature coefficient of the open-circuit voltage of silicon solar cells.

The assumption made about the independence of the short-circuit current on the temperature, although acceptable in many cases, can be removed by taking into account a temperature coefficient of the short-circuit current. Typically for silicon solar cells a value of 6.4×10^{-6} A/cm$^{2\circ}$C is frequently used.

A modification to the subcircuit of the solar cell to account for this temperature coefficient of the short-circuit current can be made and it is shown in Example 3.5.

Example 3.5

(a) Write a PSpice subcircuit with two added parameters: temperature and short-circuit current temperature coefficient.

Solution

```
*cell_4.lib
.subckt cell_4 300 303 302 params:area=1, j0=1, jsc=1, j02=1, rs=1,
+rsh=1,temp=1,isc_coef=1

girrad 300 301 value={(jsc/1000)*v(302)*area+isc_coef*(area)*(temp-25)}
d1 301 300 diode
.model diode d(is={j0*area},eg=1.17)
d2 301 300 diode2
.model diode2 d(is={j02*area}, n=2)
rs 301 303 {rs}
rsh 301 300 {rsh}
.ends cell_4
```

(b) Calculate the response of a solar cell to a temperature of 80 °C, a temperature coefficient of the short-circuit current of 6.4×10^{-6} A/cm^2°C and the same other parameters as in Example 3.4.

Solution

A '*.cir*' file has to be written:

```
*temp_2.cir
.include cell_4.lib
xcell3 0 31 32 cell_4 params:area=126.6 j0=1e-11 j02=0
+ jsc=0.0343 rs=1e-6 rsh=1000 temp=80 isc_coef=6.4e-6
vbias 31 0 dc 0
virrad 32 0 dc 1000
.plot dc i(vbias)
.dc vbias 0 0.6 0.01
.temp 80

.probe
.end
```

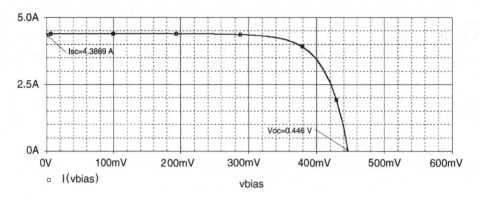

Figure 3.12   Solution to Example 3.5

The results for a temperature of 80 °C can be seen in Figure 3.12. A short circuit current of 4.38 A and an open-circuit voltage of 446 mV result from the PSpice simulation, according to the temperature coefficients used.

3.13   Effects of Space Radiation

The main effect of space radiation concerns the minority carrier lifetime degradation in the semiconductor bulk leading to an increase of the dark current density, and also a degradation of the photocurrent generated. Altogether, these effects produce a significant reduction of the maximum power of the solar cell, which is described analytically by empirical equations relating the values of the short-circuit current, open-circuit voltage and the maximum power to the fluence of a given high energy particle Φ. The units of the particle fluence are particles per unit surface and this magnitude has to be clearly differentiated from the photon flux ϕ used in Chapter 2,

$$J_{sc}(\Phi) = J_{sc}(BOL) - K_J \log\left(1 + \frac{\Phi}{\Phi_J}\right) \tag{3.53}$$

$$V_{oc}(\Phi) = V_{oc}(BOL) - K_V \log\left(1 + \frac{\Phi}{\Phi_V}\right) \tag{3.54}$$

$$P_{max}(\Phi) = P_{max}(BOL) - K_p \log\left(1 + \frac{\Phi}{\Phi_p}\right) \tag{3.55}$$

The values of the constants appearing in equations (3.53), (3.54) and (3.55) can be derived from the available data of solar cell degradation in space, which are known for a large variety of solar cells and particle type and fluences [3.3]. As an example, Table 3.7 shows the values of the constants for a silicon solar cell and a GaAs/Ge solar cell calculated from the

Table 3.7 Values of the space degradation constants for J_{sc}, V_{oc} and P_{max}

| | Silicon solar cell $2\,\Omega$ cm BSFR | | | GaAs/Ge solar cell | | |
	Short circuit (mA/cm^2)	Open circuit (mV)	Max. power (mW/cm^2)	Short circuit (mA/cm^2)	Open circuit (V)	Max. power (mW/cm^2)
K_i	5.26	42	2.91	10.9	93.5	8.79
Φ_i(cm^{-2})	3.02×10^{13}	2.99×10^{12}	5.29×10^{12}	2.51×10^{14}	1×10^{14}	1.55×10^{14}

degradation graphs in reference [3.3]. K_i and Φ_i stand for K_I for the short-circuit current, K_v for the open-circuit voltage and K_p for the maximum power.

PSpice also allows the simulation of the $I(V)$ characteristics of a solar cell after a given radiation fluence. The inputs are the known values of J_{sc}, V_{oc} and P_{max} at the beginning of life (BOL) and the output is the full $I(V)$ characteristic resulting after a given space radiation of fluence Φ. The procedure to carry out this exercise consists of several steps:

Step n^r 1: Calculate the value of J_0 in BOL conditions.

This can be accomplished recalling that in open circuit in the simplest solar cell model,

$$J = 0 = J_{sc} - J_0\left(e^{\frac{V_{oc}}{V_T}} - 1\right)$$ (3.56)

and then, aproximately, at BOL conditions

$$J_0(BOL) = J_{sc}(BOL)e^{-\frac{V_{oc}(BOL)}{V_T}}$$ (3.57)

Step n^r 2: Calculate the ideal value of the FF in BOL.

This means the FF that the solar cell with zero series resistance would have. This is given by equation (3.17) reproduced here,

$$FF_0 = \frac{v_{oc} - \ln(v_{oc} + 0.72)}{1 + v_{oc}}$$ (3.58)

where

$$v_{oc} = \frac{V_{oc}}{V_T}$$ (3.59)

is the normalized value of the open-circuit voltage.

Step n^r 3: Calculate the series resistance in BOL conditions.

This is easily accomplished by

$$FF(BOL) = FF_o(BOL)(1 - r_s) \qquad (3.60)$$

after denormalizing r_s,

$$R_s = \frac{V_{oc}}{J_{sc} \times Area} - \frac{P_{max}}{FF_o(BOL) \times J_{sc}^2 \times Area} \qquad (3.61)$$

Step n^r 4: Write the solar cell model taking into account the degradation characteristics given by the constants in Table 3.6.

$$J_0(\Phi) = J_{sc}(\Phi)e^{-\frac{V_{oc}(BOL)-K_v \log\left(1+\frac{\Phi}{\Phi_v}\right)}{V_T}} \qquad (3.62)$$

The short-circuit current can also be calculated using equation (3.53). It is assumed that the series resistance does not change after irradiation.

Example 3.6

Consider an $8\,\text{cm}^2$ silicon solar cell having been irradiated in space by a set of radiation fluences of 1×10^{10}, 1×10^{11}, 1×10^{12}, 1×10^{13}, 1×10^{14}, 1×10^{15}, 1×10^{16} 1MeV electrons/cm^2. If the BOL data are the following: $V_{oc} = 0.608$ V, $J_{sc} = 0.0436$ A/cm^2 and $P_{max} = 20.8$ mW/cm^2 simulate the PSpice $I(V)$ characteristics of the cell. Consider that the solar cell can be modelled by a single diode and has a shunt resistance of infinite value.

Before solving the problem with PSpice, we will illustrate here the steps previously described, in order to calculate some important magnitudes considering the values of the degradation constants in Table 3.7.

Step n^r 1

From equation (3.57),

$$J_0(BOL) = J_{sc}(BOL)e^{-\frac{V_{oc}(BOL)}{V_T}} = 43.6 \times 10^3 e^{-\frac{0.608}{0.026}} = 17 \times 10^{-12}\,A$$

Step n^r 2

$$v_{oc} = \frac{V_{oc}}{V_T} = \frac{0.608}{0.026} = 23.38$$

$$FF_0 = \frac{v_{oc} - \ln(v_{oc} + 0.72)}{1 + v_{oc}} = 0.828$$

Step n⁰ 3

$$R_s = \frac{V_{oc}}{J_{sc} \times Area} - \frac{P_{max}}{FF_o(BOL) \times J_{sc}^2 \times Area} = \frac{0.608}{0.0436 \times 8} - \frac{0.02}{0.828 \times (0.0436)^2 \times 8} = 0.115\,\Omega$$

Step n⁰ 4

For every fluence value, J_0 is calculated

$$J_0(\phi) = J_{sc}(\phi)e^{\frac{V_{oc}(BOL) - K_v \, \log\left(\frac{1+\phi}{\phi_v}\right)}{V_T}}$$

A PSpice subcircuit 'cell_5.lib' shown in Annex 3 is written, where it can be seen that equation (3.62) has been used to define the value of the 'is' parameter of the diode model. Moreover, the series resistance is internally calculated from equation (3.61).

The .cir file to calculate the full $I(V)$ characteristics is also shown in Annex 3 and is called 'space.cir'.

The correspondence between the names in the PSpice code and the names in the equations is the following:

Parameter in PSpice code	Parameter in equations (3.45) to (3.51)
vocbol	$V_{oc}(BOL)$
jscbol	$J_{sc}(BOL)$
fi	Φ_i
f	Φ
fv	Φ_v
kv	K_v
ki	K_i
jo	J
vocnorm	v_{oc}
uvet	V_T

The resulting $I(V)$ characteristics are shown in Figure 3.13.

The maximum power delivered by the solar cell can now be computed as 79.665 mW, that is 9.95 mW/cm^2. This value is approximately equal to the expected value of the maximum power after degradation as can be calculated using the degradation constants for the maximum power given in Table 3.7.

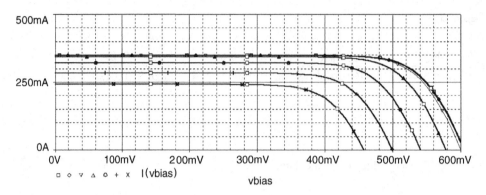

Figure 3.13 Degradation of the characteristics of the silicon solar cell of Example 3.6 for space radiation fluences of 1×10^{10}, 1×10^{11}, 1×10^{12}, 1×10^{13}, 1×10^{14}, 1×10^{15}, 1×10^{16} electrons/cm^2

3.14 Behavioural Solar Cell Model

For the purpose of modelling photovoltaic systems, what is usually required are simulations of the solar cell behaviour for changing temperature and irradiance conditions. This means that the built-in temperature analysis of PSpice is difficult to use because PSpice runs a new analysis for every value of the temperature. Usually what is available as a result is a file or a time series of irradiance and temperature values. Moreover, if solar cells are not made of one of the well-known semiconductors such as silicon, GaAs or Ge, the important physical data required to model the temperature effects of a solar cell are not easily available. Instead, most frequently the data available come from data sheets or published material by the manufacturers and concern values of electrical magnitudes from photovoltaic measurements, such as the short-circuit current, open-circuit voltage and maximum power under some standard conditions, and the temperature coefficients. If this is the case, the model of a solar cell including a diode is not very practical and is of limited use. For this reason we introduce a behavioural model, based on voltage and current sources, which is able to correctly model the behaviour of an arbitrary solar cell under arbitrary conditions of irradiance and temperature, with electrical data values as the only input.

The model assumes that the solar cell can be modelled by two current sources and a series resistance. The model is composed of a subcircuit between nodes 10 14 12 13 as shown in Figure 3.14, where two g-devices are assigned two functions as follows:

$$i(girrad) = \frac{J_{scr}A}{1000} G + \left(\frac{\mathrm{d}J_{sc}}{\mathrm{d}T}\right)(T_{cell} - T_r) \tag{3.63}$$

which returns the value of the short-circuit current at irradiance G. The derivative in equation (3.63) is the temperature coefficient of the short-circuit current and is considered constant in the range of temperatures of interest. T_r is the reference temperature which is usually considered 25 °C (in some cases 300 °K are considered). This g-device is written as:

girrad 10 11 value={(jscr/1000*v(12)*area) +coef_jsc*area*(v(17)−25)}

The second source returns the exponential term of a solar cell replacing the diode

$$i(gidiode) = \frac{Jsc(area)}{\left(e^{\frac{V_{oc}}{V_T}} - 1\right)} \left(e^{\frac{V}{V_T}} - 1\right) \tag{3.64}$$

where the value of J_{sc} is copied from the voltage at node 305 and the open-circuit voltage is copied from node 306.

The value of the temperature involved in equation (3.64) has to be the cell operating temperature T_{cell}, which is usually derived from the NOCT concept. NOCT stands for nominal operating conditions temperature and is the temparature of the cell at 800 W/m² irradiance and 20 °C of ambient temperature.

$$T_{cell} - T_a = \frac{NOCT - 20}{800} G \tag{3.65}$$

This is written in PSpice as can be seen in Figure 3.14.

The open-circuit voltage can also be written for the new temperature and irradiance values.

The temperature coefficient of the open-circuit voltage is computed for the same value of the irradiance at two temperatures,

$$\left(\frac{\partial V_{oc}}{\partial T}\right)_G (T_{cell} - T_r) = V_T \ln\left(1 + \frac{I_{scr}}{I_0}\right) - V_{ocr} \tag{3.66}$$

Adding and subtracting the value of the V_{oc} under arbitrary radiance and temperature conditions

$$\left(\frac{\partial V_{oc}}{\partial T}\right)_G (T_{cell} - T_r) = V_T \ln\left(1 + \frac{I_{scr}}{I_0}\right) - V_{ocr} + V_{oc} - V_{oc} \tag{3.67}$$

Substituting

$$V_{oc} = V_T \ln\left(1 + \frac{I_{sc}}{I_0}\right) \tag{3.68}$$

it follows,

$$V_{oc} = V_{ocr} + \left(\frac{\partial V_{oc}}{\partial T}\right)_G (T_{cell} - T_r) + V_T \ln\frac{1 + \frac{I_{sc}}{I_0}}{1 + \frac{I_{scr}}{I_0}} \tag{3.69}$$

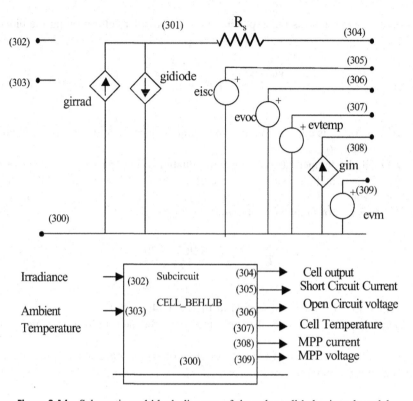

Figure 3.14 Schematic and block diagram of the solar cell behavioural model

In the general case, except when the short-circuit current is zero, unity can be neglected in the numerator and denominator of the last term in equation (3.69) and then

$$V_{oc} \approx V_{ocr} + \left(\frac{\partial V_{oc}}{\partial T}\right)_G (T_{cell} - T_r) + V_T \ln \frac{I_{sc}}{I_{scr}} \tag{3.70}$$

where V_T is calculated at the cell temperature T_{cell}.
The series resistance is calculated,

$$R_s = \frac{V_{oc}}{I_{sc}} - \frac{P_{max}}{FF_0 I_{sc}^2} \tag{3.71}$$

where FF_o is given by equation (3.17), and P_{max} is the maximum power per unit area. The PSpice code is written as,

```
rs n+ n- {vocr/(jscr*area)-pmaxr/(jscr**2*area*(vocr/0.026-log((vocr/0.026)
  +0.72))/(1+vocr/0.026))}
```

where the value is calculated at the reference temperature conditions and assumed independent of the cell temperature.

In PV systems the information about the evolution of the coordinates of the maximum power point during a given period of time is important because the electronic equipment used to connect the photovoltaic devices to loads are designed to follow this maximum power point. The PSpice code can also provide this information in two accessible nodes of the cell subcircuit. All we need are the coordinates of the maximum power point at standard conditions V_{mr} and I_{mr}. This is available from most of the data sheets provided by the manufacturers. In that case the current of the maximum power point at arbitrary conditions of irradiance and temperature can be considered to scale proportionally with the irradiance and linearly with the temperature with the temperature coefficient of the short-circuit current:

$$I_m = I_{mr} \frac{G}{G_r} + (area)Coef\_jsc(T_{cell} - T_r) \tag{3.72}$$

and from this value the voltage of the maximum power point can be calculated taking into account that

$$I_m = I_{sc} - I_o \left(e^{\frac{V_m + I_m R_s}{V_T}} - 1 \right) \tag{3.73}$$

and then

$$V_m = V_T \ln \left(1 + \frac{I_{sc} - I_m}{I_{sc}} \left(e^{\frac{V_{oc}}{V_T}} - 1 \right) \right) - I_m R_s \tag{3.74}$$

The equations described above can be implemented in a PSpice library file 'cell_beh-lib' listed in Annex 3 and schematically shown in Figure 3.14.

Example 3.7

Consider a CIGS (copper indium gallium diselenide) solar cell with the following electrical characteristics: $V_{ocr} = 0.669$, $J_{scr} = 35.7$ mA/cm^2, $P_{maxr} = 18.39$ mW/cm^2. These values have been taken from M.A. Green, K. Emery, D.L. King, S. Igari, W. Warta 'Solar cell efficiency tables (version 18)' *Progress in Photovoltaics*, pp. 287–193, July-August 2001.

The temperature coefficients are for the short-circuit current +12.5 µA/cm^2 °C and for the open-circuit voltage –3.1 mV/°C. These values have been estimated for a 1 cm^2 cell from www.siemenssolar.com/st5.html.

The PSpice model 'cell_beh.lib 'is used in the file 'cell_beh.cir' also listed in Annex 3 to plot the $I(V)$ characteristics. The result is shown in Figure 3.15, where it can be seen that the model is able to simulate the $I(V)$ characteristic.

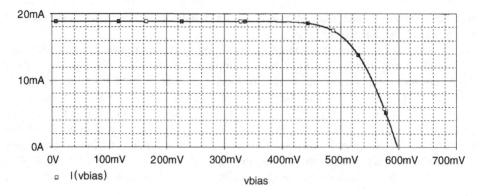

Figure 3.15   CIGS solar cell $I(V)$ characteristic simulated using the behavioural solar cell model

3.15   Use of the Behavioural Model and PWL Sources to Simulate the Response to a Time Series of Irradiance and Temperature

The behavioural model described in the previous section is aimed at simulating the response of a given solar cell to a certain profile of combined irradiance–ambient temperature time series. Although in the next chapters we will be extensively using time series of these variables it is convenient here to address two important points that help understanding of the PSpice simulation of PV systems, and not only solar cells: (a) time units and (b) variable units.

3.15.1   Time units

The operation of PV systems is meant to last over long periods of time, and the design and sizing procedures will be much assisted if long periods of time could be simulated considering arbitrary irradiance and temperature profiles. Apparently, such simulations require long CPU time, are unavailate or cumbersome. An easy way to overcome these problems is to consider two different units of time: one unit of time will be the *internal PSpice unit of time* and the second the *real PV system operating time*. This means that if we assign 1 microsecond of internal PSpice time to 1 hour of real time, a one-day simulation will be performed in a few seconds on any standard PC computer. This approach has been used in this book when necessary. In order to facilitate the understanding a warning has been placed in all figures with differences between PSpice internal time units and real time units.

3.15.2   Variable units

The use of input and output variables in the PSpice files used to simulate PV systems is required in order, for instance, to enter the values of irradiance and temperature, or to extract form the simulation results the values of variables such as the state of charge of a battery or the angular speed of a motor. PSpice is, however, a tool that only handles electrical magnitudes. This means that the only internal variable units in PSpice are restricted to electrical units. In PV systems it is obvious that non-electrical magnitudes have to be handled and this then forces the use of electrical equivalents for non-electrical magnitudes. For example, if we want to enter temperature data we have to convert the temperature to a voltage source, for example, in such a way that 1 V of the voltage source corresponds to one degree Celsius of temperature. So the internal PSpice unit is an electrical unit (volt) and the real unit is a thermal unit (°C) therefore making a domain conversion. In order to avoid confusions, warnings are issued through the book when necessary. In fact we have used such domain conversion in Chapters 1 and 2 because we have handled spectral irradiance and wavelength, but we want to stress this domain conversion issue again here due to the importance it will take from now on.

As an example of the points raised above, let us consider an example assuming we know 10 irradiance and temperature couples of values for one day and we want to know the time evolution of the main PV magnitudes of a solar cell. The values of irradiance and temperature are entered in the PSpice code as a piecewise linear voltage source where the internal unit of time is considered as one microsecond corresponding to one hour of real time.

Example 3.8

Assume the following values of temperature and irradiance

Time	Irradiance (W/m^2)	Temperature (°C)
6	0	12
8	100	12
10	200	15
11	400	18
12	450	20
13	800	22
15	750	21
17	300	17
19	200	15
20	0	13

(a) Write the PWL sources and the PSpice file to compute the time evolution of the characteristics.

The subcircuit is changed modifying the v-sources for the irradiance and temperature as follows:

```
virrad node+ node- pwl 6u 0 8u 100 10u 200 11u 400 12u 450 13u 800 15u 750 17u 300 19u 200 20u 0
    vtemp node+ node- pwl 6u 12 8u 12 10u 15 11u 18 12u 20 13u 22 15u 21 17u 17 19u 15 20u 13
```

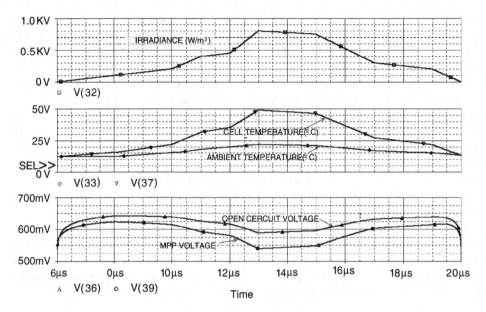

Figure 3.16 Behavioural model and PWL sources for time series simulations. Warning: The internal unit of time is the microsecond and the real time unit is the hour. The internal units of the irradiance are volts and the real units are W/m², the internal units of temperature are volts and the real units are °C

(b) Write a PSpice .cir file calling a behavioural solar cell model and performing a transient analysis covering the 10 hours of the example.

```
*cell_pwl.cir

.include cell_beh.lib
xcellbeh 0 32 33 34 35 36 37 38 39 cell_beh params:area=1, tr=25, jscr=0.0375,
+ pmaxr=0.0184, vocr=0.669, jmr = 35.52e-3, vmr=0.518, noct=47,
+ coef_jsc=12.5e-6, coef_voc=-3.1e-3

virrad 32 0 pwl 6u 1 8u 100 10u 200 11u 400 12u 450 13u 800 15u 750 17u 300 19u 200 20u 0
vtemp 33 0 pwl 6u 12 8u 12 10u 15 11u 18 12u 20 13u 22 15u 21 17u 17 19u 15 20u 13
rim 38 0 1
vbias 34 0 dc 0
.tran 0 20u 6u 0.01u
.probe
.end
```

The short circuit is included by means of an auxiliary voltage source vaux = 0 in order to be able to measure the short-circuit current. The result is shown in Figure 3.16, where it can be seen that the time evolution of irradiance and temperature can be correctly handled.

3.16 Problems

3.1   Using the single diode model of the solar cell with $r_s = 0$ and $R_{sh} = $ infinity, draw the ideal $I(V)$ characteristic of a solar cell with $J_{sc} = 0.035$ for $I_0 = 1 \times 10^{-11}$, $I_0 = 1 \times 10^{-10}$ and $I_0 = 1 \times 10^{-9}$ at irradiance $1000 \, \text{W/m}^2$. Find the values of the open-circuit voltage and short-circuit current.

3.2   One of the methods used to measure the series resistance of a solar cell is based on the comparison between the dark $I(V)$ curve and the couple of values I_{sc}–V_{oc} at several irradiance values. Write the PSpice code of a solar cell working at $27 \,^{\circ}\text{C}$ with the following parameter values: $J_{sc} = 0.0343$, $j_{01} = 1 \times 10^{-11}$, $j_{02} = 0$, $r_{sh} = 1 \times 10^6 \, \Omega$, $area = 126.6 \, \text{cm}^2$ and $r_s = 0.05 \, \Omega$. Plot the $I(V)$ characteristics for the following irradiance values: 200, 400, 600, 800, 1000, 1200 W/m^2. List the values for I_{sc} and V_{oc} for all irradiance values. Write the PSpice code of the same solar cell at the same temperature but in darkness. Plot the $I(V)$ characteristic in a semilog plot(log I vs. V). (Notice that the curve is not linear at high currents.) List the values of V in this plot for approximately the same I_{sc} values of the previous list. Verify that $(V - V_{oc})/I_{sc}$ gives the value of the series resistance.

3.3   Another method to measure the series resistance is based on two measures of the $I(V)$ characteristics of the solar cell at two irradiance values and at the same temperature. The purpose of this problem is to demonstrate this method.
      First write the PSpice code of a solar cell with the parameters $j_{sc} = 0.0343$, $j_{01} = 1 \times 10^{-11}$, $j_{02} = 0$, $r_{sh} = 1 \times 10^6 \, \Omega$, $area = 126.6 \, \text{cm}^2$ and $r_s = 0.05 \, \Omega$. Plot the $I(V)$ characteristic for two irradiance values: 600 and 1000 W/m^2. Using the cursor find a point on the curve drawn for 1000 W/m^2 where $I = I_{sc_1} - 1$ A and find the value of the voltage at this point. Call this value V_1.
      Do the same with the 600 W/m^2 plot and find the value of the voltage V_2 at the point where $I = I_{sc_2} - 1$ A. Verify that $(V_2 - V_1)/(I_{sc_1} - I_{sc_2})$ gives the value of R_s.

3.17 References

[3.1] Van Overstraeten, R.J. and Mertens, R.P. *Physics Technology and Use of Photovoltaics*, Adam Hilgher, 1986.
[3.2] Green, M.A., *Solar Cells*, Bridge Printery, Rosebery, NSW, Australia, 1992.
[3.3] Anspaugh, B.E., *Solar Cell Radiation Handbook, Addendum 1*, NASA Jet Propulsion Laboratory publication JPL 82-69, 15 February, Pasadena, California, 1989, Figures 1, 2, 3, 21, 22 and 23.

4

Solar Cell Arrays, PV Modules and PV Generators

Summary

This chapter describes the association of solar cells to form arrays, PV standard modules and PV generators. The properties of series and parallel connections of solar cells are first described and the role played by bypass diodes is illustrated. Conversion of a PV module standard characteristics to arbitrary conditions of irradiance and temperature is described and more general or behavioural PSpice models are used for modules and generators extending the solar cell models described in Chapter 3.

4.1 Introduction

Single solar cells have a limited potential to provide power at high voltage levels because, as has been shown in Chapter 2, the open circuit voltage is independent of the solar cell area and is limited by the semiconductor properties. In most photovoltaic applications, voltages greater than some tens of volts are required and, even for conventional electronics, a minimum of around one volt is common nowadays. It is then mandatory to connect solar cells in series in order to scale-up the voltage produced by a PV generator. This series connection has some peculiar properties that will be described in this chapter.

PV applications range from a few watts in portable applications to megawatts in PV plants, so it is not only required to scale-up the voltage but also the current, because the maximum solar cell area is also limited due to manufacturing and assembly procedures. This means that parallel connection of PV cells and modules is the most commonly used approach to tailor the output current of a given PV installation, taking into account all the system components and losses.

4.2 Series Connection of Solar Cells

The schematic shown in Figure 4.1 is the circuit corresponding to the series association of two solar cells. A number of cases can be distinguished, depending on the irradiance levels or internal parameter values of the different cells.

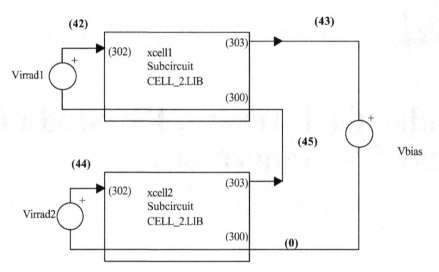

Figure 4.1   Association of two solar cells in series

4.2.1 Association of identical solar cells

The corresponding netlist for an association of two solar cells is given by:

```
*SERIES.CIR
.include cell_2.lib
xcell1 45 43 42 cell_2 params:area=126.6 j0=1e-11 j02=1E-9
+ jsc=0.0343 rs=1e-3 rsh=100000
xcell2 0 45 44 cell_2 params:area=126.6 j0=1e-11 j02=1E-9
+ jsc=0.0343 rs=1e-3 rsh=100000

vbias 43 0 dc 0
virrad1 42 45 dc 1000
virrad2 44 0 dc 1000

.dc vbias 0 1.2 0.01
.probe
.plot dc i(vbias)
.end
```

As can be seen the two solar cells have the same value as the short circuit current due to the same irradiance value and have equal values of the series and shunt resistances. It is then expected that the total $I(V)$ characteristic has the same value as the short circuit current of any of the two solar cells and that the total voltage is twice the voltage drop in one single solar cell. The output of the array is between nodes (43) and (0) and node (45) is the node common to the two cells. This is shown in Figure 4.2.

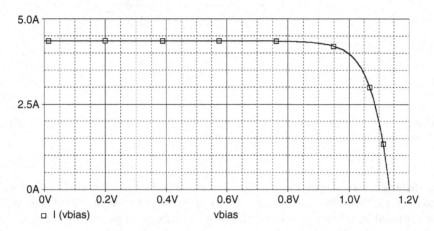

Figure 4.2   Plot of the $I(V)$ curve of the association of two identical solar cells in series

4.2.2   Association of identical solar cells with different irradiance levels: hot spot problem

Imagine that two solar cells are connected in series but the irradiance they receive is not the same. This a common situation due, for example, to the presence of dirt in one of the solar cells. The previous netlist has been modified to consider cell number 2 illuminated with an irradiance of 700 W/m^2 whereas cell number 1 receives an irradiance of 1000 W/m^2. This is achieved by modifying the value of the voltage source representing the value of the irradiance in cell number 2.

```
virrad2 44 0 dc 700
```

The result is shown in the upper plot in Figure 4.3. As can be seen the association of the two solar cells, as could be expected by the series association, generates a short circuit current equal to the short circuit current generated by the less illuminated solar cell (namely $0.0343 \times (700/1000) \times 126.6 = 3.03$ A). What also happens is that the voltage drop in the two cells is split unevenly for operating points at voltages smaller than the open circuit voltage. This is clearly seen in the bottom graph in Figure 4.3 where the voltage drop of the two cells is plotted separately. As can be seen, for instance, at short circuit, the voltage drop in cell number 1 is 533 mV (as measured using the cursor) whereas the drop in cell number 2 is -533 mV ensuring that the total voltage across the association is, of course, zero. This

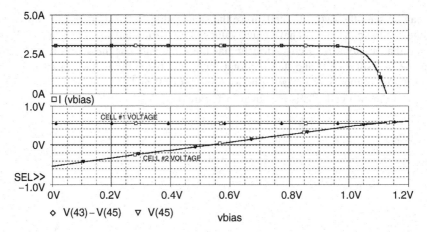

Figure 4.3 $I(V)$ curve of a two-series solar cell array with cell number one having an irradiance of 1000 W/m^2 and cell number 2 an irradiance of 700 W/m^2

means that under certain operating conditions one of the solar cells, the less illuminated, may be under reverse bias. This has relevant consequences, as can be seen by plotting not only the power delivered by the two solar cell series string, but also the power delivered individually by each solar cell as shown in Figure 4.4.

It can be seen that the power delivered by solar cell number 2, which is the less illuminated, may be negative if the total association works at an operating point below some 0.5 V. This indicates that some of the power produced by solar cell number 1 is dissipated by solar cell number 2 thereby reducing the available output power and increasing the temperature locally at cell number 2.

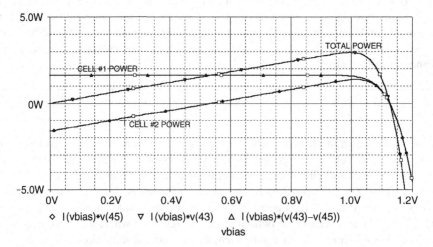

Figure 4.4 Power delivered (positive) or consumed (negative) on the two solar cells unevenly illuminated of Figure 4.3 and total power, as a function of the total voltage across the series association

This effect is called the 'hot spot' problem, which may be important in PV modules where only one of the series string of solar cells is less illuminated and which then has to dissipate some of the power generated by the rest of the cells. The analysis of this problem is extended to a PV module later in this chapter.

4.2.3   Bypass diode in series strings of solar cells

As shown above, the current available in a series connection of solar cells is limited by the current of the solar cell which is less illuminated. The extension of this behaviour for a situation in which one of the solar cells is completely in the dark, or has a catastrophic failure, converts this solar cell to an open circuit, and hence all the series string will be in open circuit.

This can be avoided by the use of bypass diodes which can be placed across every solar cell or across part of the series string. This is illustrated in the following example.

Example 4.1

Assume a series string of 12 identical solar cells. Assume that cell number six is completely shadowed. To avoid the complete loss of power generation by this string, a diode is connected across the faulty device in reverse direction as shown in Figure 4.5.

Plot the final $I(V)$ characteristic and the voltage drop at the bypass diode and the power dissipated by the bypass diode.

The corresponding netlist is shown in the file 'bypass.cir' listed in Annex 4.

The bypass diode is connected between the nodes (53) and (55) of the string. The results are shown in Figure 4.6 where a comparison is made between the $I(V)$ characterisitic of the solar cell array with all cells illuminated at 1000 W/m^2 and the same array with cell number 6 totally shadowed and bypassed by a diode. It can be seen that the total maximum voltage is 5.5 V instead of approximately 7 V.

Figure 4.7 shows the bypass diode voltage, ranging from 0.871 V to -0.532 V in the range explored, along with the power dissipation, which takes a value of 3.74 W in most of the range of voltages. These two curves allow a proper sizing of the diode.

It may be concluded that a bypass diode will save the operation of the array when a cell is in darkness, at the price of a reduced voltage.

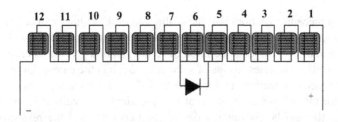

Figure 4.5   Series array of 12 solar cells with a bypass diode connected across cell number 6

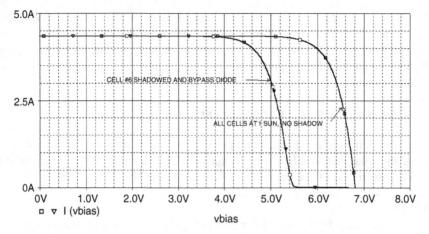

Figure 4.6   Effect of a bypass diode across a shadowed solar cell in a series array

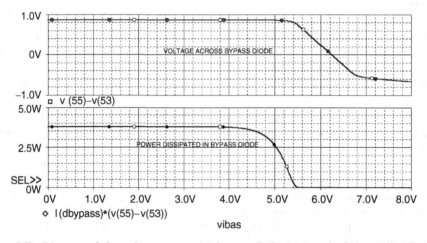

Figure 4.7   Diagram of the voltage across the bypass diode (top) and of the power dissipated (bottom)

4.3   Shunt Connection of Solar Cells

We have seen in preceding sections that the scaling-up of the voltage can be performed in PV arrays by connecting solar cells in series. Scaling of current can be achieved by scaling-up the solar cell area, or by parallel association of solar cells of a given area or, more generally, by parallel association of series strings of solar cells. Such is the case in large arrays of solar cells for outer-space applications or for terrestrial PV modules and plants.

The netlist for the parallel association of two identical solar cells is shown below where nodes (43) and (0) are the common nodes to the two cells and the respective irradiance values are set at nodes (42) and (44).

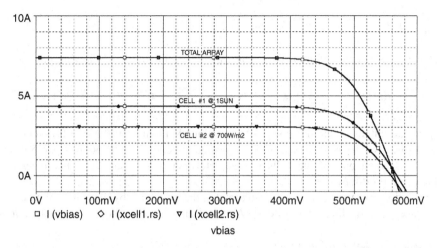

Figure 4.8 Association of two solar cells in parallel with different irradiance

```
*shunt.cir
.include cell_2.lib
xcell1 0 43 42 cell_2 params: area = 126.6 j0 = 1e - 11 j02 = 1E - 9
+ jsc = 0.0343 rs = 1e - 2 rsh = 1000
xcell2 0 43 44 cell_2 params: area = 126.6 j0 = 1e - 11 j02 = 1E - 9
+ jsc = 0.0343 rs = 1e - 2 rsh = 1000

vbias 43 0 dc 0
virrad1 42 0 dc 1000
virrad2 44 0 dc 700
.plot dc i(vbias)

.dc vbias 0 0.6 0.01
.probe
.end
```

Figure 4.8 shows the $I(V)$ characteristics of the two solar cells which are not subject to the same value of the irradiance, namely 1000 W/m^2 and 700 W/m^2 respectively. As can be seen the short circuit current is the addition of the two short circuit currents.

4.3.1 Shadow effects

The above analysis should not lead to the conclusion that the output power generated by a parallel string of two solar cells illuminated at a intensity of 50% of total irradiance is exactly the same as the power generated by just one solar cell illuminated by the full irradiance. This is due to the power losses by series and shunt resistances. The following example illustrates this case.

Example 4.2

Assume a parallel connection of two identical solar cells with the following parameters:

$$R_{sh} = 100\,\Omega,\ R_S = 0.5\,\Omega,\ \text{area} = 8\,cm^2,\ J_O = 1 \times 10^{-11},\ J_{SC} = 0.0343\,A/cm^2$$

Compare the output maximum power of this mini PV module in the two following cases: Case A, the two solar cells are half shadowed receiving an irradiance of $500\,W/m^2$, and Case B when one of the cells is completely shadowed and the other receives full irradiance of $1000\,W/m^2$.

The solution uses the netlist above and replaces the irradiance by the values of cases A and B in sequence. The files are listed in Annex 4 under the names 'example4_2.cir' and 'example4_2b.cir'.

The result is shown in Figure 4.9.

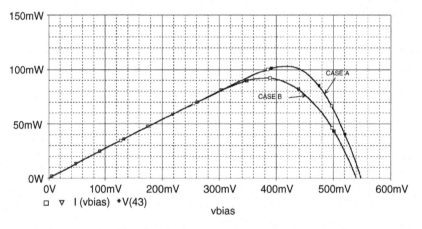

Figure 4.9   Solution of Example 4.2

As the values of the shunt resistance have been selected deliberately low to exagerate the differences, it can be seen that neither the open circuit voltage nor the maximum power are the same thereby indicating that the assumption that a shadowed solar cell can be simply eliminated does not produce in general accurate results because the associated losses to the parasitic resistances are not taken into account.

4.4   The Terrestrial PV Module

The most popular photovoltaic module is a particular case of a series string of solar cells. In terrestrial applications the PV standard modules are composed of a number of solar cells connected in series. The number is usually 33 to 36 but different associations are also available. The connections between cells are made using metal stripes. The PV module characteristic is the result of the voltage scaling of the $I(V)$ characteristic of a single solar cell. In PSpice it would be easy to scale-up a model of series string devices extending what has been illustrated in Example 4.1.

There are, however, two main reasons why a more compact formulation of a PV module is required. The first reason is that as the number of solar cells in series increases, so do the number of nodes of the circuit. Generally, educational and evaluation versions of PSpice do not allow the simulation of a circuit with more than a certain number of nodes. The second reason is that as the scaling rules of current and voltage are known and hold in general, it is simpler and more useful to develop a more compact model, based on these rules, which could be used, as a model for a single PV module, and then scaled-up to build the model of a PV plant. Consider the $I(V)$ characteristic of a single solar cell:

$$I = I_L - I_0 \left(e^{\frac{V+IR_s}{nV_T}} - 1 \right) - I_{02} \left(e^{\frac{V+IR_s}{2V_T}} - 1 \right) - \frac{V + IR_s}{R_{sh}} \tag{4.1}$$

Let us consider some simplifying assumptions, in particular that the shunt resistance, R_{sh}, of a solar cell is large and its effects can be neglected, and that the effects of the second diode are also negligible. So, assuming $I_{02} = 0$ and $R_{sh} = \infty$, equation (4.1) becomes

$$I = I_{sc} - I_0 (e^{\frac{V+IR_s}{nV_T}} - 1) \tag{4.2}$$

where $I_{sc} = I_L$ has also been assumed.

The scaling rules of voltages, currents and resistances when a matrix of $N_s \times N_p$ solar cells is considered are the following:

$$I_M = N_p I \tag{4.3}$$
$$I_{scM} = N_p I_{sc} \tag{4.4}$$
$$V_M = N_s V \tag{4.5}$$
$$V_{ocM} = N_s V_{oc} \tag{4.6}$$

where subscript M stands for 'Module' and subscripts without M stand for a single solar cell. The scaling rule of the series resistance is the same as that of a $N_s \times N_p$ association of resistors:

$$R_{sM} = \frac{N_s}{N_p} R_s \tag{4.7}$$

Substituting in equation (4.2),

$$\frac{I_M}{N_p} = \frac{I_{scM}}{N_p} - I_{01} (e^{\frac{\frac{V_M}{N_s} + \frac{I_M N_p}{N_p N_s} R_{sM}}{nV_T}} - 1) \tag{4.8}$$

$$I_M = I_{scM} - N_p I_{01} (e^{\frac{V_M + I_M R_{sM}}{nN_s V_T}} - 1) \tag{4.9}$$

Moreover, from equation (4.2) in open circuit, I_0 can be written as:

$$I_0 = \frac{I_{sc}}{\left(e^{\frac{V_{oc}}{nV_T}} - 1\right)} \tag{4.10}$$

using now equations (4.4) and (4.6)

$$I_0 = \frac{I_{scM}}{N_p\left(e^{\frac{V_{ocM}}{nV_TN_s}} - 1\right)} \tag{4.11}$$

Equation (4.11) is very useful as, in the general case, the electrical PV parameters values of a PV module are known (such as I_{scM} and V_{ocM}) rather than physical parameters such as I_0. In fact the PSpice code can be written in such a way that the value of I_0 is internally computed from the data of the open circuit voltage and short circuit current as shown below.

Substituting equation (4.11) in equation (4.9)

$$I_M = I_{scM} - I_{scM}\frac{\left(e^{\frac{V_M + I_M R_{sM}}{nV_TN_s}} - 1\right)}{\left(e^{\frac{V_{ocM}}{nV_TN_s}} - 1\right)} \tag{4.12}$$

Neglecting the unity in the two expressions between brackets,

$$I_M = I_{scM}\left(1 - e^{\frac{V_M + I_M R_{sM} - V_{ocM}}{nV_TN_s}}\right) \tag{4.13}$$

Equation (4.13) is a very compact expression of the $I(V)$ characteristics of a PV module, useful for hand calculations, in particular.

On the other hand, the value of the PV module series resistance is not normally given in the commercial technical sheets. However, the maximum power is either directly given or can be easily calculated from the conversion efficiency value. Most often the value of P_{max} is available at standard conditions. From this information the value of the module series resistance can be calculated using the same approach used in Section 3.13 and in Example 3.5. To do this, the value of the fill factor of the PV module when the series resistance is zero is required. We will assume that the fill factor of a PV module of a string of identical solar cells equals that of a single solar cell. This comes from the scaling rules shown in equations (4.3) to (4.7)

$$FF_{0M} = \frac{J_{mM}V_{mM}}{J_{scM}V_{ocM}} = \frac{J_mV_m}{J_{sc}V_{oc}} = FF_0 = \frac{v_{oc} - \ln(v_{oc} + 0.72)}{1 + v_{oc}} \tag{4.14}$$

where the normalized value v_{oc}, can be calculated either from the data of a single solar cell or from module data:

$$v_{oc} = \frac{V_{oc}}{nV_T} = \frac{V_{ocM}}{N_snV_T} \tag{4.15}$$

The series resistance is now computed from the value of the power density per unit area at the maximum power point

$$P_{\max M} = FF_M V_{ocM} J_{scM} = FF_{0M}(1 - r_{sM})V_{ocM}J_{scM} \tag{4.16}$$

and

$$R_{sM} = \frac{V_{ocM}}{I_{scM}} - \frac{P_{\max M}}{FF_{0M}I_{scM}^2} \tag{4.17}$$

Equations (4.9) and (4.10) are the basic equations of the $I(V)$ characteristics of a PV module, and are converted into PSpice code as follows corresponding to the schematics in Figure 4.10.

```
.subckt module_1 400 403 402 params:ta=1, tr=1, iscmr=1, pmaxmr=1,
+vocmr=1,
+ns=1, np=1, nd=1
girradm 400 401 value={(iscm/1000*v(402))}
d1 401 400 diode
.model diode d(is={iscm/(np*exp(vocm/(nd*ns*(8.66e-
+5*(tr+273)))))},n={nd*ns})
.func uvet() {8.66e-5*(tr+273)}
.func vocnorm() {vocm/(nd*ns*uvet)}
.func rsm() {vocm/(iscm) - pmaxm*(1+vocnorm)/(iscm**2*(vocnorm-
+log((vocnorm)+0.72)))}
rs 401 403 {rsm()}
.ends module_1
```

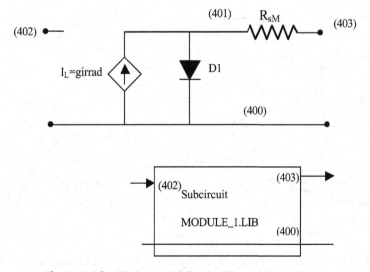

Figure 4.10 PSpice model for the PV module subcircuit

where the diode factor has been named nd in order to avoid duplicity with the name of the internal PSpice ideality factor. The above netlist includes practical values of a commercial PV module:

$$N_s = 36,\ P_{max} = 85\,W,\ I_{scm} = 5\,A,\ V_{ocm} = 22.3\,V$$

As a result of the calculations using equations (4.15) and (4.17) the total module series resistance value is $0.368\ \Omega$ and the resulting $I(V)$ characteristic can be seen in Figure 4.11, after simulation of the PSpice file 'module_1.cir' as follows,

```
*module_1.cir
.include module_1.lib
xmodule 0 43 42 module_1 params:ta=25,tr=25, iscm=5, pmaxm=85, vocm=22.3,
+ ns=36, np=1, nd=1
vbias 43 0 dc 0
virrad 42 0 dc 1000
.dc vbias 0 23 0.1
.probe
.end
```

The PSpice result for the maximum power is 84.53 W instead of the rated 85 W (which is an error of less than 0.6%).

The model described in this section is able to reproduce the whole standard AM1.5G $I(V)$ characterisitcs of a PV module from the values of the main PV magnitudes available for a commercial module: short circuit current, open circuit voltage, maximum power and the number of solar cells connected. We face, however, a similar problem to that in Chapter 3 concerning individual solar cells, that is translating the standard characterisitcs to arbitrary conditions of irradiance and temperature. The next sections will describe some models to solve this problem.

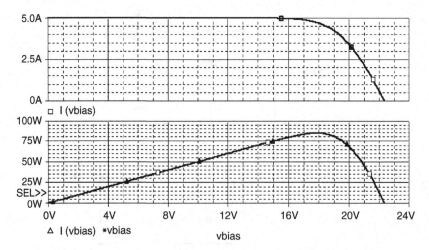

Figure 4.11 Plot of the $I(V)$ characteristic and the output power of a PV module of 85 W maximum power

4.5 Conversion of the PV Module Standard Characteristics to Arbitrary Irradiance and Temperature Values

The electrical characteristics of PV modules are rated at standard irradiance and temperature conditions. The standard conditions are for terrestrial applications, AM1.5 spectrum, 1000 W/m^2 at 25 °C cell temperature, whereas AM0 (1353 W/m^2) at 25 °C are the extraterrestrial standard conditions. Therefore, what the user knows, are the nominal values of the electrical parameters of the PV module, which are different from the values of these same parameters when the operating conditions change. The conversion of the characteristics from one set of conditions to another is a problem faced by designers and users, who want to know the output of a PV installation in average real conditions rather than in standard conditions and those only attainable at specialized laboratories.

Most of the conversion methods are based on the principles described in Chapter 3; a few important rules are summarized below:

- The short circuit current is proportional to the irradiance and has a small temperature coefficient.

- The open circuit voltage has a negative temperature coefficient and depends logarithmically on the irradiance.

Moreover, there is a significant difference between the *ambient* temperature and the cell *operating* temperature, due to packaging, heat convection and irradiance.

4.5.1 Transformation based on normalized variables (ISPRA method)

We will describe in this section a PSpice model based on the conversion method proposed by G. Blaesser [4.1] based on a long monitoring experience. The method transforms every (I, V) pair of coordinates into a (I_r, V_r) at reference conditions and vice versa, according to the following equations:

$$I_r = I \frac{G}{G_r} \qquad (4.18)$$

and

$$V_r = V + DV \qquad (4.19)$$

The temperature coefficient of the current is neglected and the temperature effects are considered in a parameter DV, defined as

$$DV = V_{ocr} + V_{oc} \qquad (4.20)$$

It is now useful to normalize the values of currents and voltages as:

$$i_r = \frac{I_r}{I_{scr}} \qquad i = \frac{I}{I_{sc}} \tag{4.21}$$

and

$$v_r = \frac{V_r}{V_{ocr}} \qquad v = \frac{V}{V_{oc}} \tag{4.22}$$

Very simple relationships between these normalized variables are obtained:

$$i_r = i \tag{4.23}$$

and

$$v = \frac{v_r - Dv}{1 - Dv} \tag{4.24}$$

where

$$Dv = \frac{DV}{V_{ocr}} \tag{4.25}$$

Equations (4.23) and (4.24) are the transformation equations of current and voltage. The transformation of the fill factor is also required, which from the above definitions is given by

$$FF = FF_r \frac{v_{mr} - Dv}{v_{mr}(1 - Dv)} \tag{4.26}$$

As can be seen Dv is an important parameter, which has been determined from many experimental measurements for crystalline silicon PV modules and can be fitted by the following expression:

$$D_v = 0.06 \ln \frac{G_r}{G} + 0.004(T_a - T_r) + 0.12 \times 10^{-3}G \tag{4.27}$$

where T_a is the ambient temperature and T_r is the operating temperature of the cells under standard conditions, i.e., 25 °C. Thus any transformation is now possible as shown in the following example.

Example 4.3

Consider the following characteristics of a commercial PV module at standard conditions (subscript r).

$N_s = 36$, $P_{maxr} = 85\,W$, $I_{scmr} = 5\,A$, $V_{ocmr} = 22.3\,V$. Calculate and plot the $I(V)$ characterisitics of this PV module at normal operating conditions ($G = 800\,W/m^2$, $T_a = 25\,°C$). The net list is shown in Annex 4 and is given the name 'module_conv.cir'.

A voltage coordinate transformation section has been added to the section calculating the characteristics under standard temperature and the new irradiance conditions and then the voltage shift transformation of equation (4.19) is implemented by substracting a DC voltage source valued DV. This is schematically shown in the circuit in Figure 4.12.

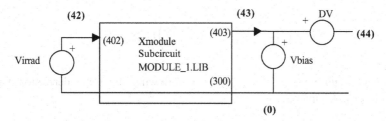

Figure 4.12 Node assignation of the circuit implemented to transform the $I(V)$ characteristics using the method based on the Dv parameter

The results provided by the corresponding PSpice simulation for an ambient temperature of 25 °C are the following:

$$V_{oc} = 19.71\,V$$

$$I_{sc} = 4\,A$$

$$P_{max} = 58.96\,W$$

4.6 Behavioral PSpice Model for a PV Module

The NOCT concept is also useful in the transformation of the $I(V)$ characterisitics of a PV module from one set of conditions to another, in the same way as was used in Chapter 3 for individual solar cells. Let us remind ourselves that NOTC stands for nominal operating cell temperature, which is defined as the real cell temperature under 800 W/m² 20 °C ambient temperature with a wind speed of 1 m/s and with the back of the cell open and exposed to the wind. This parameter helps in relating the ambient temperature to the real operating temperature of the cell. A simple empirical formula is used [4.2]:

$$T_{cell} - T_a = \frac{NOCT - 20}{800}G \tag{4.28}$$

where G is the irradiance given in W/m^2 and all the temperatures are given in °C. Of course the determination of the value of NOCT depends on the module type and sealing material. After a series of tests in a number of PV modules, an average fit to the formula in equation (4.28) is:

$$T_{cell} - T_a = 0.035\,G \tag{4.29}$$

corresponding to a NOCT $= 48\,°$C.

This simple equation (4.28) can also be used to transform the $I(V)$ characteristics by using the new value of the cell temperature in the equations of the PV module above. For an arbitrary value of the irradiance and temperature, the short-circuit current is given by:

$$I_{scM} = \frac{I_{scMr}}{1000}G + \left(\frac{dI_{scM}}{dT}\right)(T_{cell} - T_r) \tag{4.30}$$

and the open-circuit voltage can be written, following the same derivation leading to equation (3.72) in Chapter 3, as:

$$V_{ocM} \approx V_{ocMr} + \left(\frac{\partial V_{ocM}}{\partial T}\right)_G (T_{cell} - T_r) + V_T \ln\frac{I_{scM}}{I_{scMr}} \tag{4.31}$$

Using the value of the maximum power the series resistance of the module is calculated from equation (4.17).

As in Chapter 3, the available data provided by the PV manufacturers for the modules, is normally restricted to the short-circuit current, open-circuit voltage and maximum power point coordinates at standard reference conditions AM1.5G 1000 W/m^2. Module temperature coefficients for short-circuit currents and open-circuit voltages are also given. The behavioural PSpice modelling of a single solar cell described in Chapter 3 can be extended to a PV module, and it follows that the maximum power point coordinates are given by:

$$I_{mM} = I_{mMr}\frac{G}{G_r} + \left(\frac{dI_{scM}}{dT}\right)(T_{cell} - T_r) \tag{4.32}$$

$$V_{mM} = N_s V_T \ln\left(1 + \frac{I_{scM} - I_{mM}}{I_{scM}}\left(e^{\frac{V_{ocM}}{N_s V_T}} - 1\right)\right) - I_{mM}R_{sM} \tag{4.33}$$

First a subcircuit is defined for the behavioural model of the module (subckt module_beh), which is sensibly similar to the one written for a single solar cell, but adding the new parameter 'ns' to account for the number of solar cells in series. The subcircuit, shown in Figure 4.13, implements equation (4.13) in the g-device 'gidiode', and the values of the series resistance of the module.

Also two nodes generate values for the coordinates of the maximum power point to be used by MPP trackers. The complete netlist is shown in Annex 4 and is called 'module_beh.lib'.

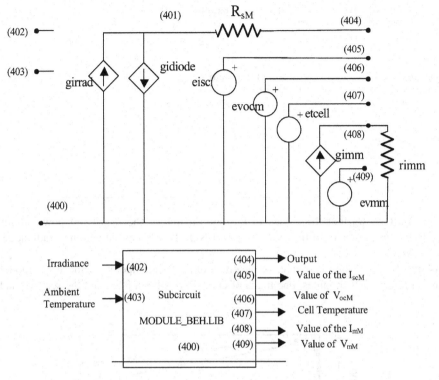

Figure 4.13 Subcircuit nodes of the behavioural model of a PV module

---

Example 4.4

Plot the $I(V)$ characteristics of a CIGS module with the following parameters: short-circuit current 0.37 A, open-circuit voltage 21.0 V, maximum power 5 W, voltage coefficient -0.1 V/°C and short-circuit current temperature coefficient $+0.13$ mA/°C. These data have been taken from www.siemenssolar.com for the ST5 CIGS module.

Solution

The file simulating the $I(V)$ characteristics is the following:

```
*module_beh.cir
*   NODES
*   (0)   REFERENCE
*   (42) INPUT, IRRADIANCE
*   (43)   INPUT, AMBIENT TEMPERATURE
*   (44) OUTPUT
*   (45)   OUTPUT, (VOLTAGE) VALUE = SHORT CIRCUIT CURRENT(A) AT
*   IRRADIANCE AND TEMPERATURE
*   (46)   OUTPUT, OPEN CIRCUIT VOLTAGE AT IRRADIANCE ANDTEMPERATURE
```

```
*    (47) OUTPUT, (VOLTAGE) VALUE = CELL OPERATING TEMPERATURE (°C)
*    (48) OUTPUT, MPP CURRENT
*    (49) OUTPUT, MPP VOLTAGE

.include module_beh.lib
xmodule 0 42 43 44 45 46 47 48 49 module_beh params: iscmr = 0.37, coef_iscm = 0.13e - 3,
+ vocmr = 21, coef_vocm = -0.1, pmaxmr = 5,
+ noct = 47, immr = 0.32, vmmr = 15.6, tr = 25, ns = 33, np = 1
vbias 44 0 dc 0
virrad 42 0 dc 1000
vtemp 43 0 dc 25
.dc vbias 0 22 0.1
.probe
.end
```

The result shows that the values of the coordinates of the maximum power point are reproduced closely (error of the order of 3%) if NOCT = 20 is used and the irradiance is set to 1000 W/m^2.

As an example of the importance of the effect of the cell temperature, the above model has been run for two PV modules, one made of crystalline silicon [4.3] and the second made of

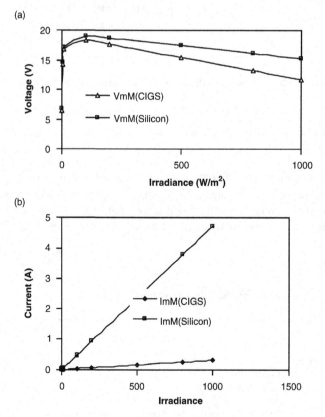

Figure 4.14 (a) Values of the voltage at the maximum power point coordinates for several irradiances and (b) values of the current at the maximum power point

CIGS [4.4]. In the simulations the ambient temperature has been maintained at 25 °C whereas the irradiance has been swept from 0 to 1000 W/m^2. The results of the values of the coordinates of the maximum power point are plotted in Figure 4.14.

As can be seen, the voltage at the maximum power point has a maximum at moderate irradiance values, thereby indicating that the irradiance coefficient of the maximum power point voltage, which will increase the value as irradiance increases, is compensated by the temperature coefficient, which produces a reduction of the voltage at higher irradiances due to the higher cell temperature involved.

4.7 Hot Spot Problem in a PV Module and Safe Operation Area (SOA)

As we have seen in Section 4.2, a shadowed solar cell can be operating in conditions forcing power dissipation instead of generation. Of course, this dissipation can be of much greater importance if only one of the cells of a PV module is completely shadowed. Dissipation of power by a single solar cell raises its operating temperature and it is common to calculate the extreme conditions under which some damage to the solar cell or to the sealing material can be introduced permanently. One way to quantify a certain safe operation area (SOA) is to calculate the power dissipated in a single solar cell (number n) by means of:

$$P_{dis} = -I_M V(n) \qquad (4.34)$$

The condition that has to be applied is that under any circumstances, this power dissipation by cell number n should be smaller than a certain limit P_{dismax}. The voltage $V(n)$ is given by:

$$V(n) = V_M - \sum_{i=1}^{i=n-1} V(i) \qquad (4.35)$$

From equation (4.34), it follows

$$-I_M V_M + I_M \sum_{i=1}^{i=n-1} V(i) < P_{dis\,max} \qquad (4.36)$$

and

$$V_M > \sum_{i=1}^{i=n-1} V(i) - \frac{P_{dis\,max}}{I_M} \qquad (4.37)$$

This equation can be easily plotted. The result for the commercial module simulated in Section 4.4 considering that $P_{dismax} = 25$ W is shown in Figure 4.15 where the axis have

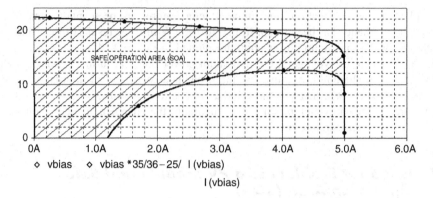

Figure 4.15   Safe operation area for a PV module. Warning: the voltage is the y-axis and the current is the x-axis

been rotated 90°: the PV module voltage is plotted by the y-axis whereas the PV current is given by the x-axis.

The safe operation area is the area between the two curves. The plot has been achieved by plotting

$$ V = \frac{35}{36} V_{bias} - \frac{25}{I(V_{bias})} $$

If ever the operating point of a solar cell falls inside the forbidden area, dissipation exceeding the limits may occur.

4.8   Photovoltaic Arrays

Photovoltaic arrays, and in general PV generators, are formed by combinations of parallel and series connections of solar cells or by parallel and series connections of PV modules. The first case is generally the case of outer-space applications where the arrays are designed especially for a given space satellite or station. In terrestrial applications arrays are formed by connecting PV modules each composed of a certain number of series-connected solar cells and eventually bypass diodes. We will concentrate in this section on the outer-space applications to illustrate the effects of shadow and discuss terrestrial applications in Chapter 5.

Generally the space arrays are composed of a series combination of parallel strings of solar cells, as shown in Figure 4.16.

The bypass diodes have the same purpose as described earlier, this is to allow the array current to flow in the right direction even if one of the strings is completely shadowed. In order to show the benefits of such structure, we will simplify the problem to an array with 18 solar cells and demonstrate the effect in Example 4.5. There are four-strings in series each string composed of three solar cells in parallel.

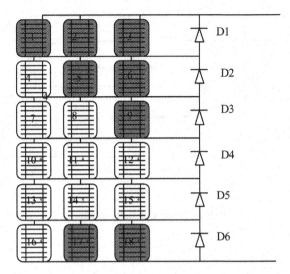

Figure 4.16   Solar cell module of 18 solar cells with bypass diodes

---

Example 4.5

Consider a space photovoltaic array composed of 18 solar cells arranged as shown in Figure 4.16. Each cell has the same characteristics:

$$\text{area} = 8\,\text{cm}^2,\ J_o = 1 \times 10^{-11},\ J_{sc} = 0.0343\,\text{A/cm}^2,\ R_{sh} = 1000\,\Omega,\ R_s = 0.1\,\Omega$$

and the array is shadowed as also shown in Figure 4.16 in such a way that there are eight solar cells completely shadowed and shown in dark in the figure. Plot the $I(V)$ characteristics of the completely unshadowed array and of the shadowed array. Plot the voltage drops across the bypass diodes and the power dissipation.

Solution

We will assume that the irradiance of the shadowed cells is zero. The corresponding netlist is an extension of the netlist in Example 4.1 but includes six bypass diodes. The file is shown in Annex 4 under the title '6x3_array.cir'. The result is shown in Figure 4.17.

As can be seen the $I(V)$ curve of the array when fully illuminated (upper plot) is severely degraded by the effect of the shadow. Taking into account that if no bypass diodes were used the total output power would be zero due to the complete shadow of one of the series strings (cells 1 to 3), there is some benefit in using bypass diodes. Of course the $I(V)$ curve is explained because some of the bypass diodes get direct bias in some region of the array operating voltage range. The same netlist allows us to plot the individual diode voltage as shown in Figure 4.18.

As can be seen diodes D4 and D5 remain in reverse bias along the array operating voltage range (0 to 3.5 V approximately) whereas diodes D3, D1, D6 and D2 become direct biased in

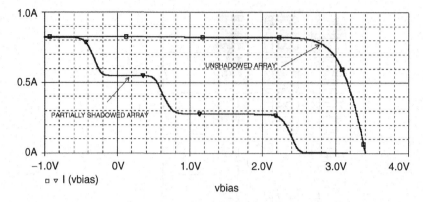

Figure 4.17   Comparison of the $I(V)$ curves of a partially shadowed array and that of the same array unshadowed

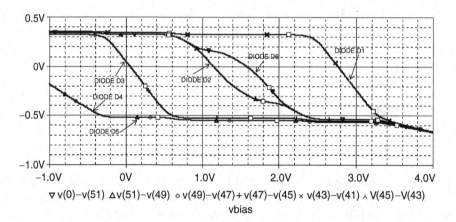

Figure 4.18   Voltage drops across the diodes in a 6×3 array

part of the operating range corresponding to the shadowed areas and depending on the severity of the shadow. These diodes dissipate significant power when they are direct biased. This calculation helps, on the one hand, in rating the correct size of the bypass diodes for a given array, and on the other hand monitors faulty devices, which can be identified by the shape of the $I(V)$ curves.

4.9   Scaling up Photovoltaic Generators and PV Plants

The association of solar cells in series allow us to build terrestrial PV modules or arrays for different applications where an standard size is not suitable for modularity or other reasons. Such is the case of satellite solar cell arrays or special size modules to be integrated in buildings.

Putting aside these special cases, where the models used are the generic solar cells model or PV series module, in the most general case a PV generator is made up of several standard PV terrestrial modules associated in a matrix $N_{sG} \times N_{pG}$ series–parallel. Scaling rules are required to relate the characteristics of the PV plant or generator to individual PV module characteristics.

The steps described in Section 4.4 can be used here to extend the scaling procedure to a PV plant. If we now consider that the unit to be used for the plant is a PV module, with the $I(V)$ parameters:

- short-circuit current: I_{scM}

- open-circuit voltage: V_{ocM}

- maximum power: P_{maxM}

- MPP current: I_{mM}

- MPP voltage: V_{mM}

- temperature coefficients for the short-circuit current and open-circuit voltage

and we build a matrix of N_{sG} modules in series and N_{pG} in parallel, the $I(V)$ characteristic of the generator scales as:

$$I_G = N_{pG}I_M$$
$$V_G = N_{sG}V_M \tag{4.38}$$

where subscripts G stands for generator and M for PV module.

Applying these rules to the generator it is straightforward to show that:

$$I_G = N_{pG}I_{scM} - \frac{N_{pG}I_{scM}}{e^{\frac{V_{ocM}}{N_sV_T}}} \left(e^{\frac{V_G+I_GR_{sM}\frac{N_{sG}}{N_{pG}}}{nN_sN_{sG}V_T}} - 1 \right) \tag{4.39}$$

where the data values, apart from the module series resistance which can be calculated by equation (4.17), are to be found in the characteristics of a single PV module which are assumed to be known. We will be using the diode ideality factor $n = 1$ because the effect of such a parameter can be taken into account, because this parameter is always multiplying N_s and then the value of N_s can be corrected if necessary.

In the same way as the coordinates of the maximum power point were calculated for a single solar cell in Chapter 3 or a single module in this chapter, they can also be calculated for a PV generator of arbitrary series–parallel size as follows,

$$I_{mG} = N_{pG}\left(I_{mMr}\frac{G}{G_r} + \left(\frac{dI_{scM}}{dT}\right)(T_{cell} - T_r) \right) \tag{4.40}$$

$$V_{mG} = N_{sG}\left(N_sV_T \ln\left(1 + \frac{I_{scM} - I_{mM}}{I_{scM}}\left(e^{\frac{V_{ocM}}{N_sV_T}} - 1 \right) \right) - I_{mM}R_{sM} \right) \tag{4.41}$$

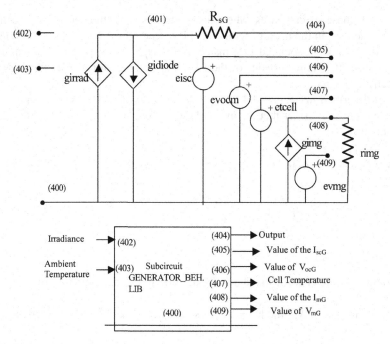

Figure 4.19 Subcircuit of a PV generator composed of $N_{sG} \times N_{pG}$ PV modules of N_s cells in series

where the value of V_{ocM} is given by equation (4.31) and

$$I_{mM} = \frac{I_{mG}}{N_{pG}} \qquad (4.42)$$

Using equations (4.38) to (4.42) a PV generator subcircuit can be written as shown in Figure 4.19 in a very similar manner as the module subcircuit.

The netlist is shown in Annex 4 and is called 'generator_beh.lib' where the subindex 'g' is used where appropriate to indicate the generator level. The file 'generator_beh.cir', also shown in Annex 4, can be used to plot the $I(V)$ characteristics of a generator of 5×2 modules. The figure is not shown because it is identical to the one of a single PV module with the voltage and currents scaled up according the scaling rules.

4.10 Problems

4.1 Simulate the $I(V)$ characteristics of a terrestrial PV module under standard conditions (AM1.5, 1000 W/m², 27 °C cell temperature) with 33, 34, 35 and 36 solar cells in series, and calculate the values of V_{oc} and P_{max}. Assume that the design specifications are such that the open-circuit voltage has to be greater than 14 V at 400 W/m² and 22 °C ambient temperature. Which designs satisfy the specifications?

4.2 In a 12-series solar cell array, as shown in Figure 4.2, two bypass diodes are used instead of one. One is placed between the − terminal and the connection between solar cells numbers 3 and 4, and the second between the + terminal and the connection between solar cells 9 and 10. Plot the $I(V)$ characteristics under standard conditions when all the cells are equally illuminated. Plot the $I(V)$ curve in the following cases: (a) cells number 1 and 10 are fully shadowed; (b) cells number 1 and 2 are fully shadowed; (c) cells number 1 and 9 are fully shadowed.

4.3 Transform the standard characteristics of a 6-series cell string to different conditions: irradiance 800 W/m^2 and ambient temperature 35 °C.

4.11 References

[4.1] Blaesser, G., 'Use of I-V extrapolation methods in PV systems analysis' in *European PV Plant Monitoring Newsletter*, 9 November, pp. 4–5, Commission of the EU Communities, Joint Research Centre, Ispra, Italy, 1993.

[4.2] Ross, R.G. and Smockler, M.I. '*Flat-Plate Solar Array Project Final Report, Volume VI: Engineering – Sciences and Reliability*, Jet Propulsion Laboratory, Publication 86–31, 1986.

[4.3] www.isofoton.es.

[4.4] www.siemenssolar.com.

5

Interfacing PV Modules to Loads and Battery Modelling

Summary

Photovoltaic systems are designed to supply either DC or AC electricity to loads. Depending on the complexity and characteristics of the final loads connected to the photovoltaic system, different elements can be found as a part of the photovoltaic system itself. These different elements may be:

- DC loads, e.g. lights, electric motors;
- batteries;
- AC loads, e.g. lamps, household appliances, electronic equipment;
- power-conditioning equipment, e.g. protection and control elements, AC–DC converters, charge regulators, DC–AC converters.

Depending on the nature of the loads, their interface to the photovoltaic system implies different complexity grades. The study of the different connection options of the photovoltaic modules to the rest of the photovoltaic system components is the objective of this chapter.

5.1 DC Loads Directly Connected to PV Modules

The simplest example of a PV system connection to a load is a direct connection of a DC load to the output of a PV module as shown in Figure 5.1. The simplest load is a resistor R.

If we want to know the coordinates of the operating point, then we have to look for the intersection between the load and the PV module $I(V)$ characteristics. Figure 5.2 shows the $I(V)$ characteristics of a PV module with a short circuit current of 4 A and an open circuit voltage of 22.1 V. The load is a 3 Ω resistor in this example and Figure 5.2 shows a plot of the current as a function of the voltage, hence the load $I(V)$ characteristic is a straight line with a slope of $1/R$.

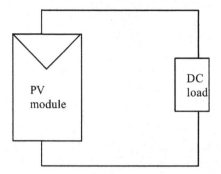

Figure 5.1   PV module directly connected to a DC load

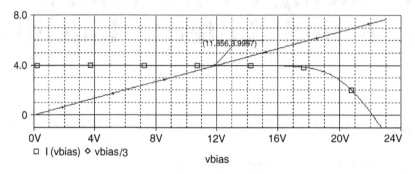

Figure 5.2   PV module and load $I(V)$ characteristics

As can be seen in Figure 5.2, for a given irradiance and temperature there is only one intersection point which sets the values of the voltage and the current. In the example shown in Figure 5.2, the current flowing into the load is 3.99 A and the voltage is set to 11.8 V. This is the voltage common to the PV module output and the resistor load.

This result means that a resistor load sets the operating point of a PV module, and by extension of a PV system, generally, at a different point than the maximum power point. As a result it can be concluded that the power transfer from the PV generator to the load is not optimized in the general case in this direct connection, unless interface circuits are inserted between the PV system and the load.

Such interface circuits are required because the PV module $I(V)$ characteristic is subject to time variations due to irradiance and temperature changes, as has been shown in Chapters 3 and 4. This will directly affect the transfer of electrical power to the load. In the simple case of a resistor as a load, analysed here, the result of a direct connection will be that the power efficiency will change as long as irradiance and temperature change.

An important effect of a direct connection to a resistor is that no power can be delivered to the load by night, which in many applications is a key specification.

5.2   Photovoltaic Pump Systems

PV pumping systems are often composed of a PV generator directly connected to a DC motor driving a water pump. Suitable models for DC motors and centrifugal pumps can be derived for PSpice simulation.

5.2.1 DC series motor PSpice circuit

In a DC motor a magnetic field is established either by a permanent magnet or by field winding, which involves a rotor with armature winding when electrical power is applied. The result is that a mechanical torque τ_e is produced, which is proportional to the magnetic flux ϕ_m, and to the armature current i_a as follows:

$$\tau_e = K_\tau \phi_m i_a \tag{5.1}$$

where K_τ is the motor torque constant. Moreover a back electromotive force(emf) is induced in the armature winding and is related to the magnetic flux and to the angular frequency ω, by

$$E_{cm} = K_{cm} \phi_m \omega \tag{5.2}$$

As in steady-state, and without losses, the mechanical power $(\omega \tau_e)$ equals the electric power $(v_a i_a)$ therefore it follows that,

$$K_{cm} = K_\tau \tag{5.3}$$

If the magnetic field is created by field winding, in a series motor, where the field and armature windings are in series, the magnetic flux is proportional to the current i_f in that winding:

$$\phi = K_f i_f \tag{5.4}$$

and hence

$$E_{cm} = K_{cm} K_f i_f \omega = K_m i_f \omega \tag{5.5}$$
$$\tau_e = K_\tau K_f i_f i_a = K_m i_f i_a \tag{5.6}$$

The dynamic equivalent circuit of a DC series motor also has to include other additional elements: resistor losses $(R_a$ and $R_f)$, inductors of the windings L_a and L_f, inertial term J and friction term F. The mechanical load torque is also included τ_L, as can be seen in Figure 5.3. The differential equations governing the dynamic operation of the DC motor are:

$$V_a = i_a R_a + L_a \frac{di_a}{dt} + E_{cm}$$
$$V_f = i_f R_f + L_f \frac{di_f}{dt} \tag{5.7}$$
$$\tau_e = J \frac{d\omega}{dt} + F\omega + \tau_L$$

where the last equation relates the mechanical magnitudes and returns the angular frequency value.

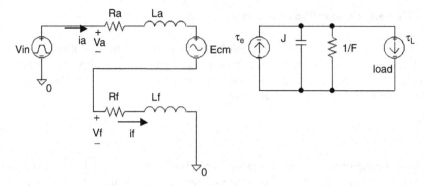

Figure 5.3 Equivalent circuit of a DC series motor with the load of a pump

All the equations shown do not consider that the core may enter into saturation. There are more complicated PSpice models that take into account these effects, however, these are beyond the scope of this book, and the reader is referred to reference [5.1] for more in-depth treatment.

5.2.2 Centrifugal pump PSpice model

When a DC motor is connected to a centrifugal pump, a load torque is applied to the mechanical port of the equivalent circuit in Figure 5.3, which is related to the angular frequency by:

$$\tau_L = A + B\omega^2 \tag{5.8}$$

where A and B are constants depending on each particular pump. As a result of this load torque an equilibrium is reached for a value of the angular speed. Moreover, the centrifugal pump has a characteristic curve relating the pumping head, H, the angular frequency and the resulting flow Q. These types of curves are widely available in commercial technical notes or web pages, and are generally given for a constant value of the angular speed. It is assumed that these curves can be approximated by a second-degree polynomial as follows:

$$H = A_1 S^2 + B_1 SQ + C_1 Q^2 \tag{5.9}$$

where S is the angular speed in rpm, and A_1, B_1 and C_1 are constants of a given pump.

A schematic circuit showing how the different parameters relate in a DC motor-pump connection is shown in Figure 5.4.

It should be noted that the input power is a function of the output angular speed due to the generator E_{cm} in the armature winding equivalent circuit. This means that the first two blocks in Figure 5.4 are tightly related by the feedback loop closing at the mechanical domain.

5.2.3 Parameter extraction

In order to supply the PSpice simulator with correct data, the main parameters in equations (5.1) to (5.9) have to be extracted. First for the pump, the data of the characteristic curve are taken and summarized in Table 5.1 from manufacturers' data taken from reference [5.1].

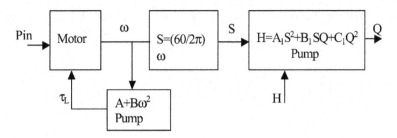

Figure 5.4 Schematic circuit showing how the different parameters relate in a DC motor-pump connection

Table 5.1   Data of the characteristic curve from reference [5.1]

Head H (metre)	Flow Q (l/s)	(W) P_{in} Power
13.4	1	480
11.5	1.5	340
5	2.3	300

By fitting equation (5.9) to the characterisitc curve at 3000 rpm, the following values of the constants A_1, B_1, C_1 have been obtained:

$$A_1 = 1.35 \times 10^{-6}$$
$$B_1 = 0.0015$$
$$C_1 = -3.32$$

Next, the values of K_m, A and B are extracted following the procedure as follows.

Step 1

Obtain the values of the main operational parameters under nominal conditions. That is the values of P_{in} and S.

Step 2

Assume a value for the input voltage if not known, and a value for the armature and field winding resistances.

Step 3

Calculate K_m.

Write the operational equations of the motor at steady-state conditions:

$$V_a = i_a R_a + E_{cm}$$
$$V_f = i_f R_f$$

(5.10)

and

$$V_{in} = i_a R_a + i_f R_f + E_{cm}$$

(5.11)

Since $i_a = i_f$, we can write the total applied voltage at nominal conditions ($\omega = \omega_{nom}$ and $i_a = i_f = i_{nom}$)

$$V_{in} = i_{nom}(R_a + R_f) + K_m i_{nom}\omega_{nom}$$

(5.12)

and finally

$$K_m = \frac{V_{in} - i_{nom}(R_a + R_f)}{i_{nom}\omega_{nom}}$$

(5.13)

Step 4

Calculate A from the starting torque.

From the mechanical equation at steady-state and at the starting conditions ($\omega = 0$)

$$\tau_e = \tau_L = A = K_m i_{min}^2$$

(5.14)

Taking into account that the motor does not start working unless a minimum power is applied P_{min}, then

$$i_{min} = \frac{P_{min}}{V_{in}}$$

(5.15)

and substituting P_{min} by its value estimated from the technical data sheets and using the nominal voltage for the voltage, we can write

$$A = K_m \frac{P_{min}^2}{V_{in}^2}$$

(5.16)

Step 5

Assume a value for F and calculate B.

The value of B is calculated from the mechanical equation at steady-state conditions at the nominal values of angular speed and input current:

$$\tau_e = F\omega_{nom} + A + B\omega_{om}^2 = K_m i_{nom}^2 \tag{5.17}$$

and

$$B = \frac{K_m i_{nom}^2 - F\omega_{nom} - A}{\omega_{om}^2} \tag{5.18}$$

***

Example 5.1

Calculate the values for A, B and K_m for the same example as in Table 5.1.

Step 1

$$H = 11.5\,m$$
$$S = 3000\,rpm$$
$$P_{in} = 340\,W$$
$$\omega = 314\,rad/s$$

Step 2

$$V_{in} = 200\,V$$
$$R_a + R_f = 0.3 \tag{5.19}$$

Step 3

$$K_m = \frac{V_{in} - i_{nom}(R_a + R_f)}{i_{nom}\omega_{nom}} = \frac{200 - 1.7 \times 0.3}{1.7 \times 3000 \times \frac{2\pi}{60}} = 0.37 \tag{5.20}$$

Step 4

$$P_{min} = 100\,W$$

$$A = K_m \frac{P_{min}^2}{V_{in}^2} = 0.37 \frac{1 \times 10^4}{200^2} = 0.0925 \tag{5.21}$$

Step 5: Calculate B

$$B = \frac{K_m i_{nom}^2 - F\omega_{nom} - A}{\omega_{om}^2} = \frac{0.37 \times 1.7^2 - 8.3 \times 10^{-4} \times 314 - 0.0925}{314^2} = 7.26 \times 10^{-6}$$

$$(5.22)$$

The values of the inertia J and of the inductances L_a and L_f do not enter the steady-state solution. However, some reasonable assumptions can be made if the dynamic response is considered. In order to calculate the small-signal dynamic performance of the motor-pump combination, the variational method to linearize equations can be used and then Laplace transforms taken. We first consider a steady-state operating point (V_{ino}, i_{ino}, w_o) and consider the instantaneous value of these three variables as a superposition of steady-state values and time-dependent increments (with subindex 1) as follows,

$$V_{in} = V_{ino} + v_1$$
$$I_{in} = I_{ino} + i_1 \qquad (5.23)$$
$$\omega = \omega_o + \omega_1$$

Substituting equation (5.23) in equations (5.7) the small-signal equations representing the dynamic operation of the motor-pump system are:

$$V_{ino} + v_1 = (i_o + i_1)(R_a + R_f) + (L_a + L_f)\frac{di_1}{dt} + K_m(i_o + i_1)(\omega_o + \omega_1)$$
$$J\frac{d\omega_1}{dt} + F(\omega_o + \omega)_1 + A + B(\omega_o + \omega_1)^2 = K_m(i_o + i_1)^2 \qquad (5.24)$$

Taking into account the steady-state equations

$$v_1 = i_1(R_a + R_f) + (L_a + L_f)\frac{di_1}{dt} + K_m(i_o\omega_1 + \omega_o i_1 + \omega_1 i_1)$$
$$J\frac{d\omega_1}{dt} + F\omega_1 + B(2\omega_o\omega_1 + \omega_1^2) = K_m(2i_o i_1 + i_1^2) \qquad (5.25)$$

Neglecting now the second-order terms $(w_1 i_1, w_1^2, i_1^2)$ it follows

$$v_1 = i_1(R_a + R_f) + (L_a + L_f)\frac{di_1}{dt} + K_m(i_o\omega_1 + \omega_o i_1)$$
$$J\frac{d\omega_1}{dt} + F\omega_1 + B2\omega_o\omega_1 = 2K_m i_o i_1 \qquad (5.26)$$

taking Laplace transforms of these linearized equations

$$v_1(s) = i_1(s)\lfloor(R_a + R_f) + s(L_a + L_f) + K_m\omega_o\rfloor + K_m i_o\omega_1(s)$$
$$(sJ + F + 2B\omega_o)\omega_1(s) = 2K_m i_o i_1(s) \qquad (5.27)$$

The mechanical dynamic response of the motor-pump association is much slower than the electrical response. We can deduce from equation (5.27) the electrical time constant assuming that the angular speed is constant while the armature current builds up. This means assuming $w_1(s) = 0$. Then the relationship between input voltage and armature current is

$$i_1(s) \approx \frac{v_1(s)}{\left[(R_a + R_f) + s(L_a + L_f) + K_m w_o\right]} = \frac{v_1(s)}{(R_a + R_f + K_m w_o)} \frac{1}{1 + s\frac{L_a + L_f}{(R_a + R_f + K_m w_o)}} \quad (5.28)$$

where the electrical time constant can be identified:

$$\tau_{ELECTR} = \frac{L_a + L_f}{(R_a + R_f + K_m w_o)} \quad (5.29)$$

Using now equation (5.27)

$$w_1(s) = v_1(s) \frac{1}{\frac{(sJ + F + 2Bw_o)}{2K_m i_o}\left[R_a + R_f + K_m w_o + s(L_a + L_f)\right] + K_m i_o} \quad (5.30)$$

that can be written as:

$$w_1(s) = \frac{v_1(s)}{K_m i_o + \frac{(F + 2Bw_o)(R_a + R_f + K_m w_o)}{2K_m i_o}} \frac{1}{1 + s\tau_M} \quad (5.31)$$

where

$$\tau_M = \frac{J}{F + 2Bw_o} \frac{1}{1 + \frac{2K_m^2 i_o^2}{(F + 2Bw_o)(R_a + R_f + K_m w_o)}} \quad (5.32)$$

Equations (5.28) and (5.32) can be used to find suitable values for the parameters L_a and L_f and J as shown in the example below.

Example 5.2

Using the results and data in Example 5.1, calculate suitable values for $(L_a + L_f)$ and J considering that the electrical time constant is 5 ms and the mechanical time constant is 5 s. Simulate the electrical and mechanical response of the motor-pump system when a small step of 1 V amplitude is applied to the input voltage while at nominal operation conditions.

Taking into account equation (5.29) and considering the same value for $(R_a + R_f) = 0.3 \, \Omega$ as in Example 5.1 it follows that

$$L_a + L_f = 1.16 \, H$$

And assuming the friction $F = 0.0083$ equation (5.32) returns $J = 0.0608$.

5.2.4 PSpice simulation of a PV array-series DC motor-centrifugal pump system

Using the parameter values derived in Examples 5.1 and 5.2 the following netlist has been written as a subcircuit of the motor-pump system, according to the schematics shown in Figure 5.3 for the motor-pump association which we will simulate.

```
.subckt pump 500 501 570 PARAMS: RA=1,LA=1,KM=1,A=1,B=1
+  F=1, J=1, RF=1, LF=1, A1=1, B1=1, C1=1, H=1

ra 501 502 {RA}
la 502 503 {LA}
econ 503 504 value={{KM}*v(508)*v(507)}
rf 504 505 {RF}
lf 505 506 {LF}
vs 506 0 dc 0
gte 0 507 value={{KM}*v(508)*v(508)}
gtl 507 0 value={A+B*v(507)*v(507)}
rdamping 507 0 {1/{F}}
cj 507 0 {J}
d2 0 507 diode
.model diode d
gif 0 508 value={(v(504)-v(505))/{RF}}
rif 508 0 1

.IC v(507)=0

*** revoluciones rpm=omega*(60/2/pi)
erpm 540 0 value={v(507)*60/6.28}
eflow 550 0 value={(-B1}*v(540) - sqrt(({B1}^2)*(v(540)^2) - 4*{C1}*
+ (A1*(v(540)^2) - {H})))/(2*{C1})}
eraiz 560 0 value={({B1}^2)*(v(540)^2)-4*{C1}*(A1*(v(540)^2)-{H})}
eflow2 570 0 value={if (v(560)>0, v(550),0)};.ends pump
```

Most of the parameters involved can be directly identified from Figure 5.3 and equations on this section. This subcircuit is used to simulate the transient response of the motor and pump, as follows:

```
*water_pump_transient

.include pump.lib

xpump 0 44 50 pump params: RA=0.15,LA=0.58,KM=0.37,A=0.0925,B=7.26e-6

+  F=0.00083, RF=0.15,LF=0.58, J=0.0608,A1=1.35e-6,B1=0.0015,C1=-3.32,H=11.5

vin 44 0 0 pulse (0,200,0,10m,10m,50,100)

.tran 0.01u 20 1e-6

.probe

.end
```

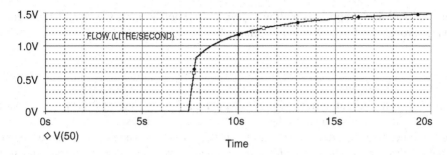

Figure 5.5 Transient of flow produced by a centrifugal pump after a voltage step of 200 V. Warning, *y*-axis is the flow (l/s)

The slow mechanical response towards the nominal conditions of operation is shown in Figure 5.5, where the evolution of the flow after the power supply has been switched ON is shown.

It is seen that the flow remains at zero until approximately 7.5 s and then rises towards the predicted steady-state value of 1.5 l/s. The electrical power required reaches the predicted 340 W in Table 5.1 for $H = 11.5$ m. The model described in this section is used in Chapter 7 to simulate long-term pumping system behaviour.

5.3   PV Modules Connected to a Battery and Load

The most commonly used connection of a PV system to a battery and a load is the one depicted in Figure 5.6 where the three components are connected in parallel. Of course, a battery is necessary to extend the load supply when there is no power generated by the PV modules in absence of irradiance, or when the power generated is smaller than required. The battery will also store energy when the load demand is smaller than the power generated by the PV modules.

A battery is an energy storage element and can be interpreted as a capacitive load connected to the PV generator output. As can be seen in Figure 5.6 the voltage Vbat is

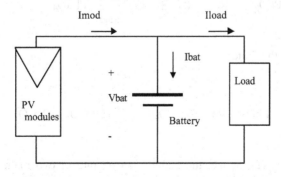

Figure 5.6 Standalone basic PV system

common to all the components and by applying Kirchhoff's current law (KCL), the current flowing through the three elements is related by:

$$I_{mod} = I_{bat} + I_{load} \tag{5.33}$$

Where I_{mod} is the output current of the PV system, I_{bat} is the current flowing to the battery and I_{load} is the current supplied to the loads. As can be seen in Figure 5.6, I_{bat} has a positive sign at the positive terminal and charges the battery, whereas when the I_{bat} sign is negative, the battery discharges. The sign of I_{bat} is determined at every time t, by the instantaneous PV system $I(V)$ characteristics, according to the irradiance and temperature values, and by the instantaneous value of the current demanded by the load.

5.3.1 Lead–acid battery characteristics

Lead–acid batteries are the most commonly used energy storage elements for standalone photovoltaic systems. The batteries have acceptable performance characteristics and life-cycle costs in PV systems. In some cases, as in PV low-power applications, nickel–cadmium batteries can be a good alternative to lead–acid batteries despite their higher cost.

Lead–acid batteries are formed by two plates, positive and negative, immersed in a dilute sulphuric acid solution. The positive plate, or anode, is made of lead dioxide (PbO_2) and the negative plate, or cathode, is made of lead (Pb).

The chemical reactions at the battery, in the charge and discharge processes are described below:

- Negative plate, anode.

$$Pb + SO_4^{2-} \underset{\text{charge}}{\overset{\text{discharge}}{\rightleftharpoons}} PbSO_4 + 2e^- \tag{5.34}$$

- Positive plate, cathode.

$$PbO_2 + SO_4^{2-} + 4H^+ + 2e^- \underset{\text{charge}}{\overset{\text{discharge}}{\rightleftharpoons}} PbSO_4 + 2H_2O \tag{5.35}$$

- Total reaction at the battery.

$$PbO_2 + Pb + 2H_2SO_4 \underset{\text{charge}}{\overset{\text{discharge}}{\rightleftharpoons}} 2PbSO_4 + 2H_2O \tag{5.36}$$

As has been discussed above, the battery can operate in two main modes: charge or discharge, depending on the I_{bat} sign.

While in charge mode, the current I_{bat} flows into the battery at the positive terminal, and it is well known that the battery voltage V_{bat} increases slowly and the charge stored increases. On the contrary, while in discharge mode, the current flows out of the positive terminal, the battery voltage, V_{bat}, decreases and the charge stored decreases supplying charge to the load.

In addition to the two main operation modes, the complex battery behaviour is better modelled if two additional operating modes are considered: undercharge and over-charge.

When the battery charge is close to the recommended minimum value and the circuit conditions force the battery to further discharge, the undercharge state is reached, characterized by a strong decrease of the electrolyte internal density, causing sedimentation at the bottom of the battery elements. This process strongly reduces the total battery capacity and, if the battery remains in this mode for a long time, irreversible damage can be caused.

The last mode of operation, or overcharge mode, is reached when the battery charge exceeds the maximum recommended value. At this point two different effects appear: gassing and corrosion of the positive battery electrode material. In this mode the battery shows an effective reduction of the available battery capacity. If the overcharge remains the battery enters into saturation and no more charge will be stored. Figure 5.7 shows a typical evolution of the battery voltage along the modes of operation described above for a 2 V battery element. As can be seen, after a slow increase of the voltage, overcharge and saturation regions produce a stagnation of the element voltage. As soon as the discharge is started, the battery voltage drops quite sharply and then tends to smaller values until the underdischarge zone is reached when a drop to zero volts occurs.

The main parameters that usually define and rate a battery are (a) the nominal capacity, C_x, for a rate of discharge of x hours, (b) the charge/discharge rate and (c) the state of charge, SOC.

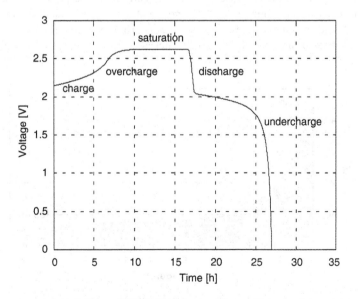

Figure 5.7   Battery modes of operation

(a) Nominal capacity

The nominal capacity is defined as the total charge that can be stored. This parameter is given for certain measurement conditions by the manufacturers, usually by measuring the charge delivered by the battery in a given period of time at a given discharge rate and temperature.

Depending on the length of time considered for the discharge, different nominal capacities can be defined. Most standard time lengths provided by the manufacturers are 5 h, 10 h and 100 h. According to these time lengths, nominal capacities, C_5, C_{10} and C_{100} are defined and given in units of Ah, where the subscript index indicates the discharge time length.

For a given battery, the capacity also depends on the charge/discharge rate, Figure 5.8 illustrates an example of the available battery capacity, considering different constant discharge current rates, for a $C_{20} = 100$ Ah battery.

Sometimes the capacity is given in units of energy, indicating the total energy that can be supplied by the battery from full charge to cut-off voltage, and is related to the capacity expressed in Ah by,

$$C_x(\text{Wh}) = C_x(\text{Ah})Vbat\,(\text{V}) \tag{5.37}$$

where x is the discharge time length and $Vbat$ is the battery voltage.

(b) Charge/discharge rates

The charge or discharge rates are defined as the relationship between the nominal capacity of the battery and the charge/discharge current values. In the case of a discharge the discharge rate is the time length needed for the battery to discharge at a constant current. The discharge rate coincides with the capacity subscript index considered in the previous section.

The battery capacity is a function of the charge and discharge rates. The battery capacity increases for longer discharge rates, this effect appearing as a consequence of the deeper electrolyte penetration into the battery plate material.

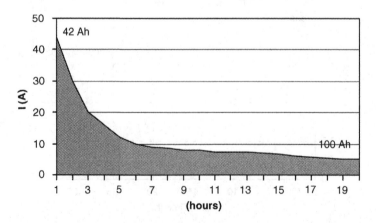

Figure 5.8   Battery capacity variation as a function of discharge rate

(c)   Battery state of charge, SOC

The state of charge is an important parameter present in most of the available battery models and is related to the electric charge stored by the battery at a given time. The SOC value is the ratio of the available charge at a given time divided by the maximum capacity and is written as follows:

$$SOC = \left(1 - \frac{Q}{C}\right)$$ (5.38)

$$0 \leq SOC \leq 1$$ (5.39)

where C (Ah), is the battery capacity and Q (Ah) is the charge already delivered by the battery at the time of interest.

As can be seen in equation (5.38), the SOC is a function of the battery capacity, C, which strongly depends, apart from on the electrolyte temperature, on the charge/discharge rate history as discussed above. This dependence has to be taken into account for correct modelling of the battery behaviour.

A parameter which is defined as the complement to unity of the SOC, is called depth of discharge, DOD, it represents the fraction of discharge reached by the battery, and is given by,

$$DOD = 1 - SOC$$ (5.40)

Sometimes in the technical literature, the SOC is given in units of energy and hence equals the value of the remaining energy at a given time. We will be using within the battery model the following parameters for the SOC:

SOC_1	Initial battery state of charge in %
SOC_m	Maximum battery energy (Wh)
SOC_n (t) (%)	Normalized value of the remaining energy to SOC_m

5.3.2   Lead–acid battery PSpice model

Lead–acid batteries are difficult to model, and the estimation of the battery state of charge value is recognized as one of the most complex tasks [5.3], [5.4]. A lead–acid battery PSpice model, implementing a dynamic estimation of the SOC (Wh), is described in this section allowing an accurate simulation of the battery behaviour.

The model is based on the equations described by Lasnier and Tang in reference [5.5] and in an earlier PSpice version [5.6], which has since been revised. The model has the following input parameters:

- Initial state of charge: SOC_1 (%), indicating available charge.

- Maximum state of charge: SOC_m (Wh), maximum battery capacity.

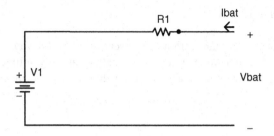

Figure 5.9   Battery model

- Number of 2 V series cells: n_s.
- Two empirical constants depending on the battery characteristics:
  - K (adimensional): charge/discharge battery efficiency;
  - D (h^{-1}): battery self-discharge rate.

The electrical battery model is composed of a voltage source V_1 in series with a resistor R_1, as shown in Figure 5.9. The values of V_1 and R_1 depend on the battery operation mode at a given time.

Although all modes of operation can be taken into account in the model we restrict ourselves here to the two main modes of operation: charge and discharge, while remembering that most PV systems include overcharge and undercharge protections to avoid a reduced battery life and extra cost to PV system maintenance and part replacements.

The implementation of the model consists basically of assigning different expressions to the values of V_1 and R_1 in each different mode as follows:

Charge mode

$$V_1 = V_{ch} = (2 + 0.148\beta)n_s \tag{5.41}$$

$$R_1 = R_{ch} = \frac{0.758 + \frac{0.1309}{(1.06 - \beta)}}{SOC_m}n_s \tag{5.42}$$

with

$$\beta = \frac{SOC}{SOC_m} \tag{5.43}$$

where the subindex '*ch*' indicates charge mode and, as defined above, SOC_m is the maximum value of battery state of charge, i.e. one of the input model parameters, and SOC is the battery state of charge at a given time, calculated internally by the model, both in Wh units. The battery voltage is given by

$$V_{bat} = V_{ch} + I_{bat}R_{ch} \tag{5.44}$$

Discharge mode

$$V_1 = V_{dch} = (1.926 + 0.124\beta)n_s \qquad (5.45)$$

$$R_1 = R_{dch} = \frac{0.19 + \frac{0.1037}{(\beta - 0.14)}}{SOC_m}n_s \qquad (5.46)$$

where the subindex '*dch*' indicates discharge mode. The output battery voltage is given by

$$V_{bat} = V_{ch} + I_{bat}R_{dch} \qquad (5.47)$$

where the battery current (I_{bat}), defined as shown in Figure 5.7, takes negative values.

The battery model is represented by the PSpice equivalent circuit shown in Figure 5.10, where controlled voltage sources '*evch*' and '*evdch*' implement the voltages V_{ch} and V_{dch}, respectively.

The PSpice implementation of the migration of the battery mode from charge to discharge or vice versa is achieved using the two current controlled switches, '*Wch*' and '*Wdch*' shown in Figure 5.10.

As can be seen, these switches connect the battery output nodes to the corresponding internal voltage sources and resistors depending on the operation mode, which is identified by the sign of the current I_{bat}: positive sign corresponding to charge mode and negative sign to discharge mode, sensed at the 'Vcurent' source in the figure.

An important part of the PSpice model is related to the estimation of the instantaneous value of the *SOC* (Wh). The estimation is performed as described by the following equation:

$$SOC(t + dt) = SOC(t)\left(1 - \frac{D}{3600}dt\right) + k\left(\frac{V_{bat}I_{bat} - R_1 I_{bat}^2}{3600}\right)dt \qquad (5.48)$$

where all the parameters have already been defined. Equation (5.48) basically is the energy balance equation computing the value of the *SOC* increment as the energy increment in a differential of time taking into account self-discharge and charge discharge efficiency. As time has units of seconds, some terms have to be divided by 3600 so that *SOC* is in Wh.

Equation (5.48) can be simplified substituting V_{bat} as a function of V_1, see Figure 5.9,

$$SOC(t + dt) = SOC(t)\left(1 - \frac{D}{3600}dt\right) + \left(\frac{kV_1 I_{bat}}{3600}\right)dt \qquad (5.49)$$

and finally,

$$\frac{SOC(t + dt) - SOC(t)}{dt} = \frac{(kV_1 I_{bat})}{3600} - \frac{DSOC(t)}{3600} \qquad (5.50)$$

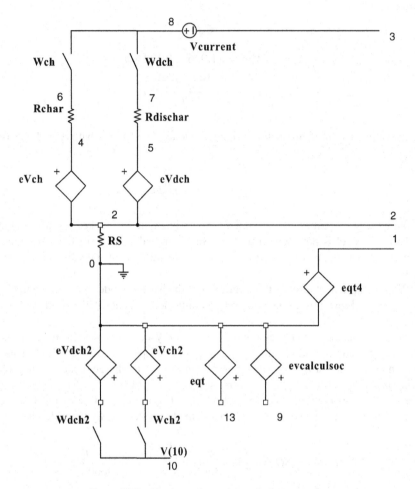

Figure 5.10   Battery schematic equivalent circuit

Equation (5.50) can be easily solved using the 'sdt' PSpice function which is the time integral operation, as shown in the subcircuit netlist, where the voltage controlled source 'eqt' evaluates the *SOC* value as follows:

$$SOCn(t) = SOC1 + \frac{1}{SOCm}\int \left(\frac{kV_1 I_{bat}}{3600}\right) - \frac{D\,SOCn(t-\tau)SOCm}{3600}\,dt \qquad (5.51)$$

where SOC_1 is the initial battery state of charge in %, SOC_m is the maximum battery state of charge in Wh units, and $SOC_n(t)$ is the *SOC* normalized to the maximum battery state of charge SOC_m, then SOC_n is the *SOC* in %. As can be seen, to simplify the numerical resolution of equation (5.50) an approximation is used by equation (5.51), where the term $SOC_n(t-\tau)$ is introduced in substituting $SOC(t)$, τ is the internal time step used by PSpice in the simulation.

The controlled voltage sources: 'evcalculsoc', 'eqt', 'eqt4', 'evch2', 'evdch2' and switches 'Wch2' and 'Wdch2' are used to evaluate the $SOC_n(t)$ evolution, as shown in Figure 5.10 and also in the following netlist of 'bat.cir'. Finally the controlled voltage source 'eqt4', of the following netlist also shown in Figure 5.10, limits the SOC_n value at the battery model output, battery subcircuit node 1. This is done as follows:

$$\text{eqt4 1 0 value} = \{\text{limit } (v(13),\ 0,\ 1)\}$$

where $v(13)$ is the evaluated SOC_n value, following equation (5.51), by the controlled voltage source 'eqt' of the battery model.

The battery model subcircuit implemented by this netlist has three access nodes, shown at Figure 5.11:

- $\pm$Vbat, as shown, nodes 3 and 2 in the netlist.

- SOC_n, node 1. The voltage of this node corresponds to the normalized $SOC(t)$ value, calculated from equation (5.51), and limited between 0 and 1.

```
* lead-acid battery model *
* Bat.cir
.subckt bat 3 2 1 PARAMS: ns=1, SOCm=1, k=1, D=1, SOC1=1

evch 4 2 value={(2+(0.148*v(1)))*ns}
evdch 5 2 value={(1.926+(0.124*v(1)))*ns}
rs 2 0 0.000001
rch 4 6 {rchar}
.func rchar() {(0.758+(0.1309/(1.06-SOC1)))*ns/SOCm}
rdch 5 7 {rdischar}
.func rdischar() {(0.19+(0.1037/(SOC1-0.14)))*ns/SOCm}
vcurrent 3 8 dc 0
Wch 6 8 vcurrent sw1mod
.model sw1mod iswitch (ioff=-10e-3, ion=10e-3, Roff=1.0e+8, Ron=0.01)
Wdiscar 7 8 vcurrent sw2mod
.model sw2mod iswitch (ioff=10e-3, ion=-10e-3, Roff=1.0e+8, Ron=0.01)

* SOCn evolution
eqt 13 0 value={SOC1+(sdt(v(9))/SOCm)}
eqt4 1 0 value=limit (v(13), 0, 1)}
evcalculsoc 9 0 value={(k*v(10)*i(vcurrent)/3600)-(D*SOCm*v(13)/3600)}

evch2 11 0 value={(2+(0.148*v(1)))*ns}
evdch2 12 0 value={(1.926+(0.124*v(1)))*ns}
Wdch2 12 10 vcurrent sw2mod
Wch2 11 10 vcurrent sw1mod
.ends bat
```

Despite the limitations of this model, restricted to charge and discharge modes of operation, and hence restricting the use of the model within the safe zone of *SOC* values, typically comprised between 30% and 80% of the maximum, the results provide good estimates in general, and with the advantage of faster computation times and usability with

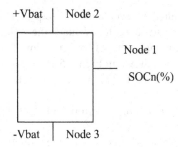

Figure 5.11   Schematic diagram of bat.cir showing access nodes

student or evaluation PSpice versions. This is why the effects of temperature have not been considered here for the battery although they can be easily included, see [5.5] and [5.6].

The following Example 5.3 shows the main features of the model.

Example 5.3

Write a PSpice netlist to simulate the evolution of the battery voltage and of the state of charge considering a battery driven by a sinusoidal current source of 20 A amplitude and a 1 kHz frequency, as shown in Figure 5.12. Consider a 12-element, 2 V nominal voltage, maximum $SOC = 1200$ Wh, charge/discharge battery efficiency $k = 0.8$, battery self-discharge rate $D = 0.001\,\mathrm{h}^{-1}$, and an initial value of the normalized state of charge, $SOC_1 = 50\%$.

Solution

The circuit is depicted in Figure 5.12. The PSpice netlist is the following:

```
* Example 5.3. cir
.inc bat.cir
.temp = 27
xbat1 3 2 1 bat params: ns = 12, SOCm = 1200, k = 0.8, D = 1e-3, SOC1 = 0.5
isin 2 3 sin (0 20 0.001 0 0 0); sinusoidal current source
r44 1 0 1000000
.plot dc v(3)
.tran 1s 2000s
.probe
.end
```

Figure 5.13 shows the current flowing in and out of the battery and the battery voltage values as a function of time. The two different battery modes of operation, charge and discharge, can be clearly identified by the sharp voltage changes at the current zero-crossings.

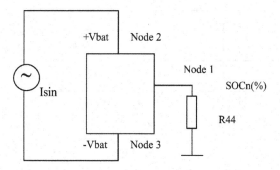

Figure 5.12 Schematic diagram for Example 5.3

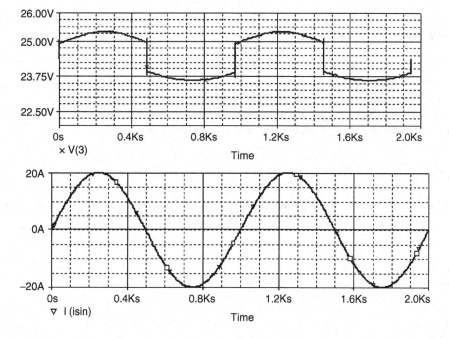

Figure 5.13 Battery voltage, top, and current, bottom, Example 5.3

The values of the normalized state of charge are shown in Figure 5.14, where it can be seen that the cycle starts at 0.5 (which is the initial SOC_1 value) and increases to a maximum value of 0.53 at one-quarter of the period of the sinusoidal source. The SOC_n has a periodical evolution according to the battery current forced by this simulation.

5.3.3 Adjusting the PSpice model to commercial batteries

The battery model discussed so far can be adjusted to improve PSpice simulation results of PV systems involving any given type of commercial battery, either by taking into account the data provided by the manufacturer or by deriving empirically the values of the model parameters. This is illustrated in Example 5.4.

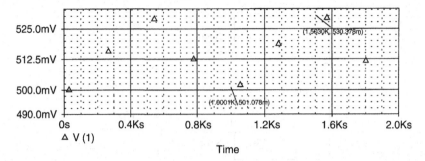

Figure 5.14 SOC_n evolution. Warning the y-axis is the value of the normalized SOC(adimensional)

Example 5.4

Consider a commercial battery model 7TSE 70 by ATERSA in reference [5.7] with the following characteristics: C_{10} of 784 Ah, $SOC_1 = 0.9$, $V_{bat} = 2\,\text{V}$. Consider also a charge/discharge efficiency of 100%, $k = 1$, battery self-discharge rate: $D = 1\text{e} - 5\ (\text{h}^{-1})$.

Compute the values of the internal source V_1 and R_1 from the equations (5.41) to (5.46) and write a modified subcircuit netlist for this specific battery.

Solution

After calculation the resulting parameter values are included in the modified file as follows:

```
*7TSE.cir
.subckt bat 3 2 1 PARAMS: ns = 1, SOCm = 1568, k = 1, D = 1e-5, SOC1 = 0.9
evch 4 2 value = {(2+(0.16*v(1)))*ns}
evdch 5 2 value = {(1.926+(0.248*v(1)))*ns}
rs 2 0 0.000001
rch 4 6 {rchar}
.func rchar() {(0.758+(0.1309/(1.06- SOC1)))*ns*100/SOCm}
rdch 5 7 {rdch}
.func rdch() {(0.19+(0.1037/(SOC1-0.14)))*ns*250/SOCm}
vcurrent 3 8 dc 0
Wch 6 8 vcurrent sw1mod
.model sw1mod iswitch (ioff = -10e-3, ion = 10e-3, Roff = 1.0e +8, Ron = 0.01)
Wdch 7 8 vcurrent sw2mod
.model sw2mod iswitch (ioff = 10e-3, ion = -10e-3, Roff = 1.0e +8, Ron = 0.01)
* SOCn
eqt 13 0 value = {SOC1 +(sdt(v(9))/SOCm)}
eqt4 1 0 value = {limit (v(13), 0, 1)}
evcalculsoc 9 0 value = {(k*v(10)*i(vcurrent)/3600)-(D*SOCm*v(13)/3600)}
evch2 11 0 value = {(2+(0.16*v(1)))*ns}
evdch2 12 0 value = {(1.926+(0.248*v(1)))*ns}
Wdch2 12 10 vcurrent sw2mod
Wch2 11 10 vcurrent sw1mod
.ends bat
```

5.3.4 Battery model behaviour under realistic PV system conditions

Models are aimed at reproducing the behaviour of the system under realistic operation conditions. In a real PV system, the irradiance and temperature values evolve during the observation time according to meteorological conditions and site location. Although the irradiance changes are not very sudden in general, they are of random nature and differ from the laboratory conditions under which the parameters of the PV components have been measured. This raises the question as to whether the performance of the battery model will accurately reproduce the dynamics of the system operation. This section describes a simple procedure which can be used to compare the simulation with real-system operation in those cases where system monitoring data are available.

PSpice can easily handle monitoring data provided they are in electronic format and can be transformed in a 'stimulus' PSpice file. The corresponding format to stimulus files is as shown below.

Syntax for stimulus files

* file_name.stl

.stimulus <vname> <iname> pwl

+ <time value> <variable value>

where a name starting with a 'v' is to be used for a voltage source or a name starting by an 'i' is to be used for a current source. Following the stimulus pwl source definition, all lines must list the values: time and variable; every line has to start with a + sign at the first column.

Once this file has been written, it can be included in a. cir file as follows:

. include file_name.stl

vxx   node+ node−   stimulus   vname

in case it is a voltage source. Vname in this sentence has to be the same as in the stimulus sentence within the stimulus file. If it is a current source, then ixx should be used instead.

This is illustrated below where part of files exportmed3v.stl and expormedI.stl are listed. After the first line, which corresponds to the title, the file is defined as a stimulus by the *.stimulus* sentence, and given a name: Vgg and Igg. The pwl command indicates the stimulus type: piecewise lineal waveform.

```
***exportmed3V  Vbat(V)          *** exportmedI  Ibat(Ic − Id)
.stimulus Vgg pwl                 .stimulus Igg pwl
+    0     26.414                  +    0     0.42916811
+   120   26.44                    +   120   0.55016811
+   240   26.462                   +   240   0.57716811
+   360   26.483                   +   360   0.62513536
+   480   26.506                   +   480   0.65013536
+   600   26.518                   +   600   0.64816811
+   720   26.527                   +   720   0.62516811
+   840   26.539                   +   840   0.62513536
```

Finally the data is organized into three columns, using a text file format where the column separator is a tabulator. The first column must be filled by + signs as indicated above, the second one corresponds to the *x*-axis, time, and the third column corresponds to data: currents, voltages, etc. This is included as:

.inc exportmed3v.stl

where exportmed3v.stl is the stimulus file included in the simulation.

Finally the included stimulus must be associated to a current or voltage source, for example, the following sentence in the netlist:

vmeasured 4 2 stimulus Vgg

will create a voltage source, vmeasured, between nodes 4 and 2 associated to the voltage data included in file exportmed3v.stl, because the stimulus name Vgg associated to vmeasured source is the same as the stimulus name fixed at this file.

In Example 5.6, the two previously shown stimulus files are used.

Example 5.5

Write a PSpice netlist of a standalone PV system composed of 8 PV modules ($I_{sc} = 4.4\,\text{A}, V_{oc} = 20.5\,\text{V}$), 4 parallel rows of 2-series PV modules, connected to 12 ATERSA 7TSE 70 battery elements. The battery is also connected to a controlled load as shown in Figure 5.15.

The PV system has been monitored for a period of five consecutive days in Barcelona, Spain and the battery voltage (*Vbat*) and battery current (*Ibat*) have been measured, as well as voltages and currents at the loads and at the output of the PV modules. Some incandescent lamps have been used as load.

The measured battery current over these five days, *Ibat*, has been included in the PSpice netlist as a current source, defined as a pwl current stimulus in the sexpotmedI.stl file. Figure 5.16 shows the battery current, *Ibat*, evolution over the five days.

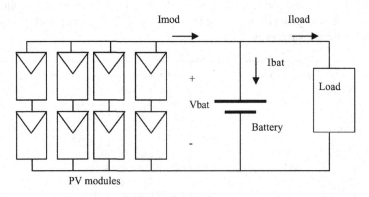

Figure 5.15 PV system description

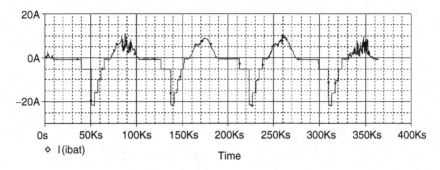

Figure 5.16   Battery current, I_{bat}, evolution

As can be seen the load is connected to the battery at night and the power demand of the load is not constant. The netlist is the following:

```
* Example 5.5.cir
.temp = 27
.inc bat.cir
xbat1 3 2 1 bat params: ns = 12, SOCm = 13200, k = .8, D = 1e-5, SOC1 = 0.9
.inc exportmed3V.stl
vmeasured 4 2 stimulus Vgg
.inc exportmedI.stl
ibat 2 3 stimulus igg
r44 4 2 1000000
.plot dc v(3)
.tran 1s 364200s
.probe
.end
```

The measured battery voltage over these five days, vmeasured, has been included in the PSpice netlist as a voltage source, defined as a pwl voltage stimulus in the expotmed3V.stl file. This source is included in the netlist to allow a final comparison with the battery voltage obtained by the PSpice simulation as shown in Figure 5.17. In this figure v(4) is the real measured battery voltage, and v(3) is the simulation result obtained for the same node voltage.

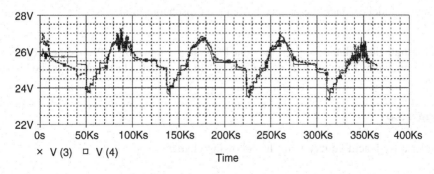

Figure 5.17   Battery voltage evolution

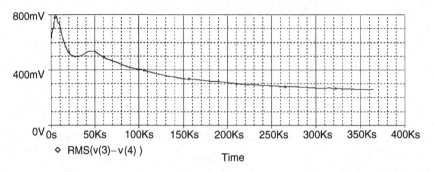

Figure 5.18 Error evaluation

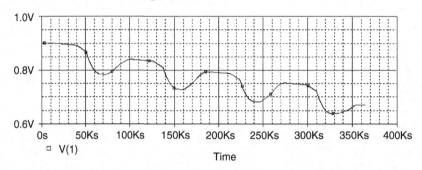

Figure 5.19 SOC_1 simulation results. Warning, the y-axis is the numerical value of the SOC_n

Figure 5.18 shows the value of the error between simulated, v(3), and measured, v(4), calculated using the RMS PSpice function.

Finallly, the battery normalized state of charge evolution, $SOCn$, obtained as result of this PSpice simulation is shown in Figure 5.19.

As can be seen the value of SOC_n ranges from the initial value of 0.9 to a final value of 0.668 at the end of the five simulated days.

As explained earlier in this chapter, the battery model does not work very well for SOC values over 0.8, because overcharge is not considered by the model. This explains the error evolution shown in Figure 5.18, where the error is larger at the begining of the simulation, corresponding to high values of SOC. As the same time as SOC decreases along the simulation, the error decreases too, as expected when the battery behaviour is more accurately modelled once the battery has entered discharge mode.

To illustrate the simulation of a standalone PV system formed by a PV module, a battery and a load, as shown by Figure 5.6, using the PSpice battery model, a last example is shown below.

Example 5.6

Consider a lead–acid battery with the following characteristics:

- Initial state of charge: $SOC_1 = 0.85$

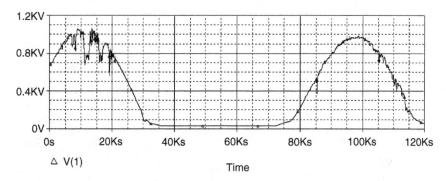

Figure 5.20 Irradiance profile for Example 5.6

- Maximum state of charge: $SOCm = 576$ Wh
- Number of 2 V series cells: $ns = 6$
- $K = 0.7$, charge/discharge battery efficiency
- $D = 1e - 3(\text{h}^{-1})$, Battery self-discharge rate

The battery is connected to a PV module formed by 36 solar cells in series. The module characteristics are:

- short-circuit current, $iscmr = 5$ A
- open-circuit voltage, $vocmr = 22.3$ V
- maximum output power, $pmaxr = 85$ W

Finally a 4 Ω DC load is connected permanently to the battery.

Consider the irradiance profile shown in Figure 5.20 as input data for the simulation of the PV system behaviour.

(a) Write the PSpice netlist for the simulation of the above-described PV system. Include the irradiance profile as Irrad2.stl file. The circuit is shown in Figure 5.21.

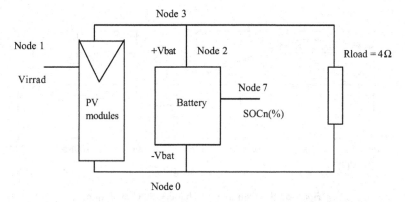

Figure 5.21 Schematic diagram for Example 5.6

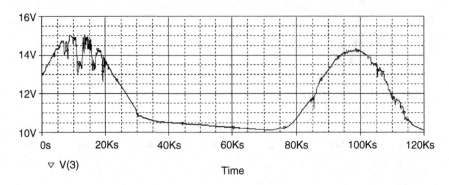

Figure 5.22   Example 5.6, battery voltage evolution

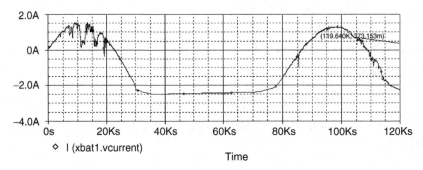

Figure 5.23   Example 5.6, battery current evolution

The netlist is included in Annex 5 as 'example 5.6' netlist.

(b) Run a Pspice simulation and obtain the battery voltage and current evolution. The simulation results are shown in Figures 5.22 and 5.23.

(c) Plot the battery *SOC* evolution and obtain its final value. The *SOC* evolution is shown in Figure 5.24.

The battery *SOC* value at the end of the two simulated days is 0.37.

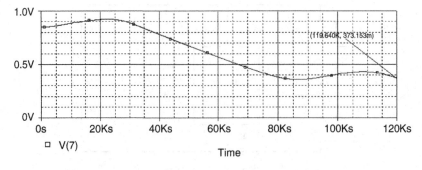

Figure 5.24   Example 5.6, battery SOC evolution

5.3.5 Simplified PSpice battery model

The battery model presented in the previous section can be further simplified by considering that the resistance R_1 of the battery model, shown in Figure 5.9, has the same value when battery is either in charge or in discharge mode. This can reduce simulation accuracy in some cases, but by selecting a correct value for a specific battery, good simulation results are generally obtained.

This model simplification results in a new battery model which avoids some of the switches in the model in Figure 5.10. Elimination of these switches reduce the probability of convergence problems using PSpice. The resulting simplified battery model netlist is shown below. Instead of switches, IF sentences are used to obtain a more compact model. We strongly suggest the use of the simplified model if convergence problems arise in a specific simulation due to the switches.

Figure 5.25 shows an schematic plot corresponding to the netlist associated with this simplified battery model. The connection nodes of this new model are the same as those used in the previous battery model shown in Figure 5.11.

The corresponding netlist is as follows:

```
************ Simplified battery model
*batstd.cir
.subckt batst 3 2 1 PARAMS: ns=1, SOCm=1, k=1, D=1, SOC1=1
evch 4 2 value={(2+(0.16*v(1)))*ns}
evdch 5 2 value={(1.926+(0.248*v(1)))*ns}
rserie 8 88 {rs}
.func rs() {(0.7+(0.1/(abs(SOC1-0.2))))*ns/SOCm}
ebat 88 2 value={IF (i(vcurrent)>0, v(4), v(5))}
vcurrent 3 8 dc 0
rsc 2 0 0.0001
eqt 13 2 value={SOC1+(sdt(v(9))/SOCm)}
eqt4 1 2 value= {limit (v(13), 0, 1)}
evcalculsoc 9 2 value={(k*v(10)*i(vcurrent)/3600)-(D*SOCm*v(13)/3600)}
ecoch 10 2 value={IF (i(vcurrent)>0, v(4), v(5))}
.ends batst
```

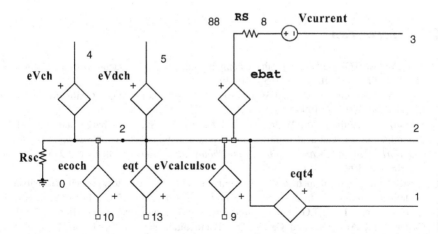

Figure 5.25 Schematic diagram of the simplified PSpice battery model

5.4 Problems

5.1 Simulate a standalone PV system formed by a PV module, a battery and two loads in parallel with the following characteristics:
  - PV module: formed by 36 solar cells in series, $ns = 36$, maximum power: $pmaxmr = 85\,W$, short-circuit current: $iscmr = 5\,A$, open-circuit voltage: $vocmr = 22.3\,V$.
  - Battery: 12 V, six elements of 2 V in series, $ns = 6$. Battery capacity: $SOCm = 576\,Wh$, initial state of charge: $SOC1 = 0.75$.
  - Load: two resistive loads of 90 Ω and 8 Ω in parallel.

  Consider as input data the irradiance profile included in file irrad2d.stl shown in Figure 5.20 and a constant temperature of 20 °C. Using the PSpice models presented in this chapter for PV module, module_1.lib and batteries, bat.cir, run the simulation of this standalone PV system and obtain: final battery SOC, power delivered by the PV module to battery and loads and battery current evolution.

5.2 Simulate a standalone PV system formed by a PV generator, a battery and a resistive load with the following characteristics:
  - PV generator: four, 85 Wpeak, PV modules, 2 series x 2 parallel: $nsg = 2$ $npg = 2$. Each module formed by 36 solar cells in series: short-circuit current: $iscmr = 5.2\,A$, open-circuit voltage: $vocmr = 21.2\,V$. The maximum power point of the PV module in standard conditions has the following coordinates: $immr = 4.9\,A$, $vmmr = 17.3\,V$.
  - Battery: 24 V, 12 elements of 2 V in series, $ns = 12$. Battery capacity: $SOCm = 984\,Wh$, initial state of charge: $SOC1 = 0.6$.
  - Load: a resistive load of 6 Ω permanently connected to the battery.

  Consider as input data the irradiance profile included in file irrad2d.stl shown in figure 5.20 and a constant temperature of 22 °C. Using the PSpice model presented in this chapter for batteries, bat.cir and the generator_beh.lib file for the PV generator, run the simulation of this standalone PV system and obtain: final battery SOC, power delivered by the PV module to battery and loads and battery current evolution.

5.5 References

[5.1] Yang, Y.C., and Gosbell, V.J., 'Realistic computer model of a DC machine for CADA topology on SPICE2', *PESC'88*, IEEE pp. 765–771, 1988.

[5.2] Wenham, S.R., Green, M., and Watt, M.E., 'Applied Photovoltaics Centre for Photovoltaic Devices and Systems, 1994.

[5.3] Schmitz, C., Rothert, M., Willer, B., and Knorr, R., 'State of charge and state of health determination for lead-acid batteries in PV power supply systems', *Proceedings of the IEEE 2nd World Conference and Exhibition on Photovoltaic Energy Conversion*, 2157–64. Vienna, Austria, July 1998.

[5.4] Casacca, M.A., and Salameh, Z.M., 'Determination of lead-acid battery capacity via mathematical modelling techniques', *IEEE Transactions on Energy Conversion*, **7** (3), 442–6, September 1992.

[5.5] Lasnier, F., and Tang, T.G., *Photovoltaic Engineering Handbook*, Adam Hilguer, 1990.

[5.6] Castañer, L., Aloy, R., and Carles, D., 'Photovoltaic system simulation using a standard electronic circuit simulator, *Progress in Photovoltaics*, **3**, 239–52, 1995.

6

Power Conditioning and Inverter Modelling

Summary

Power-conditioning equipment for protection or control such as DC-DC converters, charge regulators or DC-AC converters are part of a PV system in most applications. This chapter is focused on the description of models for these elements in order to simplify the PSpice simulation for a wide range of PV system architectures incorporating power-conditioning equipment. The models explained in this chapter have different degrees of complexity and simulation quality.

6.1 Introduction

The energy supplied by the PV modules to the system is subject to variations depending on the operating conditions, especially the irradiance and temperature values, as well as the load demand profile. Power-conditioning devices and circuits are included in the PV system with the purpose either of protection or of losses reduction, allowing the PV generator to work as close as possible to the maximum power point, thereby optimizing the energy transfer, resulting in a more efficient system.

Charge regulator elements can also be introduced into the system to prevent undesired operation conditions and to protect the battery from entering overcharge and undercharge states.

This chapter describes some of these power-conditioning elements as well as charge regulation modules and the operation features in a PV system.

6.2 Blocking Diodes

In absence of irradiance the PV module short-circuit current, I_{sc}, is zero and the module $I(V)$ characteristic becomes similar to that of a diode as shown in Figure 6.1, where an ideal module with an equivalent infinite shunt resistance has been considered.

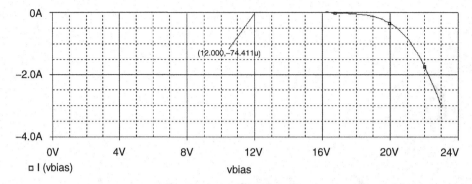

Figure 6.1 PV generator characteristics in darkness

Under these circumstances, the PV generator acts as an additional load to the battery. Taking into account the PV systems current relationship in equation (5.35) in Chapter 5, part of the battery current, I_{bat}, could be derived through the PV modules. This effect is represented in Figure 6.1, where this part of the battery current can be identified as I(vbias). The cursor shows the value of current that can be derived from the PV module considering the module connection to a 12 V battery in the absence of irradiance. This effect can be prevented by including a blocking diode in the system. Figure 6.2 shows a PV system where a blocking diode has been connected in series with the PV array.

Some considerations have to be taken into account:

- The presence of a diode in series implies a diode voltage drop, which can be of the order of 1 V depending on the diode model and power rating.

- The magnitude of the current loss, I_{mod}, shown in Figure 6.1, depends strongly on the PV equivalent shunt resistance. Despite this dependence, considering typical equivalent shunt resistance values of real PV modules, the observed values of I_{mod} can be very small.

According to the above considerations, the presence of a blocking diode in the PV system has to be carefully considered as a safety protection, however it may not be necessary in some cases, especially if charge regulation elements are included. In this case the protection

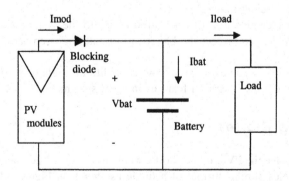

Figure 6.2 Block diagram of a PV system with blocking diode

function of the blocking diode will be covered by the regulation elements described in the sections below.

6.3 Charge Regulation

A key issue in any power conversion system is efficiency. As the energy flow from input to output strongly depends, not only on the irradiance and temperature, but also on the working conditions of all other system components, an efficient management of the flow becomes important in order to prevent the battery from entering overcharge and undercharge states.

Two basic types of electronic regulators are usually used for the battery charge management: parallel or shunt regulators and series regulators.

6.3.1 Parallel regulation

Shunt regulators are connected in parallel to the PV generator with the aim of diverting the excess energy generated, which occurs the battery reaches the overcharge mode. The input signal controlling the action of the regulator is the battery voltage, in particular when it reaches the onset of the overcharge mode, a shunt path for the current is activated. Figure 6.3 shows a typical configuration of this kind of regulator.

The transistor is OFF and I_{shunt} equals zero while $V_{control}$ remains smaller than the overcharge onset voltage, which must be known. When $V_{control}$ reaches this onset voltage the transistor turns ON and I_{shunt} becomes positive, diverting part of I_{mod} across the shunt branch, limiting I_{bat} and therefore the battery voltage V_{bat}.

In Figure 6.4, $V_{control}$ is generated by a simple reference comparator, which takes V_{bat} and the onset reference value (in the figure V_{ref}) at the input and issues a signal with a negative value, close to the negative bias comparator supply, when the battery voltage is below the reference value, and a positive voltage, close to the positive bias comparator supply, when the battery voltage is greater than the reference value.

The shunt regulation branch, composed of a transistor and a resistance, should be rated in order to dissipate the excess power generated by the PV modules and not delivered to the battery.

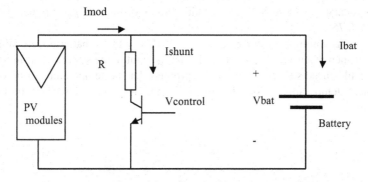

Figure 6.3 Typical shunt regulator configuration

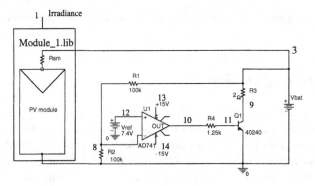

Figure 6.4   PV system with shunt regulation elements

As an example of a PSpice simulation of shunt regulation, consider the PV system shown in Figure 6.4.

This system includes a PV module, a battery and a shunt regulator circuit composed of a voltage comparator and a npn bipolar transistor as the main components, and four resistors. In particular, R_1 and R_2 are voltage dividers generating a voltage proportional to the battery voltage at the negative input of the comparator, R_3 and R_4 are power resistors in series with the collector circuit of the transistor and R_3 is in series with the base of the transistor. These two resistors limit the current through the transistor terminals.

Example 6.1 shows a simulation corresponding to the circuit shown in Figure 6.4. PSpice models for operational amplifiers and for transistors can be easily found in the PSpice library.

Example 6.1

Consider the circuit shown in Figure 6.4, write a netlist of the circuit and simulate the operation considering that the PV module is composed of 36 solar cells connected in series, with a short-circuit current of 5 A, an open voltage of 22.3 V and a total power of 85 W under standard AM1.5G 1 kW/m^2 irradiance and 25 °C operating temperature.

The battery shown in the circuit is formed by an association of six 2 V elements in series with a total capacity of 1840 Wh and the initial value of the SOC_n at the beginning of the simulation is 0.75.

The battery has a nominal voltage value $V_{bat} = 12$ V, and the battery overcharge onset is assumed to happen when $V_{bat} = 14.8$ V. Consider as input to the system an irradiance profile given by the file 'irrad.stl'. To simplify the problem let the temperature to be constant at 25 °C ambient temperature. Consider the PV module model described in Chapter 4, 'module_1.lib'.

Solution

We start writing the netlist

```
******** Example 6.1 netlist
****Module, shunt regulator and battery connection
xmodule 0 3 1 module_1 params:ta=25, iscmr=5, tr=25,
+vocmr=22.3,ns=36,np=1,nd=1,pmaxmr=85
.inc module_1.lib
xbat1 3 0 7 bat params: ns=6, SOCm=1240, k=.8, D=1e-5, SOC1=0.75
.inc bat.cir
R1 3 12 100000
R2 12 0 100000
R3 3 9 2
Q_Q1 9 11 0 Q40240
.model Q40240 npn
vref 8 0 dc 7.4
x741 12 8 13 14 10 ad741
.inc opamp.lib
R4 10 11 1250
vcc 13 0 dc 15
vee 14 0 dc -15
.inc irrad.stl
vmesur 1 0 stimulus Vgg
.tran 1s 130000s
.probe
.end
```

The irradiance profile for this example has been taken from real monitored data and has been included in the netlist by means of 'irrad.stl'. Figure 6.5 show the irradiance profile corresponding to two days of April in Barcelona (Spain).

Figure 6.6 shows the resulting battery voltage waveform. As can be seen, the battery voltage V(3), is limited to the value of 14.8 V by the shunt regulator as set by the circuit. The PV module output voltage shown, V1(xmodule.d1), is the voltage at node before the series resistance module Rsm (see Figure 6.4).

As can be seen in Figure 6.4 the operational amplifier compares $V_{bat}/2$ to a constant reference fixed voltage, $V_{ref} = 7.4$ V. When V_{bat} reaches the value of 14.8 V, i.e. entering into overcharge, the output of the comparator becomes positive enabling the shunt branch

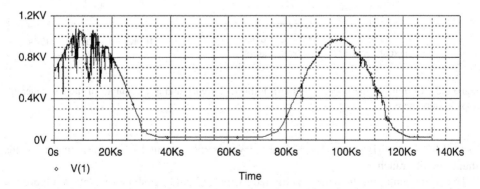

Figure 6.5   Irradiance profile in Example 6.1

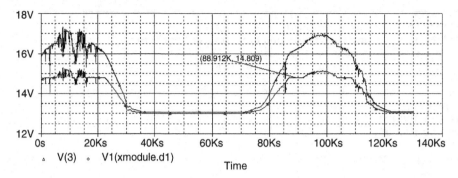

Figure 6.6 Simulation results for battery voltage, V(3) (bottom), and PV module voltage (top), V1(xmodule.d1), evolution

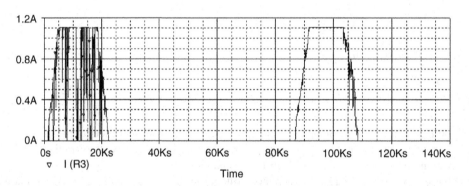

Figure 6.7 Current evolution across the shunt circuit branch, I(R3)

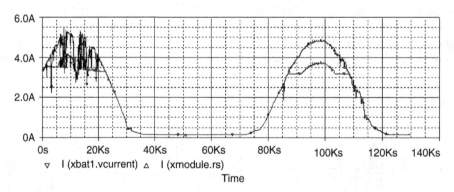

Figure 6.8 Currents at the battery (bottom), I(xbat1.vcurrent), and at the output of the PV generator (top), I(xmodule.rs)

formed by the transistor and R_3. Figure 6.7 shows the current I(R3) evolution across this shunt circuit branch.

The current entering the battery is limited by the action of the shunt regulation. Figure 6.8 shows the PV module output current, I(xmodule.rs), and the current at the battery,

I(xbat1.vcurrent). The difference between these two currents is the current derived by the regulation branch, formed by the transistor and R3, shown in figure 6.8.

6.3.2 Series regulation

The second regulation approach is known as series regulation. Today most of the charge regulators in photovoltaic applications use this kind of regulation [6.1]. Figure 6.9 shows a block diagram of a standalone PV system including a series charge regulator element.

As described above, the battery has a recommended voltage window between the low ($V_{\min}$) and high ($V_{\max}$) where it operates at rated capacity and efficiency. If the battery is forced to work outside of this window, it may be irreversibly damaged or operate incorrectly. The series charge regulators prevent the battery from working out of this voltage window.

Basically, the way the series charge controller works, is by opening the load when the battery reaches $V_{\min}$ and connecting the load circuit when the battery has been recharged enough so that its output voltage recovers. On the other hand, the charge regulator disconnects the battery from the PV array when full charge is achieved, this means when the battery voltage reaches $V_{\max}$, and resets the connection as soon as the battery has been discharged enough.

This series regulation can be easily implemented using standard electromechanic controlled switches or relays as shown in Figure 6.10. Table 6.1 illustrates an example for

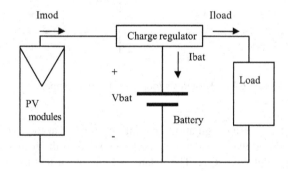

Figure 6.9 PV system including a series charge regulator

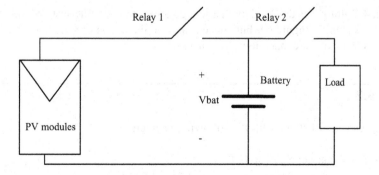

Figure 6.10 Schematic representation of charge series regulation

Table 6.1   State of switches

Relay 1 Battery-PVpanel		Relay 2 Battery-load	
CLOSED PV modules connected	OPEN PV modules disconnected	OPEN Load disconnected	CLOSED Load connected
$V_{bat} = 12.8$ V	$V_{max} = 13.9$ V	$V_{min} = 11$ V	$V_{bat} = 12$ V

a battery with a nominal voltage of 12 V, showing the desired battery voltages for switching relays 1 and 2; in this case the battery voltage window operation is limited by $V_{max} = 13.9$ V and $V_{min} = 11$ V.

The signal controlling the relay switching can be easily made using two comparator circuits with the hysteresis loops shown by Figure 6.11.

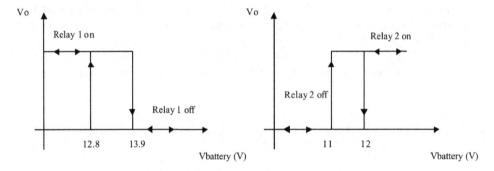

Figure 6.11   Hysteresis loops for the two comparators

By the same nature as series regulation, the only power lost at the regulator circuit is the losses at the relays themselves, which are small compared to the losses at the resistor and transistor used in parallel regulation. The balance of energy suffers to some extent because part of the energy generated by the PV modules is lost when the input to the battery switch is open.

Example 6.2 illustrates this kind of charge regulation strategy. The system simulated is the same system shown in Figure 6.9 but concentrates on the action of switch 1, which governs the battery–PV modules connection/disconnection.

Example 6.2

Consider a lead–acid battery with the following characteristics:

- Initial state of charge: $SOC_1 = 0.45$
- Maximum state of charge: $SOC_m = 1000$ Wh

- Number of 2 V series cells: $n_s = 6$

- $K = 0.8$, charge/discharge battery efficiency

- $D = 1 \times 10^{-5}$, battery self-discharge rate

- Battery voltage for PV module's disconnection: 13.9 V

- Battery voltage for PV module's reconnection: 13.4 V

The battery is connected to a Pv module formed by 36 solar cells in series. The module characteristics are:

- Short-circuit current, $i_{scmr} = 5\,A$

- Open-circuit voltage, $v_{ocmr} = 22.3\,V$

- Maximum output power, $p_{maxr} = 85\,W$

Finally an 800 Ω DC load is connected permanently to the battery.

Consider the irradiance profile shown in Figure 6.12 as input data for the simulation of the PV system behaviour.

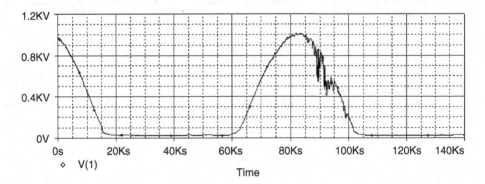

Figure 6.12 Irradiance profile

(a) Write the PSpice netlist for the simulation of the above PV system including a series charge regulator to control the PV module/battery switching. Use a non-inverting Schmitt trigger to implement the switching control. Include the irradiance profile as 'Irrad.stl file'. The netlist is the following:

```
** Example 6.2.cir
***** module, charge regulator and battery connection
xmodule 0 3 1 module_1 params: iscmr =5, tr=25,
+vocmr = 22.3,ns = 36,np = 1,nd = 1,pmaxmr = 85, ta = 25
.inc module_1.lib
.inc irrad.stl
vmesur 1 0 stimulus Virrad
```

```
xbat1 4 0 7 batstd params: ns = 6, SOCm = 1000, k = .8, D = 1e-5, SOC1 = .45
.inc batstd.cir
Rbat 4 0 800
R1 4 8 10000
R2 8 0 10000
x741 12 8 13 14 10 ad741
.inc opamp.lib
Vcc 13 0 dc 15
Vee 14 0 dc -15
vref 16 0 dc 6.8
R6 12 16 4100
R7 12 100 220000
Wch 3 4 vcurrent swlmod
.model swlmod iswitch (ioff = -10e-5, ion = 10e-6, Roff = 1.0e+8, Ron = 0.01)
vcurrent 10 100 dc 0
.tran 1s 140000s
.probe
.end
```

(b) Plot the PV modules output voltage evolution

The result is reached by ploting node V(3) which is the output node of the PV array as shown in Figure 6.13.

(c) Plot the battery voltage evolution, showing the connection and disconnection voltage levels.

The result is shown in Figure 6.14.
As can be seen the voltage drops once V_{max} is reached and the PV array remains disconnected from the battery until the reconnection voltage is reached.

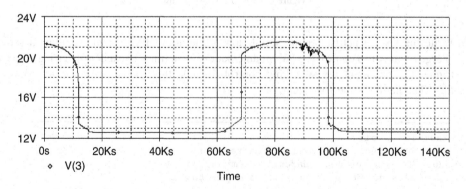

Figure 6.13 PV generator output voltage (note that the minimum voltage corresponds to the battery voltage at night)

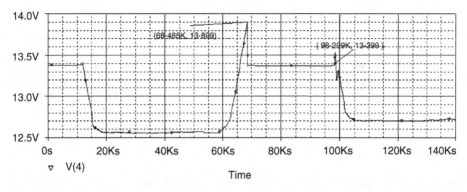

Figure 6.14   Battery voltage

6.4 Maximum Power Point Trackers (MPPTs)

In order to achieve an optimal power transfer, from generator to load, it is imperative to maintain both the PV generator and the load at their respective optimum operating conditions. Of course, the maximum power point is the target for the operating point of the PV generator and this is the main task of the MPPT circuits.

These circuits are especially useful in applications where loads are DC motors for daytime operation, and also in applications where the life of the load can be strongly reduced when forced to work in anomalous or extreme conditions, as can be the case of pumps in pumping applications.

Maximum power point trackers are basically DC/DC converters and can be represented as shown in Figure 6.15. The input power at the DC/DC converter is:

$$P_i = V_i I_i \qquad (6.1)$$

And the power supplied by the converter at the output is:

$$P_o = V_o I_o \qquad (6.2)$$

The relationship between input and output power defines the DC/DC converter efficiency, η:

$$\eta = \frac{P_o}{P_i} \qquad (6.3)$$

Figure 6.15   DC/DC converter schematic diagram

In a DC/DC tracker, the input impedance of the DC/DC converter must be adapted in order to force the PV generator to work at the maximum power point. The DC/DC converter output must also be adapted to the specific load characteristics. The DC/DC converter can give a variable DC output voltage from a nominally fixed DC input voltage. Depending on the load requirements the output voltage, V_o, can be lower or higher than the input voltage, V_i. For this reason DC/DC converters can be divided into two main basic categories:

- Voltage reductors, also called step-down or buck converters.

- Voltage elevators, also called step-up or boost converters.

Other types of DC-DC converters are full-bridge DC-DC converters, CúK DC-DC converters and buck-boost converters, which can be used in applications where special requirements are needed [6.2–6.5], mainly negative polarity output, output voltage higher or lower than input voltage or DC motor drives. In the past few years power electronics has produced high performance converters, which can be used in PV applications.

Figure 6.16 shows a PV system with a DC/DC converter between the PV generator and the load.

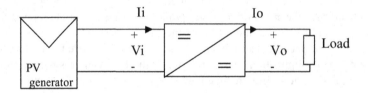

Figure 6.16   PV system block diagram showing the DC/DC converter connection

6.4.1   MPPT based on a DC-DC buck converter

The basic circuit for a buck converter is shown in Figure 6.17. The switch opens and closes the input circuit and then the voltage at the cathode of the diode D1 has a rectangular voltage waveform, which is later filtered by the LC high efficiency filter providing a quasi-continuous voltage at the output node equal to the average value of the rectangular waveform. The diode D1 ensures a return path for the inductor current when the switch is open forcing a low positive voltage at its cathode terminal. The average value of the rectangular waveform can be adjusted to control the length of the ON and OFF (t_{on} and t_{off})

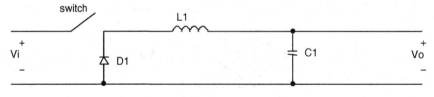

Figure 6.17   Buck converter basic circuit

states of the switch. The control signal of the switch usually has a constant period T and the duty cycle, Dc defined as:

$$Dc = \frac{t_{on}}{T}$$

(6.4)

is the variable. When the switch is ON, the current flows through the inductor and the diode is in reverse bias state; the inductor then stores energy. When the switch is turned OFF, the current still flowing through the inductor forces the diode ON during the interval t_{off}.

Considering an ideal switch, a constant instantaneous input voltage, V_{in}, and a pure resistive load at the output of the converter, the average output voltage, V_o, is given by:

$$V_o = \frac{1}{T}\int_0^T v_o(t)\mathrm{d}t = \frac{1}{T}\int_0^{t_{on}} V_i(t)\mathrm{d}t = Dc \; V_i$$

(6.5)

As the duty cycle is smaller than the unity, the result in equation (6.5) means that $V_o < V_i$.

The output voltage of the converter can be set to a given value by proper election of the duty cycle value. Considering an ideal DC/DC converter with no power losses associated, with an efficiency $\eta = 1$, the input power must be equal to the output power:

$$P_i = P_o$$

(6.6)

$$V_i I_i = V_o I_o$$

(6.7)

$$\frac{V_o}{V_i} = Dc = \frac{I_i}{I_o}$$

(6.8)

In photovoltaic applications DC/DC buck converters are usually used as DC power supplies, where the input voltage V_i is variable and depends on the temperature and irradiance conditions, but the output voltage V_o remains constant. The duty cycle can be made variable in such a way that the input voltage to the DC-DC converter is always set to the PV generators maximum power point voltage of operation.

6.4.2   MPPT based on a DC-DC boost converter

DC-DC converters as voltage elevators are also used in PV applications, especially in battery charging. Conceptually, the basic circuit for a step-up converter is shown in Figure 6.18. As can be seen, the circuit has the same components as the step-down converter but they are arranged differently. The switch state is also governed by a control signal with a constant period T and a variable duty cycle.

In this converter the output voltage V_o is always greater than the input voltage V_i. When the switch is ON the diode is reverse biased, and only the capacitor is connected to the output. The inductor stores energy from the input, and then,

$$V_i = L\frac{\mathrm{d}i_L}{\mathrm{d}t}$$

(6.9)

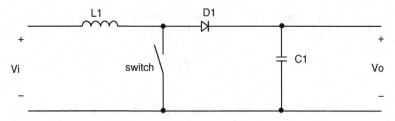

Figure 6.18   Boost converter basic circuit

where i_L is the inductor current. Usually these converters use a clock frequency much greater than the frequency at which the input and output voltage change, so time integration of equation (6.9) leads to,

$$i_L(t_{on}) - i_L(0) = t_{on}\frac{V_i}{L} \tag{6.10}$$

When the switch is ON the diode enters into conduction and the energy is transferred to the capacitor and to the output from the inductor and the input source V_i. Now,

$$V_i - V_o = L\frac{di_L}{dt} \tag{6.11}$$

and

$$i_L(T) - i_L(t_{on}) = t_{off}\frac{V_i - V_o}{L} \tag{6.12}$$

Applying the continuity and periodicity conditions to the inductor current:

$$\frac{V_i}{L}t_{on} = \frac{V_i - V_o}{L}t_{off} \tag{6.13}$$

we finally find the relationship between input and output voltage as a function of the duty cycle, Dc.

$$\frac{V_o}{V_i} = \frac{T}{t_{off}} = \frac{1}{1 - Dc} \tag{6.14}$$

Considering an ideal DC/DC converter with no power losses associated, and hence with an efficiency $\eta = 1$, the input power must be equal to the output power:

$$P_i = P_o \tag{6.15}$$
$$V_i I_i = V_o I_o \tag{6.16}$$
$$\frac{I_o}{I_i} = (1 - Dc) \tag{6.17}$$

As in the case of voltage reductors, it is possible to adjust the input voltage V_i at the maximum power point of PV modules according to the required load voltage V_o, by appropiate election of the duty cycle.

When a DC-DC converter is used as a maximum power point tracker, buck or boost, an adequate control signal must be generated to control the switch state. Usually this signal is generated using pulse width modulation (PWM), techniques, and adding a control circuit to the converter in order to generate the switch control signal [6.5]. The control circuit must adjust the duty cycle of the switch control waveform for maximum power point tracking, as a function of the evolution of the power input at the DC-DC converter.

6.4.3   Behavioural MPPT PSpice model

The basic operation of the DC-DC switching converters described in previous sections, show that a detailed simulation of one period of the clock signal, provides an accurate and low loss design of the electronics, but has little use in the operation of PV systems where the PV engineer is more concerned with the long-term behaviour of the system.

This is the reason why the simple and behavioural PSpice model proposed to simulate a DC-DC converter, buck or boost, is not based on the electronic topologies of the converters, but instead, focuses on the input–output relationships and on the working principles, therefore realizing a correct maximum power point tracking for the PV generator and an optimization of the power load demands.

In Figure 6.19, a plot of the $I(V)$ characteristics of a PV module is shown along with three iso-power curves:

- curve (A), which is tangent to the $I(V)$ curve at the maximum power point and hence corresponds to the iso-power $P_{\max}$ curve;

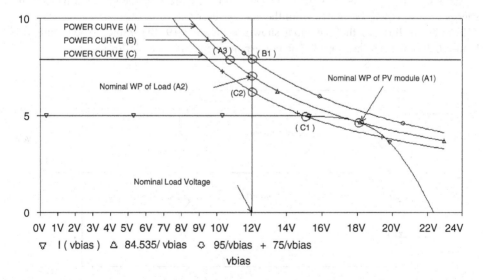

Figure 6.19   Illustration of the working points of a PV module and the load

- curve (B) is the iso-power curve for a power value greater than P_{max}; and

- curve (C) is the iso-power curve for a power value smaller than P_{max}.

The nominal operating conditions, assuming a 100% efficient MPP tracker, are for the PV module, the working point (A1), which is the maximum power point, and the point (A2) for the load, corresponding to a given value of the nominal load voltage.

If the system is only composed of a PV module, a DC/DC converter and a load, the energy balance of the circuit for a 100% efficient DC/DC converter tells us that the nominal conditions can only be sustained along curve A for a given irradiance.

Imagine now that for the same value of irradiance and temperature, a change is produced in the load power that is now greater than the PV module maximum power P_{max}. What happens is that the working point of the load shifts to point (A3), if we want to deliver all the current demanded; in turn the output voltage cannot be maintained and takes a smaller value than the nominal value (the opposite strategy can also be implemented).

If, in contrast, the power demanded by the load takes a smaller value than P_{max} (curve (C)) then the maximum power point bias cannot be sustained and the working point shifts to point (C1) and the load working point is located at point (C2).

In practice the behaviour shown in Figure 6.19 will only happen in direct connection of a PV generator to a load by means of a DC/DC MPP tracker, such as for instance a water pumping system. In other PV systems with an energy reservoir such as a standalone (battery) or grid-connected (the grid itself) system, the working point of the load and of the PV module would be:

- case B corresponding to curve (B), point (B1) for the load and (A1) for the module; and

- case C corresponding to curve (C), point (C2) for the load and point (A1) for the PV module.

In both cases the DC/DC converter efficiency is the nominal and the extra energy produced by the PV module in case C will be injected into the battery or the grid and the deficit of energy in case B will be drawn from the energy reservoir.

In order to illustrate the behaviour shown in Figure 6.19, the PSpice model outlined in Figure 6.20 has been developed. The corresponding netlist is as follows:

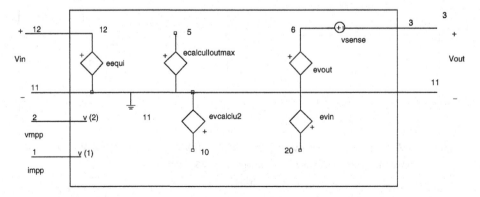

Figure 6.20   Schematic representation of a PSpice DC/DC converter model

```
**** dcdcf.cir
**** DC/CD converter
.subckt dcdcf 11 12 1 2 3 params: n = 1, vo1 = 1
eequi 12 11 value = {v(20)}
evin 20 11 value = {IF (v(10)-v(5)>0, v(2), vo1*v(10)/(n*v(1)))}
**** power control
ecalculIoutmax 5 11 value = {n*v(1)*v(2)/vo1}
vsense 6 3 dc 0
evcalcul2 10 11 value = {i(vsense)}
evout 6 11 value = {IF (v(10)-v(5)>0, (n*v(1)*v(2)/v(10)), vo1)}
.ends dcdcf
```

As can be seen, the converter is defined as a subcircuit with two input parameters: n and v_{o1} defined as follows:

- n is the converter efficiency, as defined in equation (6.3);

- v_{o1}, is the desired converter output voltage.

The subcircuit has five connecting nodes, as can be seen in Figures 6.20 and 6.21:

- Node 11: ground.

- Node 12: Converter input.

- Node 1: input, value of the current of the maximum power point of the PV generator i_{mmp}.

- Node 2: input, value of the voltage coordinate of the maximum power point.

- Node 3: Converter output.

Nodes 1 and 2 give information to the DC-DC converter about the instantaneous values of the coordinates of the maximum power point of the PV generator. From these data, the DC/DC converter behavioural model takes the appropriate action of MPP tracking by connecting

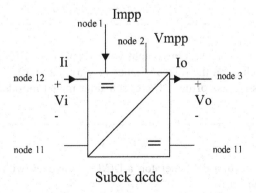

Subck dcdc

Figure 6.21   Connecting nodes of the PSpice DC/DC converter model

to the generator output, node 12, which is the converter input, a voltage-controlled voltage source: 'eequi'. This voltage source is directly controlled by node 20. Voltage at node 20 is determined by the controlled voltage source 'evin', according to the power demanded at the DC/DC output and to the converter efficiency following the rules described in Figure 6.19 and either shifting the working point of the PV module to (C2) or keeping (A1) depending on the case, as described above.

At the converter model output another voltage-controlled voltage source is placed: 'evout', subcircuit node 6.

Between node 6 and the converter output, node 3, a voltage source: 'vsense', is placed to sense the current supplied to the output, 'i(vsense)'.

In order to implement this feature the DC-DC behavioural model calculates the maximum available current at the output taking into account:

- the converter efficiency: n;

- the desired output voltage value, v_{o1}; and

- the available power at the input, which is assumed at the MPP: $P_i = V_{mM} I_{mM}$.

This calculation is made by the e-device 'ecalcuIoutmax', included in the subcircuit, node 5. A comparison is made by an IF statement in such a way that if the current demanded by the load is smaller than the maximum output current calculated by the e-device, then, the converter output voltage is set to the desired value v_{o1}. In the opposite case, this voltage is limited to the value:

$$\frac{I_{mpp} V_{mpp}}{I_{load}} n \qquad (6.18)$$

This load power control strategy is defined in the PSpice netlist as follows:

```
ecalcuIoutmax  5 11 value = {n*(v(1)/1000)*v(2)/vol}; calculates the maximum
*available output current
vsense 6 3 dc 0; zero voltage ancillary voltage source to sense the load current value
evcalcul2 10 11 value = {i(vsense)}
evout 6 11 value = {IF (v(10)-v(5)>0, (n*v(1)*v(2)/v(10)), vol)}
```

where the last line of the code sets the output voltage either to v_{o1} or to the value given by equation (6.18) depending on the relative values of v(5) and v(10).

Example 6.3 shows the use of the DC/DC converter model introduced in this section.

Example 6.3

Consider a PV system with a PV generator, a DC/DC converter, with MPP tracking, and a load as shown in Figure 6.16.

The PV module is composed of 36 solar cells in series. The module characteristics are:

- short-circuit current, $i_{scmr} = 5.2\,\text{A}$

- open-circuit voltage, $v_{ocmr} = 21.2\,\text{V}$

- Maximum output power, $p_{maxr} = 85\,\text{W}$

- Maximum power point coordinates at standard conditions: $I_{mppr} = 4.9\,\text{A}$ and $V_{mppr} = 17.3\,\text{V}$.

The DC/DC converter is considered ideal, with an efficiency of 100%, and the desired output voltage is $V_{o1} = 12\,\text{V}$. Finally a 40 Ω DC load is connected permanently to the DC/DC converter output. Consider the irradiance profile shown in Figure 6.22, 'irrad.stl file', and a constant temperature of 25 °C as input data for the simulation of the PV system behaviour.

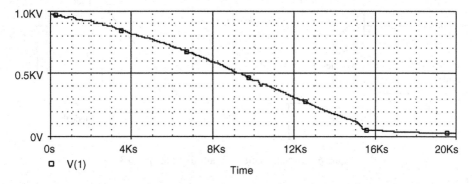

Figure 6.22 Irradiance profile

(a) Write the PSpice netlist for the above PV system, using the corresponding subcircuits described in this chapter for the PV modules and the DC/DC converter.

The netlist is the following:

```
* ppm.cir
.inc generator_beh.lib
xgen 0 1 2 3 405 406 407 20 30 generator_beh params:
+iscmr = 5.2, coef_iscm = 0.13e-3, vocmr = 21.2, coef_vocm = -0.1, pmaxmr = 85,
+noct = 47, immr = 4.9, vmmr = 17.3, tr = 25, ns = 36, nsg = 1, npg = 1
vtemp 2 0 dc 25
.inc irrad.stl
vmesur 1 0 stimulus Virrad
.inc dcdcf.cir
xconv 0 3 20 30 40 dcdcf params: n = 1, vo1 = 12
Rload 40 0 5
.tran 1s 20000s
.probe
.end
```

where the subcircuit model generator_beh.lib of Chapter 4 has been used with a constant ambient temperature of 25 °C and a DC/DC converter with 100% efficiency and output nominal voltage of 12 V.

(b) Plot a schematic diagram of the PV system associated with the above netlist showing the connection of the different blocks.

The schematic diagram is shown in Figure 6.23.

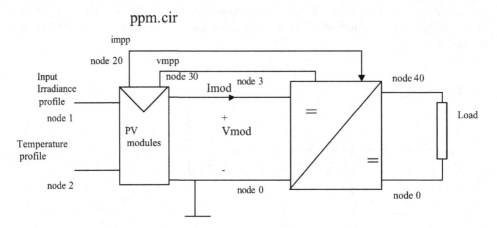

Figure 6.23   Schematic diagram for Example 6.3

(c) Plot the simulation results obtained for:

   - PV module output voltage (node 3).
   - PV module output current, current entering into DC/DC converter (node 3). Compare these results with the simulation results, with:
   - v_{mpp}, PV module maximum power point voltage (node 30).
   - i_{mpp}, PV module maximum power point current v(20).

The results are the following shown in Figure 6.24 and 6.25.

As can be seen in the figures the DC/DC converter ensures correct tracking of the maximum power point of operation of the PV modules forcing them to work at this point when necessary, that is, when the power demand at the output of the converter is large enough.

(d) Plot the values of P_i and P_o that is the input power to the DC/DC converter and output power delivered to the load.

The plot is shown in Figure 6.26, where it can be seen, that the DC/DC converter supplies the constant power load when the input power is sufficient. When the power demanded by the load exceeds the available power, the output power is limited by the converter. This effect can also be observed by plotting the values of the current supplied to the load and the DC/DC converter voltage output, as shown in Figure 6.27.

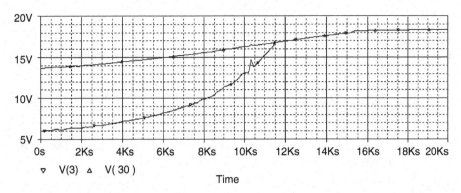

Figure 6.24   Evolution of voltages at the input of the DC/DC converter (output of the PV module, v(3))(bottom), and at the maximum power point of the PV module (v(30)) (top)

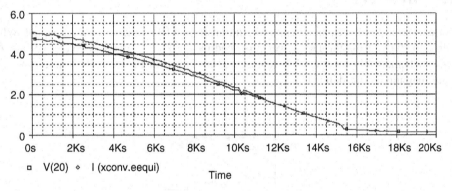

Figure 6.25   Evolution of currents at the input of the DC/DC converter (output of the PV module (top), I(xconv.eequi), and at the maximum power point of the PV module (v(20)) (bottom)

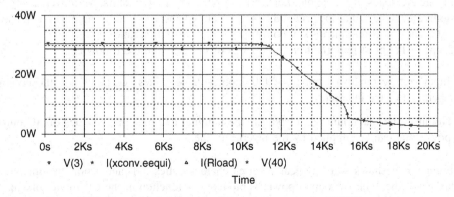

Figure 6.26   Input (top) and output (bottom) power at the DC/DC converter

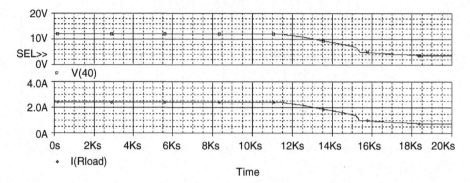

Figure 6.27   Voltage (top) and current (bottom) at the load connected to the output of the DC/DC converter

6.5   Inverters

Inverters are power electronic circuits which transform a DC signal into an AC signal. When applied to PV systems the DC power produced by the PV generator is converted into AC power. Schematically inverters are represented as shown in Figure 6.28, where input and output magnitudes are defined.

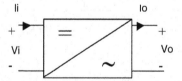

Figure 6.28   Inverter or DC/AC converter schematic diagram

AC power is necessary in PV systems because of the majority preferred conventional loads are AC loads, for example, household appliances or AC motors. Moreover, many PV systems are connected to the grid and hence the output energy of the complete system has to be AC. The inverter is then a key component of the overall system.

Some of the most important characteristics of inverters are the following:

- Input voltage range: the range of DC input voltage values acceptable for the nominal operation of the inverter.

- Nominal and maximum output power: the nominal and maximum values of AC output power that can be supplied by the inverter to the loads.

Figure 6.29 shows some typical inverter characteristics, adapted from manufacturers' datasheets, where the AC output power is plotted as a function of the DC input voltage. As can be seen the output power increases according to the DC input voltage increase until the

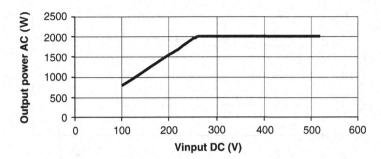

Figure 6.29   Typical input–output characteristics of an inverter

maximum output power is reached and then the output power of the inverter remains constant despite further increase of the DC input.

As the output power saturation is reached the input reaction of the characteristics' is such that the DC input current decreases, as shown in Figure 6.30, indicating that the DC power at the inverter input remains approximately constant.

Total harmonic distortion (THD) is defined as the ratio of the root mean square (rms) value of the signal composed of all harmonics but the first, to the rms value of the first harmonic:

$$THD = \frac{\sqrt{\sum_{n=2}^{\infty} V_n^2}}{V_1} \qquad (6.19)$$

where V_1 is the rms value of the first harmonic, and V_n are the rms values of the nth harmonic.

Inverter efficiency is defined as the ratio of the rms value of the output power to the value of the DC input power, as:

$$\eta = \frac{P_{AC}}{P_{DC}} \qquad (6.20)$$

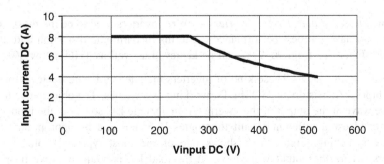

Figure 6.30   Typical input characteristics of an inverter

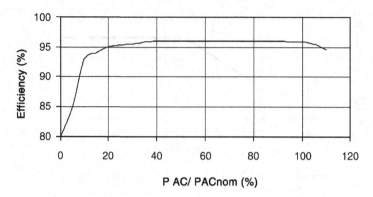

Figure 6.31   Typical efficiency characteristics of an inverter

Figure 6.31 shows a typical plot of an inverter efficiency, adapted from commercial data sheets, as a function of the AC power at the inverter output. The AC output power is commonly expressed in %, the AC output power is normalized to the nominal AC output power of the inverter.

Inverters have to meet very high standards; public utility bodies have set narrow limits for tolerable voltage and frequency deviations from the nominal. As a result, most of the existing inverters in the marketplace exhibit levels of total harmonic distortion below 3% and efficiencies ranging from 90 to 96%.

Many types of inverters are available with different characteristics, properties and performances. Inverters can be classified in several ways:

- Inverter output: single-phase or three-phase inverters.

- Switching operating principles: line-commutated inverters. Inverters where the internal switch devices, such as thyristors, are controlled by the mains, or self-commutated inverters that use self-control of power switches.

- Inverter power rating.

A more general inverter classification can be made according to the inverter's internal topology. Following this criteria, inverters can be classified into three main categories:

- *Pulse-width modulated (PWM) inverters* – PWM inverters used to be the most expensive, but they do have the best performances in terms of harmonic distortion and efficiency. Typical efficiencies from 90–97% can be achieved as well as THD below 3%.

- *Square wave inverters* – These kind of inverters include solid-state devices, for example power bipolar transistors or MOSFETs, working as switches. The objective is to produce a square wave at the output of the inverter from a single DC source at the input. This can be achieved using different circuit topologies [6.5–6.6] and by appropriate control of alternate switching. Figure 6.32 shows one of the most typical topologies of these inverters, where the output voltage V_o, at the load R_1, is a square wave if appropriate control of the transistor base voltages is performed.

Most of these inverters have a transformer at the output in order to increase the amplitude to the desired level. Square wave inverters used to be the cheapest because of their simplicity, with good efficiencies ranging from 75–95%, but they suffer from high THD, up to 45%.

- *Modified sine wave inverters* – These inverters combine the previous switching strategy with the use of resonant circuits to provide an inverter output waveform closer to a sine wave; they have intermediate performance between the two other types. Efficiencies ranging from 70–80% are commonly achieved and THD levels are around 5%.

6.5.1 Inverter topological PSpice model

A variety of PSpice models for inverters can be found in the literature [6.5]–[6.6]. These models have been developed according to the internal inverter architecture and can be associated with the above inverter categories. In order to describe the internal operation of a circuit inverter we will simulate as an example one of the most popular topologies, shown in Figure 6.32, which is reported as a 'classical bridge' in most power electronics books. The reader is encouraged to consult reference [6.4], for an example of a PSpice netlist for this type of inverter. We have adapted the values of the components of this classical bridge to values adequate for PV applications. In particular, the next example shows a low frequency inverter, working at the grid frequency, in order to analyse the inverter output signals, without transformers at the inverter output. In practice this approach is the origin of many important losses in converter efficiency and directly forces the size of both inductive and capacitive components to be very large in volume and weight. To avoid these problems, to improve efficiency and to reduce the size of the inverters, in today's designs the internal frequency is often selected to be several orders of magnitude greater than the grid frequency.

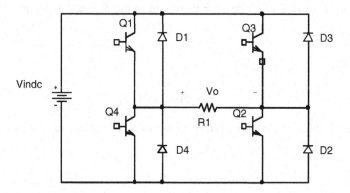

Figure 6.32 Conventional schematic of a square wave inverter

Example 6.4

In the circuit shown in Figure 6.33, four power bipolar transistors Q2N3055 are used. The model is available in the PSpice evaluation library. Four diodes with the default model *d*,

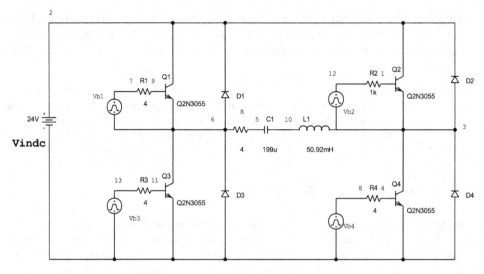

Figure 6.33   Inverter schematics with an RLC load

with transit time $tt = 1\,\mu s$ are also connected. Imagine that the load is composed of a $4\,\Omega$ resistor and an inductor and a capacitor in series and that the input DC voltage is 24 V.

(a) Calculate the value of the inductor L if the quality factor of the RLC circuit is specified to be $Q = 6.35$ and the resonant frequency is set to 50 Hz.

Solution

As the Q factor is related to the resistor value and to the resonant frequency by:

$$Q = L\frac{2\pi fr}{R}$$

it follows that:

$$L = 50.92\,\text{mH}$$

(b) Calculate the value of the capacitor, from the value of the resonant frequency.

Solution

As the resonant frequency is related to the values of the inductor and the capacitor by:

$$fr = \frac{1}{2\pi\sqrt{LC}}$$

it follows that

$$C = 199\,\mu F$$

(c) Write the PSpice netlist including a transient analysis for 1 s and a Fourier analysis for the output voltage and for the output current up to the 12th harmonic considering a fundamental frequency of 50 Hz.

Solution

Taking into account that the syntax for the Fourier analysis is:

Syntax for Fourier analysis

.FOUR fund_freq harm_number variable_1 variable_2

where:

fund_freq is the fundamental armonic frequency;
harm_number is the number of harmonics to be calculated;
variable_1 etc. are the variables for which the Fourier analysis is to be performed.

The netlist can be written as follows:

```
*** inverter 1.cir
vindc  2  0  24v
q1  2  9  6  q2n3055
d1  6  2  diode
r1  9  7  4
vb1  7  6  pulse 0 14  0  1ps 1ps 9.9ms  20ms

q2  2  1  3  q2n3055
d2  3  2  diode
r2  12  1  4
vb2  12  3  pulse 0 14 10ms 1ps 1ps 9.9ms 20ms

q3  6  11  0  q2n3055
r3  11  13  4
d3  0  6  diode
vb3  13  0  pulse 0 14 10ms 1ps 1ps 9.9ms 20ms

q4  3  4  0  q2n3055
d4  0  3  diode
r4  8  4  4
vb4  8  0  pulse 0 14 0 1ps 1ps 9.9ms 20ms

r  6  5  4
c  10  5  199uf
l  3  10  50.92mh
.model diode d(t=1e-6)
.model q2n3055 npn
.tran 1ms 1
.four  50hz  12  i(r)  v(6,3)
.options abstol=0.5ma reltol=0.01 vntol=0.001v
.probe
.end
```

Figure 6.33 reproduces the schematic representation for the above netlist. The sources Vb1 to Vb4 are voltage sources of pulse type controlling the switching of the transistors Q1 to Q4. Q1 and Q4 are turned ON for one-half of the period while Q2 and Q3 are OFF. The opposite happens during the second half of the period where Q2 and Q3 are forced to be ON while Q1 and Q4 remain OFF.

(d) Plot the output voltage waveform.

Solution

Figure 6.34 shows the output voltage, Vo = V(3)−V(6), waveform. As can be seen it is a square wave with a period of 20 ms and an amplitude of 23.9 V. This is due to the switching action of the transistors that act as short circuits when they are ON and as open circuits when they are OFF.

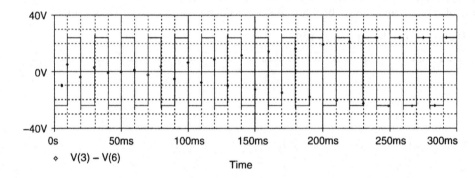

Figure 6.34   Inverter output voltage in Example 6.4

(e) Compute the total harmonic distortion of the voltage waveform.

Solution

The result is directly available in the output file of the simulation as follows:

```
FOURIER COMPONENTS OF TRANSIENT RESPONSE V(6,3)
DC COMPONENT = −3.177885E − 05
```

HARMONIC NO	FREQUENCY (HZ)	FOURIER COMPONENT	NORMALIZED COMPONENT	PHASE (DEG)	NORMALIZED PHASE (DEG)
1	5.000E + 01	3.048E + 01	1.000E + 00	1.969E + 00	0.000E + 00
2	1.000E + 02	6.319E − 05	2.073E − 06	−8.596E + 01	−8.793E + 01
3	1.500E + 02	1.019E + 01	3.343E − 01	5.825E + 00	3.856E + 00
4	2.000E + 02	5.935E − 05	1.947E − 06	−8.905E + 01	−9.102E + 01
5	2.500E + 02	6.139E + 00	2.014E − 01	9.652E + 00	7.683E + 00
6	3.000E + 02	6.413E − 05	2.104E − 06	−9.104E + 01	−9.300E + 01

7	3.500E+02	4.411E+00	1.447E-01	1.349E+01	1.152E+01
8	4.000E+02	6.280E-05	2.060E-06	-9.070E+01	-9.267E+01
9	4.500E+02	3.457E+00	1.134E-01	1.730E+01	1.533E+01
10	5.000E+02	6.461E-05	2.120E-06	-8.828E+01	-9.025E+01
11	5.500E+02	2.854E+00	9.364E-02	2.111E+01	1.914E+01
12	6.000E+02	6.397E-05	2.099E-06	-8.992E+01	-9.189E+01

TOTAL HARMONIC DISTORTION = 4.414693E+01 PERCENT

Where the 12 harmonic frequency, amplitude and phase and its normalized values are listed. As can be seen, and as expected, the odd harmonics are the most important here, and a total harmonic distortion of 44.14% is found. It can also be seen that the DC component of the output voltage is not zero, although the value is in the range of tens of microvolts. As a general rule of inverter output requirement this DC component value must be as close to zero as possible.

(f) The load impedance in the circuit analysed is an RLC series circuit providing a filtered load current waveform. Plot the output current.

Solution

The inverter load current is plotted and the result is shown in Figure 6.35, which is close to a sinusoidal waveform after the initial transient.

(g) Calculate the rms value of the current waveform.

Solution

This can be directly obtained using a built-in PSpice function 'rms' at the probe utility and the result is shown in Figure 6.36; the rms value is found to be at the end of several waveform cycles, $I_o(\text{rms}) = 4.92$ A.

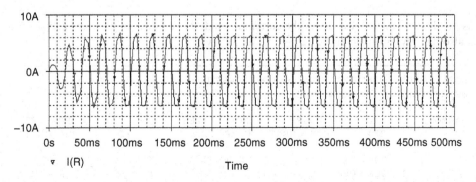

Figure 6.35   Inverter load current in Example 6.4

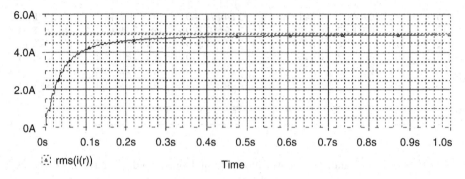

Figure 6.36   rms value of the inverter output current in Example 6.4

The rms value of the output voltage can be theoretically calculated, considering a perfect square waveform of 24 V amplitude, as:

$$V_{0(rms)} = \sqrt{\frac{1}{T}\int_0^T V_0^2 \, dt} = \sqrt{\frac{2}{20\times 10^{-3}}\int_0^{10\times 10^{-3}} 24^2 \, dt} = 24 \text{ V}$$

and then

$$I_{0(rms)} = \frac{V_{0(rms)}}{R} = \frac{24}{4} = 6$$

The value obtained by the simulation is different but of the same order of magnitude than the theoretical value due to the rise and fall times of the real waveform.

(h) Perform a Fourier analysis for the output current waveform.

Solution

As the .FOUR statement includes the current through the resistor, the Fourier analysis result is also directly available at the output file as follows:

```
FOURIER COMPONENTS OF TRANSIENT RESPONSE I(R)
DC COMPONENT = -2.248941E-05
```

HARMONIC NO	FREQUENCY (HZ)	FOURIER COMPONENT	NORMALIZED COMPONENT	PHASE (DEG)	NORMALIZED PHASE (DEG)
1	5.000E+01	6.911E+00	1.000E+00	-2.547E+01	0.000E+00
2	1.000E+02	7.918E-04	1.146E-04	-1.716E+02	-1.461E+02
3	1.500E+02	6.700E-01	9.695E-02	-6.633E+01	-4.087E+01

4	2.000E + 02	9.830E − 04	1.423E − 04	−7.670E + 01	−5.124E + 01
5	2.500E + 02	1.523E − 01	2.203E − 02	3.973E + 01	6.519E + 01
6	3.000E + 02	6.035E − 04	8.733E − 05	3.394E + 01	5.941E + 01
7	3.500E + 02	8.244E − 02	1.193E − 02	−1.083E + 02	−8.280E + 01
8	4.000E + 02	3.718E − 04	5.380E − 05	2.701E + 01	5.247E + 01
9	4.500E + 02	7.000E − 02	1.013E − 02	1.720E + 02	1.975E + 02
10	5.000E + 02	5.207E − 04	7.534E − 05	1.402E + 02	1.657E + 02
11	5.500E + 02	5.669E − 02	8.203E − 03	−5.334E + 01	−2.787E + 01
12	6.000E + 02	1.389E − 04	2.011E − 05	1.529E + 02	1.784E + 02

TOTAL HARMONIC DISTORTION = 1.009803E + 01 PERCENT

which gives a total harmonic distortion for the output current of 10%.

(i) Plot the inverter efficiency.

Solution

The inverter instantaneous power efficiency can be evaluated from PSpice simulation results as follows:

$$\eta = \frac{P_o}{P_i} = \frac{R I_o^2}{V_i I_i} = \frac{4\, rms(i(R))^2}{AVG(V_{indc}\, ABS(I(V_{indc})))}$$

The inverter efficiency can also be easily plotted from the simulation results as shown in Figure 6.37, where an efficiency of 90% is obtained for the above inverter, once the permanent regime has been reached.

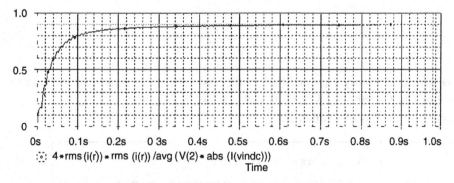

Figure 6.37 Inverter efficiency in Example 6.4

6.5.2 Behavioural PSpice inverter model for direct PV generator–inverter connection

The topological inverter PSpice model described in the previous section is useful in the design of the internal architecture of inverters, although accurate results are only available if accurate models for the switching power devices are used.

In addition to the difficulty of accurate modelling of power electronic components, the photovoltaic systems engineer is more interested in the behaviour of the inverter in the long term rather than in a few internal frequency cycles. For this reason, and in the same way than we used for the power regulators, we will describe in this section a behavioural model for a generic inverter. This allows us to explore some features of the connection of PV generators to inverters.

The input parameters of the inverter behavioural model are:

- Inverter efficiency: nf. The model considers a constant value for this parameter. As shown in Figure 6.31, the efficiency of most inverters is non-constant but the value of this parameter remains reasonably constant for a wide range of inverter output power.

- Maximum output power: P_m. The maximum power that the inverter can supply to the load.

- Pload: Load power demand.

The inverter model is assumed to be similar to an ideal voltage source, with zero DC component and zero total harmonic distortion at the output voltage.

The behavioural model is schematically described in Figure 6.38. As can be seen, the inverter output supplies an ideal sinusoidal voltage while the input is biased by a voltage source, 'emod', forcing the inverter input voltage, which can be either the output of the PV generator or the output of the DC/DC converter, to be

$$Vi_{DC} = \frac{V_{OAC} I_{OAC}}{nf\ I_{iDC}} \tag{6.21}$$

where:

$$nf = \frac{P_{AC}}{P_{DC}} = \frac{V_{OAC} I_{OAC}}{V_{iDC} I_{iDC}} \tag{6.22}$$

is the inverter efficiency, V_{oAC} and I_{oAC} are the rms values of the output voltage and current, and V_{iDC} and I_{iDC} are the DC inverter input voltage and current.

In PSpice code this is written as:

$$\boxed{\text{emod 1 3 value} = \{(\text{v}(11) * 220/(\text{nf} * \text{v}(5))) - 0.8\}} \tag{6.23}$$

where:

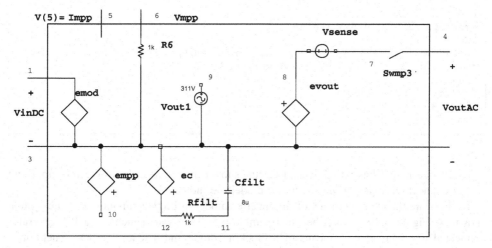

Figure 6.38   Schematic of the PSpice file Inverter1.cir

- v(11) is the rms value of the inverter output current, obtained after filtering. As can also be seen in the complete netlist below, the DC source 'vsense' is connected at the output of the inverter to sense the inverter output current. The 'ec' controlled source extracts the absolute value of the current across 'vsense' and finally, the low pass filter composed of 'Rfilt' and 'Cfilt' obtains the rms value of the inverter output current as a voltage at node 11, v(11).

- The rms value of the inverter output voltage is 220 V.

- v(5) is the value of the PV generator maximum power point current.

- A correction factor of 0.8 V has also been included in equation (6.23) to take into account the voltage drop in a blocking diode.

Finally two protection mechanisms have been considered in the model:

- When the power demanded by the load exceeds the maximum inverter output power defined by the model parameter P_m, the inverter output voltage is forced to zero.

- When the power at the inverter input multiplied by the inverter efficiency, decreases below the power demanded by the inverter loads, P_{load}, the switch 'SWmp3' disconnects the load of the inverter output.

The netlist of this inverter model is described below.

```
********** Inverter, behavioural model
********** Inverter2.cir
.subckt inverter2 1 3 4 5 6 PARAMS: nf = 1, Pm = 1, Pload = 1
eVout 8 3 value = {IF (Pload < Pm, v(9), v(3))}
vsense 7 8 dc 0
Vout1 9 3 sin (0 311 50Hz)
```

```
r6 6 3 1000
empp 10 3 value = {v(5)*v(6)*nf/(Pload)}
SWmp3 4 7 10 3 sw3mod
.model sw3mod vswitch (Roff = 1e + 9, Ron = 0.0001, voff = 0.991, von = 1.01}
ecorriente 12 3 value = {abs(i(vsense))}
Rfiltro 11 12 10000
Cfiltro 11 3 0.000008
emod 1 3 value = {(v(11)*220/(nf*v(5))) - 0.8}
.ends inverter2
```

The above netlist has an equivalent circuit shown in Figure 6.39, where the input and output connection nodes of the subcircuit model are indicated.

To illustrate the characteristics of the inverter model, a PSpice simulation of the system shown by Figure 6.40 is described in Example 6.5. The system includes a PV generator connected to an inverter by means of a blocking diode and a load. The PV generator is modelled by 'generator_beh.lib', from Chapter 4.

inverter1.cir

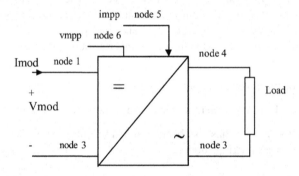

Figure 6.39   Schematic representation of the inverter

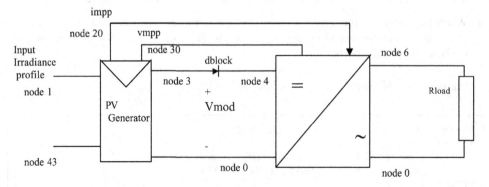

Figure 6.40   Schematic block diagram for inverter model simulation

Example 6.5

Imagine a system such the one depicted in Figure 6.40 in which the PV generator is composed of six modules in series. Each of the PV modules is made of 36 solar cells connected in series and at standard conditions a short-circuit current of $i_{scmr} = 5.2$ A and an open-circuit voltage; $v_{ocmr} = 21.2$ V. The module maximum power point coordinates are $i_{mppr} = 4.7$ A, $v_{mppr} = 17.3$ V and the maximum power is $p_{maxr} = 510$ W.

The inverter parameters are, efficiency, $nf = 0.9$, maximum inverter output power $P_m = 420$ W, expected power load demand (nominal load power), $P_{load} = 107.5$ W. Finally a resistive load of 450 Ω is connected to the output inverter.

(a) Write the PSpice netlist considering an irradiance provided by file irrad3.stl and that the ambient temperature is constant at 25 °C.

Solution:

```
******* Inverter simulation
******* Example 6.5.cir
.inc generator_beh.lib
xgenerator 0 1 43 3 45 46 47 20 30 generator_beh params:
+iscmr = 5.2, coef_iscm = 0.13e - 3, vocmr = 21.2, coef_vocm = 1, pmaxmr = 85,
+noct = 47, immr = 4.9, vmmr = 17.3, tr = 25, ns = 36, nsg = 6, npg = 1
.inc irradprueba2.stl
vmesur 1 0 stimulus Virrad
vtemp 43 0 dc 25
dblock 3 4 diode
.model diode d(n = 1)
.inc inverter2.cir
xinvert 4 0 6 20 30 inverter2 params: nf = 0.9, Pm = 600, Pload = 107.5
Rload 6 0 450
.options stepgmin
.tran 0.1ps 2.80s
.probe
.end
```

The file irrad2.stl includes the irradiance profile shown in Figure 6.41, where the internal time is in units of seconds. We have considered a rather unrealistic irradiance profile with very fast changes in order to study the PV system behaviour without the need of a very long simulation time. As can be seen the time scale ends at 2.64 seconds.

(b) Run a simulation and plot the output voltage of the inverter for the first few cycles.

Solution

Figure 6.42 shows the results for the inverter output voltage between 600 ms and 900 ms. As can be seen the output is sinusoidal of the desired frequency.

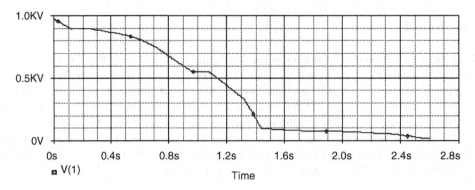

Figure 6.41   Irradiance profile data included in file irrad2.stl

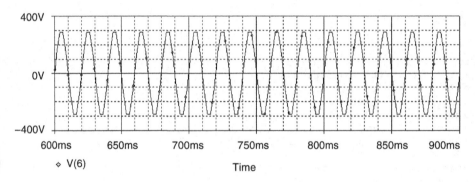

Figure 6.42   Inverter output voltage

(c) Show the action of the load disconnection when the power at the input is smaller than the load demand.

Solution

This is seen by looking at the output current close to the internal time value of 1.43 s. When the power at the inverter input decreases below the expected power demanded by the inverter loads, 107.5 W, due to the irradiance profile, the switch SWmp3 disconnects the load from the inverter output. This effect can be seen in Figures 6.43 and 6.44.

Comparing the simulation results shown in Figures 6.43 and 6.44 the effect of the load disconnection can be clearly identified. When the power at the inverter input decreases as a result of the irradiance profile, and becomes smaller than the power demanded by the load at the inverter output, the inverter disconnects the load and the inverter output voltage drops to zero.

(d) Show the time waveform for the inverter input voltage and input current.

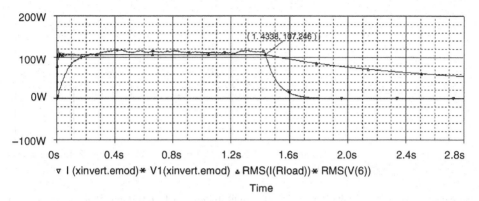

Figure 6.43 Power at the inverter input (bottom): V1(xinvert.emod)* I(xinvert.emod) and power supplied to the load (top): rms(I(Rload)*rms(V(6))

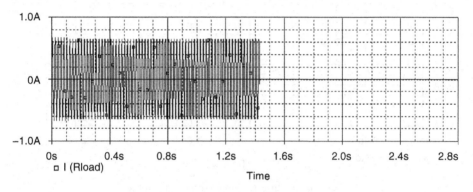

Figure 6.44 Inverter output current waveform

Solution

Figure 6.45 shows the inverter input voltage waveform, which increases at the beginning of the observation period due to the fact that the irradiance value forces the PV output current to decrease as shown in Figure 6.46. This effect is a consequence of the inverter characteristic shown in Figure 6.30. The total power at the inverter input remains constant, because it is equal to the output power divided by the efficiency.

The disconnection effect can also be observed, when the inverter input power is not enough to satisfy the power load demand, the inverter disconnects the load and the inverter input voltage decreases strongly, leaving the PV modules practically in an open circuit.

6.5.3 Behavioural PSpice inverter model for battery–inverter connection

In some applications inverters are not directly connected to the PV modules but through a charge regulator or to a battery. The interface requirements between a battery and an inverter force changes to the behavioural model of the inverter, which based on the model described in the previous section, includes:

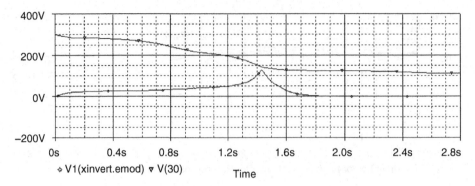

Figure 6.45   Evolution of voltage at the inverter input (bottom): V1(xinvert.emod) and evolution of the voltage maximum power point of the PV modules (top), v(30)

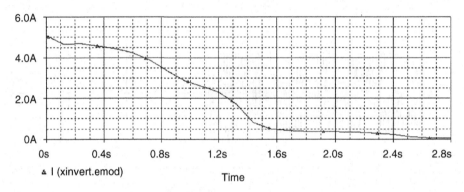

Figure 6.46   Inverter input current waveform

- The power control at the inverter input is based on the battery normalized state of charge *SOC*. If the *SOC* decreases below 30% the switch SWmp3 disconnects the load from the inverter output. The battery SOC value is tested at inverter nodes 5 and 6.

- Inverter input: at the inverter input the bias is set by a current source, substituting the voltage source of the previous model. This new g-device, 'gindc' in the netlist shown in Example 6.6, forces a current at the input of the inverter, which is related to the power demanded by the load, the inverter efficiency and the input voltage of the inverter, as follows:

$$Ii_{DC} = \frac{V_{OAC}I_{OAC}}{nf \, V_{iDC}} \tag{6.24}$$

which is implemented in PSpice code as follows:

```
gindc 1 3 value = {(v(11) * 220/(nf * v(1)))}
```
(6.25)

Where v(11) is the rms value of the inverter output current, v(1) is the inverter DC input voltage, nf is the inverter efficiency and a rms value of 220 has been considered for the inverter output voltage.

Examples 6.6 and 6.7 illustrate how this new model works.

Example 6.6

Write a subcircuit model for the new behavioural model.

Solution

The netlist is shown below where the new sentences have been highlighted.

```
********** Inverter2 model 1.0
** example 6.6.cir
.subckt inverter2 1 3 4 5 6 PARAMS: nf = 1, Pm = 1, Pload = 1
eVout 8 3 value = {IF (Pload < Pm, v(9), v(3))}
vsense 7 8 dc 0
Vout1 9 3 sin (0 300 50Hz)
r6 6 3 1000
empp 10 3 value = {v(5)/(0.3)}
SWmp3 4 7 10 3 sw3mod
.model sw3mod vswitch (Roff = 1e + 9, Ron = 0.001, voff = 0.981, von = 1.1}
ec 12 3 value = {abs(i(vsense))}
Rfil 11 12 10000
Cfilt 11 3 0.000008
gindc 1 3 value = {(v(11)*220/(nf*v(1)))}
.ends inverter2
```

Example 6.7

Simulate the PV system shown in the block diagram in Figure 6.47.
 This PV system is composed of:

- The PV generator: 216 solar cells, series connected, resulting in a PV generator with a short-circuit current, $i_{scmr} = 5$ A, an open-circuit voltage, $v_{ocmr} = 130$ V; the maximum output power is $p_{maxr} = 480$ W, and the maximum power point coordinates at standard conditions are: $i_{mppr} = 4.7$ A, $v_{mppr} = 102$ V. The PV generator model we are using in this example is different from the ones described earlier in the sense that it is a simplification aimed at speeding up the simulation speed. This is required due to the number of non-linear devices involved in the whole system (the PV generator itself, battery, diode, DC/DC and DC/AC converters).

The following netlist describes the PV generator:

```
************pv generator
*********** moduleppt.cir
.subckt moduleppt 10 12 13 14 15 17   params: iscmr=1, ta=1, tempr=1,
+vocmr=1, ns=1, np=1, nd=1,
+pmaxr=1, imppr=1, vmppr=1
girrad 10 11 value={(iscmr/1000*v(12))}
d1 11 10 diode
.model diode d(is={iscmr/(np*exp(vocmr/(N*uvet)))}, n={N})
.func N() {ns*nd}
Rirrad 12 10 100000
Rtemp 13 10 1000000
vtemp 13 10 dc {tempr}
rsm 11 14 {RSM}
.func RSM() {vocmr/iscmr-pmaxr/(ffom*iscmr^2)}
.func ffom() {(vocmnorm-log(vocmnorm+0.72))/(1+vocmnorm)}
.func uvet() {8.66e-5*(tempr+273)}
.func vocmnorm() {vocmr/(nd*ns*uvet)}
**** MPP calculation
gimout 10 15 value={iscmr*(v(12)/1000)}
evdc 16 10 value={0.004*(v(13)-tempr)+(0.12e-3*v(12))+0.06*log
+(1000/v(12))*vocmr}
evmppout 17 10 value={vmppr-v(16)}
.ends moduleppt
```

The characteristics of the other system elements are:

- DC/DC converter: with an efficiency of 100% and an output voltage of 28 V.

- Battery: battery voltage = 24 V, battery capacity $C = 471$ Ah, charge/discharge efficiency: 0.91, autodischarge rate, $D = 1 \times 10^{-9}$ h^{-1} and initial state of charge, $SOC_1 = 0.65\%$.

- An inverter: with an efficiency, nf = 0.9 and the maximum AC output power, $P_m = 500$.

- Load: total power demanded by the load, *Pload* = 350 W, *Rload* = 138 Ω.

- Two blocking diodes: d2 and d3.

Example 5/5.cir

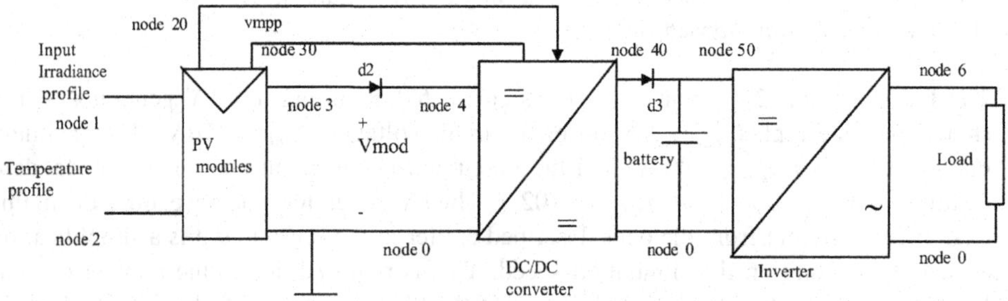

Figure 6.47   Schematic block diagram, Example 6.6

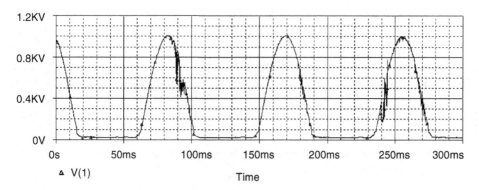

Figure 6.48 Irradiance profile

Consider the irradiance profile shown in Figure 6.48, corresponding to four days of April in Barcelona, file: 'april.stl'. Note that the time scale has units of ms; the irradiance profile has been included in this file, scaling the time by a factor of 10^{-6}, to improve simulation speed.

(a) Write the netlist for the simulation of the PV system in Figure 6.47, considering the PV modules model shown above in moduleppm.cir netlist.

Solution

The PSpice equivalent circuit netlist to simulate the above PV system is as follows:

```
****** Example 6.7.cir
.inc moduleppt.cir
xmodule 0 1 2 3 20 30 moduleppt params: iscmr = 5, tempr = 25, vocmr = 130,
+ns = 216, np = 1, nd = 1, pmaxr = 480, imppr = 4.7, vmppr = 102, ta = 25
.inc april.stl
vmesur 1 0 stimulus Virrad
r111 20 0 1
d2 3 4 diode
.model diode d(n = 1)
.inc dcdcf.cir
xconv 0 4 20 30 40 dcdcf params: n = 1, vo1 = 28
d3 40 50 diode
.inc batstdif.cir
xbat 50 0 100 batstdif PARAMS: ns = 12, SOCm = 11304, k = 0.91, D = 1e - 9, SOC1 = 0.65
.inc inverter2.cir
xinvert 50 0 6 100 100 inverter2 params: nf = 0.9, Pm = 500, Pload = 350
rload 6 0 138
.options abstol = 0.1mA Vntol = 0.001
.tran 0.1s 0.3s
.probe
.end
```

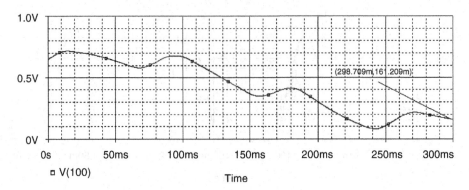

Figure 6.49 SOC evolution, Example 6.6

(b) Obtain the battery state of charge, SOC, evolution.

Solution

Figure 6.49 shows the battery state of charge evolution, starting with a SOC = 0.65% with a final SOC = 0.16%.

(c) Detect the inverter disconnection to the load.

Solution

Figure 6.50 shows the inverter output voltage. As can be seen, when the battery SOC is lower than 30%, the inverter output is disconnected from the load. The frequency of the inverter output voltage is 50 Hz, while the full time scale explored is 0.3 seconds representing in real time 83.3 h as mentioned before. The frequency at the inverter output has not been scaled in

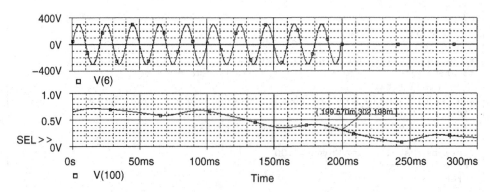

Figure 6.50 Inverter output voltage, v(6) on the top, and SOC, v(100) at the bottom

the same way as the time scale in order to reduce the simulation time. The disconnection occurs at 199.57 ms of internal PSpice time which corresponds to 55.43 hours of system operation; obviously the number of cycles of the inverter output voltage in Figure 6.50 does not correspond to the real behaviour of this signal.

(d) Plot the results for the battery current waveform and for the current supplied by the PV generator.

Solution

This is shown in Figure 6.51, I(xbat.vcurrent), where positive values indicate that the battery is charging. This figure also shows the PV generator output current, I(xmodule.rsm).

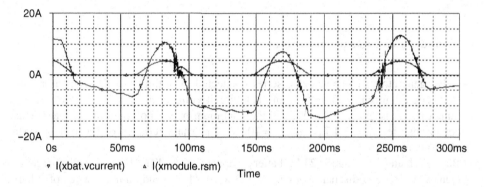

Figure 6.51   Battery current (I(xbat.vcurrent)) and PV modules output current (I(xmodule.rsm)) waveforms

6.6 Problems

6.1 Consider a PV system with a block diagram as shown in Figure 6.47. The characteristics of the elements involved are the following:

Lead–acid battery

 – Initial state of charge: $SOC_1 = 0.78$
 – Maximum state of charge: $SOC_m = 6720\,\text{Wh}$
 – Number of 2 V series cells: $ns = 6$
 – $K = 0.75$, charge/discharge battery efficiency
 – $D = 1 \times 10^{-4}\,h^{-1}$, battery self-discharge rate.

PV module: formed by 33×2 solar cells: $ns = 33$, $np = 2$

- Short-circuit current, $i_{scmr} = 6.54$ A
- Open circuit voltage, $v_{ocmr} = 19.8$ V
- Maximum output power, $p_{maxr} = 94$ W
- Vmaximum power point $= 16$ V.
- Imaximum power point $= 5.88$ A.

DC/DC converter: Efficiency $= 0.8$, $V_{o1} = 14$ V.

Inverter: Efficiency $= 0.9$, $P_{max} = 500$ W.

Load: 180 Ω DC load is connected permanently to the battery.

Consider the irradiance profile shown in Figure 6.48, Example 6.7, as input data for the simulation of the PV system behaviour. Perform a PSpice simulation of the PV system behaviour and obtain the battery state of charge at the end of the simulated period of time.

6.2 Consider a PV system composed of:

- A PV generator: 216 solar cells, series connected, resulting in a PV generator with a short-circuit current, $i_{scmr} = 5$ A, an open-circuit voltage, $v_{ocmr} = 130$ V; the maximum output power is: $p_{maxr} = 480$ W, and the maximum power point coordinates at standard conditions are $i_{mppr} = 4.7$ A, $v_{mppr} = 102$ V. The characteristics of the rest of system elements are:
- Battery: battery voltage $= 24$ V, battery capacity $C = 471$ Ah, charge/discharge efficiency, 0.91, autodischarge rate, $D = 3 \times 10^{-9}$ h^{-1}, and initial state of charge, $SOC_1 = 0.8\%$.
- Inverter: efficiency, $nf = 0.9$ and the maximum AC output power, $P_m = 500$.
- Load: total power demanded by the load, $P_{load} = 37$ W, $R_{load} = 1300\,\Omega$.
- Two blocking diodes: d1 between the PV modules and the battery, and d2 between the battery and the inverter.

Consider the irradiance profile shown by Figure 6.48, corresponding to four days of April in Barcelona, file: 'april.stl'. Note that the time scale has units of ms, the irradiance profile has been included in this file, scaling the time by a factor of 10^{-6} to improve simulation speed.

Using the netlists presented previously in this chapter, 'moduleppt.cir', 'inverter3.cir' and 'batstdif.cir' for the PV generator, inverter and battery modelling, obtain:

(a) the corresponding netlist for PSpice simulation of the system;
(b) the battery SOC final value;
(c) the battery voltage at the end of the simulation.

6.7   References

[6.1] Woodworth, J.R., Thomas, M.G., Stevens, J.W., Harrington, S.R., Dunlop, J.P., and Swamy, M.R., 'Evaluation of the batteries and charge controllers in small stand-alone photovoltaic systems', *IEEE First World Conference on Photovoltaic Energy Conversion*, **1**, pp. 933–45. Hawaii, December 1994.

[6.2] Wong Siu-Chung, and Lee Yim-Shu, 'PSpice modeling and simulation of hysteric current-controlled Cúk converter', *IEEE Transactions on Power Electronics*, **8**(4), October 1993.

[6.3] Lee, Y.-S., Cheng, D.K.W., and Wong, S.C., 'A new approach to the modeling of converters for PSPICE simulation', *IEEE Transactions on Power Electronics*, **7**(4), 741–53, October 1992.

[6.4] Mohan, N., Undeland, T.M., and Robbins, W.P., *Power Electronics*, Wiley, 1995.

[6.5] Hui, S.Y., Shrivastava, Yash, Shatiakumar, S., Tse, K.K., and Chung, Henry, Shu-Hung, 'A comparison of non deterministic and deterministic switching methods for dc-dc power converters', *IEEE Transactions on Power Electronics*, **13**(6), 1046–55, November 1998.

[6.6] Rashid, Muhammad, H., *PSpice Simulations of Power Electronics, selected Readings*', IEEE, 1996.

[6.7] Rashid, Muhammad, H., '*Power Electronics Laboratory using PSpice*, IEEE, 1996.

7

Standalone PV Systems

Summary

Standalone photovoltaic systems are described and PSpice models developed to illustrate the basic concepts of energy balance, energy mismatch, loss of load probability and short and long-term simulations and comparison with measured waveforms. Analytical and simple sizing procedures are described and combined with long-term simulations using synthetic solar radiation time series.

7.1 Standalone Photovoltaic Systems

Standalone PV systems are the most popular systems used worldwide despite the more recent interest of the market in grid-connected systems. A standalone PV system should provide enough energy to a totally mains-isolated application. The standard configuration of this system is depicted in Figure 7.1, where it can be seen that the PV generator is connected to the storage battery (this connection may include protection and power-conditioning devices), and to the load. In some specific cases, such as water pumping, the generator can be connected directly to the motor.

We have described in previous chapters, the models available for the various elements encountered in PV applications, such as the PV modules, power-conditioning devices, inverters and batteries and specific operating properties. This chapter describes simple procedures to size a PV system for a given application and illustrates several important concepts:

- Equivalent length of a standard day or peak solar hours (PSH) concept.

- Generation and load energy mismatch.

- Energy balance in a PV system.

- Loss of load probability (LLP).

- System-level simulation and comparison with monitoring data.

- Long-term simulation using stochastically generated radiation and temperature series.

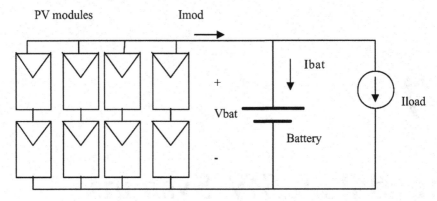

Figure 7.1   Standalone PV system

7.2   The Concept of the Equivalent Peak Solar Hours (PSH)

It is useful to introduce the concept of the equivalent peak solar hours (PSH) defined as the length of an equivalent day in which the irradiance is 1000 W/m^2, T_{cell} = 25 °C, in such a way that the radiation (time integral of the irradiance over the day) is the same in one sun-equivalent day as in a real day.

Let us consider that the real irradiance profile in a given day is $G(t)$ and that the irradiance of the equivalent day is by definition 1 kW/m^2 during a time of length PSH hours. If we write that the total daily radiation in the real day has the same value as in the equivalent day, then

$$\int_{day} G(t)\mathrm{d}t = 1 \cdot PSH \tag{7.1}$$

As can be seen, if the units of the irradiance are kW/m^2, then the numerical value of the daily sun radiation equals the numerical value of the parameter *PSH*.

The definition of PSH is further illustrated in the following example.

Example 7.1

Consider the average irradiance data for a given location, namely Barcelona (41.318° North, Spain), 16 January. (a) Write a PSpice .cir file to calculate the total radiation received that day.

We first write a subcircuit containing the hourly radiation data for that day [7.1], arranged as a PWL source 'virradiance' in the file 'irrad_jan_16' as follows,

```
*irrad_jan_16.lib
.subckt irrad_jan_16 12 10
virradiance 12 10 pwl 0u 0,1u 0,2u 0,3u   0,4u 0,5u 0,6u 0,7u 0
+8u 0,9u 76,10u 191,11u 317,12u 418,13u 469,14u 469,15u 418,16u 317
+17u 191,18u 76,19u 0,20u 0,21u 0,22u 0,23u 0,24u 0
.ends irrad_jan_16
```

These irradiance values are available at the subcircuit node (12).

A .cir file is written to compute the time integral as:

```
*irrad_jan_16.cir
xrad 72 0 irrad_jan_16
.include irrad_jan_16.lib
erad 70 0 value = {sdt(v(72)*1e6)}
vpsh 73 0 pulse (0 1000 11u 0 0 2.941u 24u)
epsh 74 0 value = {sdt(v(63)*1e6)}
.tran 0.001u 24u
.probe
.end
```

where node (70) returns the value of the daily radiation.

The result is shown in Figure 7.2 where it can be seen that the time integral gives a value of the total daily radiation of 2.94 kWh/m²_day. According to equation (7.1) this means that the value of the parameter *PSH* is 2.94 hours.

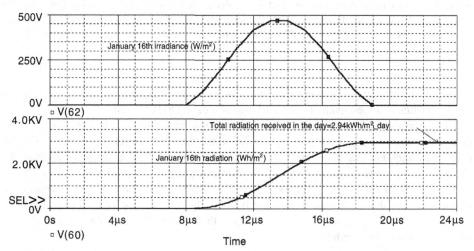

Figure 7.2 (a) Irradiance profile in Barcelona, 16 January inclination 40°. *Warning*: y-axis is the irradiance in W/m². (b) Time integral of irradiance. *Warning*: y-axis is the radiation in kWh/m². In both graphs, the x-axis internal PSpice unit is a microsecond and real unit is time in hours (1 μs = 1 hour)

(b) Plot the irradiance profile and radiation of the equivalent day. According to the definition we choose a 2.94 hour day length at 1 kW/m² (as the definition is independent of the initial and final time values of the irradiance pulse, we choose 11:00 as the initial time of the equivalent day and 13.94 hours the final time). This is shown in Figure 7.3.

The source 'vpsh' is a pulse-type voltage source simulating the equivalent one sun day. The time integral is available at node (74).

The utility of the PSH concept comes from the fact that the technical information usually available on PV commercial modules only includes standard characteristics. Even though these can be translated into arbitrary irradiance and temperature conditions as described in Chapter 4, in order to have a first-order hand calculation model of the energy potential of a

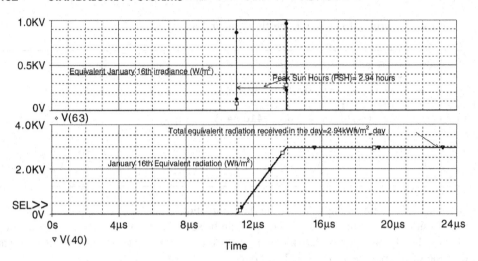

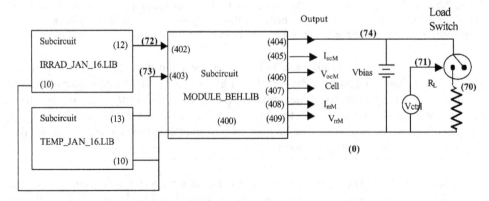

Figure 7.3 (top) Equivalent one sun irradiance profile in Barcelona, 16 January inclination 40°. *Warning*: y-axis is the irradiance in W/m². (bottom) Time integral of equivalent one sun irradiance. *Warning*: y-axis is the radiation in kWh/m². In both graphs, the x-axis internal PSpice unit is a microsecond and real unit is time in hours (1 μs = 1 hour)

given PV plant, the PSH concept allows us to work with the values of the standard characteristics if the equivalent length of one-sun day (PSH) is used.

This approximation is, of course, subject to some inaccuracy in the evaluation of the energy produced by a given PV generator due to the fact that effects of temperature are not taken into account. An assessment of the accuracy of the PSH concept is illustrated in Example 7.2. We will consider in the next examples and subsections a PV system as depicted in Figure 7.4, which shows a block diagram of a PV module driven by irradiance and temperature profiles connected to an ideal battery Vbat and a resistive load in series with a switch. As we will consider three cases of load, this switch allows an easy modification of the load connection schedule.

Figure 7.4 Block diagram for the circuit used to show power mismatch, night-time load and day-time load

Example 7.2

Consider a simple installation such as the one depicted in Figure 7.4 where the switch is open and the battery voltage is set to zero. We will use the two outputs of the maximum power point coordinates in this example.

Consider the irradiance of 16 January used in Example 7.1, and the temperature profile given in Annex 7 by the file 'temp_jan_16.lib'.

Calculate the total energy generated at the maximum power point by this PV module.

Solution

We have to use the time integral of the values of the maximum power delivered by the PV module over that day to compute the total energy supplied by running the following PSpice file 'psh.cir':

```
*psh.cir

.include irrad_jan_16.lib
.include temp_jan_16.lib
.include module_beh.lib
xtemp 73 0 temp_jan_16
xirrad 72 0 irrad_jan_16
xmodule 0 72 73 74 75 76 77 78 79 module_beh params: iscmr = 5,coef_iscm = 9.94e-6,
+vocmr = 22.3, coef_vocm = - 0.0828, pmaxmr = 85,noct = 47,immr = 4.726, vmmr = 17.89,
+tr = 25, ns = 36, np = 1

vaux 74 0 dc 0
.tran 0.01u 24u
.probe
.end
```

As v(78) and v(79) are the values of the current and of the voltage respectively at the maximum power point, the energy generated that day is given by

$$\int_{day} P_{max}(t)dt = \int_{day} I_{mM}(t)V_{mM}(t)dt = 263\,\text{Wh/day} \tag{7.2}$$

after running the 'psh.cir' file.

The value of the equivalent-day total energy delivered by this same module (note that the standard characteristics provide 85 W at 1 sun) during a day length of PSH, is

$$85\,\text{W} \times PSH = 85\,\text{W} \times 2.94\,\text{h} = 249.9\,\text{Wh\_day} \tag{7.3}$$

where the value of PSH = 2.94 is the value of the total radiation received on 16 January in Barcelona at a 40° tilted surface expressed in kWh/m^2 as calculated in Example 7.1.

Comparing the results obtained in equations (7.2) and (7.3) in Example 7.2, it becomes clear that the PSH concept underestimates the total energy delivered by the PV module in a January day by approximately 5%. This is due to the fact that the ambient temperature data in January at that location takes an average value close to 10 °C and as the irradiance values are also low, the cell operating temperature is lower than the value of the reference standard temperature leading to larger power generated by the PV module. Problem 7.1 addresses the same issue in June where both irradiance and temperature are higher, leading to a different conclusion.

7.3 Energy Balance in a PV System: Simplified PV Array Sizing Procedure

Sizing standalone PV systems is not an easy task due to the random nature of the sun's radiation at a particular place, the effects of the horizon, the albedo reflection of the surroundings, the orientation of the collecting surface (both azimuth and inclination) and to the unreliable data on the energy demand by the user.

All sizing methods are then subject to the same inaccuracy as the method we are describing here, which is a simple procedure that considers the energy balance of the photovoltaic system and the PSH concept. This procedure allows a clear description of how the different parameters enter the problem. It has also been shown [7.2] that the method has practical value by comparing the results of more rigorous procedures [7.3].

The energy balance in a PV system is established by a general equation stating that the energy consumed in a given period of time equals the energy generated by the PV system in the same period of time. This general equation in practical terms has to be established for the most suitable period of time for a given application. For example, if the application is a summer water pumping system, we are interested in the energy balance in the summer season, whereas if the application is a year-round standalone plant, we will write the energy balance for a period of one year. In any of these cases, the most conservative criteria usually considers, for the period of time analysed, the month having less solar radiation (worst-case design), while another, less conservative approach considers the average monthly radiation value (average design).

This energy balance method uses the peak solar hours (PSH) concept to write the energy balance equation in a given day, as:

$$P_{\max Gr} PSH = L \tag{7.4}$$

where $P_{\max Gr}$ is the nominal output power of the PV generator at standard conditions (1 sun AM1.5 and 25 °C operating temperature), PSH is the value of the peak solar hours (which is numerically equal to the global in-plane radiation in $kWh/m^2 day$), and L is the energy consumed by the load over this average (or worst) day.

Equation (7.4) can now be written for the two design scenarios:
(a) Worst-case design:

$$P_{\max Gr} (PSH)_{\min} = L \tag{7.5}$$

where $(PSH)_{\min}$ is the value of PSH in the worst month.

(b) Average design:

$$P_{\max Gr}(\overline{PSH}) = L \tag{7.6}$$

where $(\overline{PSH})$ is the average value of the 12 monthly PSH values.

We continue using an average design. Replacing the nominal maximum power of the PV generator by its value:

$$V_{mGr}I_{mGr}(\overline{PSH}) = L \tag{7.7}$$

Considering that a PV generator is composed of N_{sG} series string of N_{pG} parallel PV modules, it follows

$$N_{sG}V_{mMr}N_{pG}\,I_{mMr}(\overline{PSH}) = L \tag{7.8}$$

where V_{mMr} and I_{mMr} are the voltage and current coordinates of the maximum power point of one PV module under standard 1Sun AM1.5.

From equation (7.8) the basic design equation can be drawn:

$$N_{sG}\,N_{pG} = \frac{L}{V_{mMr}\,I_{mMr}(\overline{PSH})} \tag{7.9}$$

Usually the loads in a standalone PV system are connected to a DC voltage, namely V_{cc}. The load L can now be written as:

$$L = 24V_{cc}\,I_{eq} \tag{7.10}$$

where I_{eq} is the equivalent DC current drawn by the load over the whole day. Replacing in equation (7.9):

$$N_{sG}V_{mMr}\,N_{pG}\,I_{mMr}(\overline{PSH}) = 24V_{cc}\,I_{eq} \tag{7.11}$$

There are a number of reasons why the real energy generation of the PV system might be different than the value given by the left-hand side in equation (7.11). One of them, as has been mentioned earlier, is that the peak solar hour concept underestimates or overestimates the energy generation. The other reasons are related to the energy efficiency of the several interfaces between the PV generator and the load, as for example, the efficiency of the MPP tracker, the efficiency of the charge–discharge cycle of the battery itself, wiring losses, and DC/AC converter efficiency to supply AC loads. This means that the design process has to be deliberately oversized to account for these energy losses. It is very useful to introduce here an engineering oversizing factor or 'safety factor' (SF) in such a way that equation (7.11) becomes:

$$N_{sG}V_{mMr}\,N_{pG}\,I_{mMr}(\overline{PSH}) = (SF)24V_{cc}\,I_{eq} \tag{7.12}$$

Finally equation (7.12) can be split in two,

$$N_{sG} = (VSF)\frac{V_{cc}}{V_{mMr}} \tag{7.13}$$

$$N_{pG} = (CSF)\frac{24I_{eq}}{I_{mMr}(\overline{PSH})} \tag{7.14}$$

where independent equations are found to size the number of PV modules in series (7.13) and in parallel (7.14). As can be seen the array oversize safety factor (SF) has also been divided into a voltage safety factor (VSF) and a current safety factor (CSF).

In PV literature the PV generator size is often normalized to the daily load L as:

$$C_a = \frac{N_{sG} \times N_{pG} \times P_{\max Gr} \times (\overline{PSH})}{L} \tag{7.15}$$

Example 7.3

Calculate the values of N_{sG} and N_{pG} for a PV system having to supply a DC load of 2000 Wh/day at 24 V in Sacramento (USA). The PV surface is tilted 40° and faces south. The solar radiation data available for Sacramento [7.4] are shown in Table 7.1 in kWh/m² month. The PV modules which will be used in the system have the following characteristics at standard conditions (1000 W/m², AM1.5, 25 °C):

$$P_{maxr} = 85 \text{ W}, \ V_{ocr} = 22.3 \text{ V}, \ I_{scr} = 5 \text{ A}, \ I_{mMr} = 4.72 \text{ A}, \ V_{mMr} = 18 \text{ V}$$

We first compute the value of I_{eq}

$$L = 24I_{eq}V_{cc} = 2000 \text{ Wh/day} \tag{7.16}$$

$$I_{eq} = \frac{2000}{24 \times 24} = 3.46 \text{ A} \tag{7.17}$$

Table 7.1 Radiation data for Sacramento (CA)

Month	Monthly radiation (kWh/m² month)
Jan	88
Feb	113
Mar	159
Apr	185
May	208
Jun	208
Jul	222
Aug	217
Sep	199
Oct	169
Nov	107
Dec	83

The value of PSH is now required and it is computed by calculating the average daily radiation value from Table 7.1, taking into account that January, March, May, July, August, October and December have 31 days each; that April, June, September and November have 30 and that February has 29 days. The daily radiation values for every month are given in Table 7.2.

Table 7.2   Daily radiation

Month	Daily radiation (kWh/m² day)
Jan	2.83
Feb	3.89
Mar	5.12
Apr	6.16
May	6.70
Jun	6.93
Jul	7.4
Aug	7
Sep	6.63
Oct	5.45
Nov	3.56
Dec	2.67

The average value of the average monthly daily radiation data in Table 7.2 is 5.35 kWh/m² day.

Considering the values of the parameters of a single PV module in equations (7.13) and (7.14) and assuming that the safety factors are unity, it follows,

$$N_{sG} = 1.33$$

$$N_{pG} = 3.28$$

The recommended size of the PV array will be 4×2 which is the closest practical value. As can be seen in practice the system becomes oversized by the factors: $VSF = 1.5$ and $CSF = 1.21$ due to rounding to the closest higher integer.

7.4   Daily Energy Balance in a PV System

The procedure described in Section 7.3 to size the PV array for a given application can be extended to calculate the size of the battery required. In a standalone PV system, the battery has in general to fulfil several tasks, among them:

(a) Cover the difference between the instantaneous values of load demand and of the power generation (power mismatch) at a given time of the day.

(b) Cover the load energy requirements during night time (night time load).

(c) If the system is to be fully autonomous in a yearly basis, the battery also has to cover the seasonal deficits produced.

Of course, the battery size compromises the reliability of the electricity supply in a given application. Prior to the battery sizing procedure we will illustrate details of the energy balance of the system in three scenarios for the load profile in a time frame restricted to one day.

7.4.1 Instantaneous power mismatch

If we now focus on one-day dynamics, we often face situations in which the power demanded by the load does not exactly match the power being generated in this same time by the PV generator. The excess of energy produced during these time windows can be stored by the battery, provided it is not yet full. If the battery is fully charged at that time, most of the PV system managing electronics opens the circuit of the PV array to avoid battery overcharge. However, during the time windows where the load is demanding more power than currently available at that time, then the battery has to provide this extra energy.

At a given time the power available at the output of the PV generator, $P_G(t)$ depends on the irradiance and temperature values and also on the peculiarities of the connections to the load, such as protection and power conditioning equipment.

If the value of the instantaneous power demanded by the load at time t is $P_L(t)$, the energy balance at a given time is:

$$E(t) = \int_0^t [P_G - P_L(t)]\mathrm{d}t \tag{7.18}$$

where the integral in equation (7.18) can take positive or negative values during the time frame of analysis.

In order to illustrate this concept, the following example evaluates the energy excess and deficit during an average day in January in Barcelona using the same data as in Example 7.1 for the hourly radiation and temperature data in file 'temp_jan_16.lib' file in Annex 7.

Example 7.4

Imagine a PV generator such as the one depicted in Figure 7.4 composed of a single PV module supplying power to a load at 12 V. For illustration purposes we will consider that the total load energy demanded during one day equals the total energy supplied by one PV module with the same characteristics described in Section 7.1. In this example we will assume that the load consumes a constant power at all times, this means that the load switch is permanently closed. No maximum power point tracker is used. As will be shown below as the PV generator is biased by the battery at 12 V, the energy produced during the day is 176.5 W.

(a) Calculate the value of the load current
The load current at 12 V is given by

$$I_L = \frac{\int_0^{24} P_L\, dt}{V_{cc}24} = \frac{176.5}{12 \times 24} = 0.612 \text{ A} \tag{7.19}$$

(b) Calculate the value of the equivalent load resistance

$$R_L = \frac{V_{cc}}{I_L} = 19.6\,\Omega \tag{7.20}$$

(c) Write the PSpice netlist. The netlist is the following:

```
*mismatch.cir
.include irrad_jan_16.lib
.include temp_jan_16.lib
.include module_beh.lib
xtemp 43 0 temp_jan_16
xirrad 42 0 irrad_jan_16

xmodule 0 42 43 44 45 46 47 48 49 module_beh params: iscmr = 5,coef_iscm = 9.94e-6,
+vocmr = 22.3, coef_vocm =- 0.0828, pmaxmr = 85,noct = 47,immr = 4.726, vmmr = 17.89,
+tr = 25, ns = 36, np = 1

vbat 44 0 dc 12
rload 44 0 19.6
.tran 0.01u 24u 0 0.01u
.option stepgmin
.probe
.end
```

(d) Run a PSpice simulation and calculate the maximum and minimum value of the energy
IN-OUT of the battery. The result is shown in Figure 7.5.

As can be seen, in the upper graph in Figure 7.5, the current at the battery branch is plotted
against time. It can be seen that before sunrise and after sunset, the current is negative
indicating that there is a flow of current from the battery to the load as there is no generation
by the PV module. In the bottom graph in Figure 7.5, the value of the integral of the current–
voltage product at the battery gives the energy in or out of the battery at a time t.

$$E_{bat} = \int_0^t I_{bat} V_{cc}\, dt \tag{7.21}$$

Using the cursor it can be measured that the maximum value of the energy is $+42.99$ Wh
and the minimum value is -64.35 Wh. As can also be seen, and as designed, the system is
fully balanced because the resulting net flow at the end of the day is zero. It can be easily
verified that the time integral of the output current of the PV generator (Ixmodule.rsm)

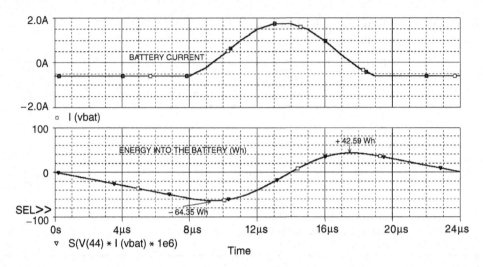

Figure 7.5 Result of the mismatch.cir file in Example 7.4. Battery current (upper graph) and energy into the battery (bottom graph). *Warning*: x-axis units are internal PSpice units (microseconds) corresponding to real time units of hours (1 µs corresponds to 1 hour)

multiplied by 12 V gives 176.5 W, which is the value used in equation (7.19) to balance the load to the energy generated that day.

The results shown in Example 7.4 indicate that the size of the battery should be at least 64.35 Wh + 42.99 Wh = 107.34 Wh, which is known as the *daily cycling* of the battery.

7.4.2 Night-time load

It is obvious that not all the loads are consuming energy all day long as was the case of the previous example. In fact in many standalone PV systems the load is consuming most of the energy at night. One extreme case happens when all the energy is consumed by the load during night-time periods. If we use the same data in Example 7.4 but changing the load to a night load (from 19:00 to 23:00) but keeping the total energy consumed equal to 176.5 Wh_day as in the previous example, we introduce the corresponding modifications, in particular, the new value of the load resistor is:

$$I_L = \frac{\int_0^{24} P_G \, dt}{V_{cc}4} = \frac{176.5}{12 \times 4} = 3.677 \, \text{A} \tag{7.22}$$

and

$$R_L = \frac{V_{cc}}{I_L} = 3.263 \, \Omega \tag{7.23}$$

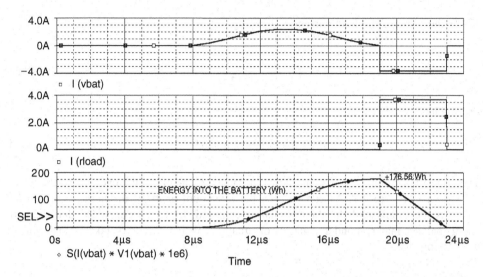

Figure 7.6 Evolution of the energy balance in a night-time load scenario, current in or out of the battery (top graph), load current (middle graph) and energy into the battery (bottom graph). *Warning*: *x*-axis units are internal PSpice units (microseconds) corresponding to real time units of hours (1 µs corresponds to 1 hour)

Some other modifications to the PSpice file are needed, mainly to switch the load resistor on only during the night-time period; we now use the switch in series with the load shown in Figure 7.4 and drive it with a controlling signal. This is included in the netlist by adding the following lines:

```
rload 80 0 3.263
sload 74 80 81 0 switch
.model switch vswitch roff = 1e8 ron = 0.01 voff = 0 von = 5
vctrl 81 0 0 pulse (0,5,19u,0,0,4u,24u)
```

where it can also be seen that the switch has a model with a very low ON resistance (0.01 Ω) instead of the default value of 1 Ω. This is due to the low values we have to use for the load, which represent a certain amount of power demanded by the load. In that case the simulation results are shown in Figure 7.6 after running the file 'nightload.cir' described in Annex 7.

It can be seen that the battery receives energy during the day (176.5 Wh) and delivers it to the load during the four hours of night consumption.

The size of the battery required in that case would be 176.5 Wh.

7.4.3 Day-time load

We could consider also a third case (see Problem 7.1) where the load consumes power during the day time. Assuming a total length of the consumption time to be 8 hours from 8:00 to

17:00, the value of the load current is given by,

$$I_L = \frac{\int_0^{24} P_G \, dt}{V_{cc}8} = \frac{176.5}{12 \times 8} = 1.838 \, \text{A} \tag{7.24}$$

and the load resistance is,

$$R_L = \frac{V_{cc}}{I_L} = 6.528 \, \Omega \tag{7.25}$$

Then, the PSpice file 'dayload.cir' shown in Annex 7, calculates the energy and the results are shown in Figure 7.7. As can be seen the maximum depth of discharge is −41.1 Wh, which is the smallest of the three scenarios analysed. It can be concluded that the daily energy balance may respond to a variety of situations, which have to be taken into account at the design stage.

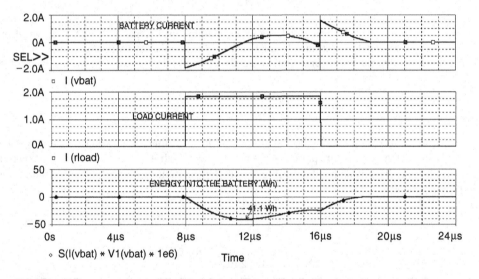

Figure 7.7 Profiles of the battery current (top graph), load current (middle graph) and energy into the battery (bottom graph) for the day-time load scenario. *Warning*: x-axis units are internal PSpice units (microseconds) corresponding to real time units of hours (1 μs corresponds to 1 hour)

7.5 Seasonal Energy Balance in a PV System

The majority of stand-alone PV systems are intended to be self-sufficient during the year. This adds several requirements to the battery size due to the seasonal variations of the energy stored in the accumulator. In order to analyse this problem using PSpice, we will consider a PWL source for the irradiance, which is composed of the values of the monthly radiation at the plane of the PV array (tilted surface). Taking into account the definition of peak solar hours (PSH) in Section 7.2, the values of the Table 7.3 represent the values of the PSH per month in Sacramento (USA), as an example.

Table 7.3 PSH values per month

Month	Monthly radiation (kWh/m^2 month)
Jan	88
Feb	113
Mar	159
Apr	185
May	208
Jun	208
Jul	222
Aug	217
Sep	199
Oct	169
Nov	107
Dec	83

This is transformed into a PWL source as follows:

```
.subckt monthly_radiation 12 10
Vradiation 12 10 pwl 0u 0,0.99u 0,1u 88,1.99u 88,2u 113,2.99u 113,3u
+159,3.99u 159, 4u 185,4.99u 185, 5u 208,5.99u 208,6u 208,6.99u 208, 7u
+222,7.99u 222, 8u 217,8.99u 217,9u 199,9.99u 199, 10u 169,10.99u 169,
+11u 107,11.99u 107, 12u 83,12.99u 83,13u 0
.ends monthly_radiation
```

where January lasts from 1u to 1.99u and so on to preserve monthly values of the radiation. The monthly values of the energy generated are calculated by:

$$(E_{gen})_{month} = N_{sG} \times N_{pG} \times P_{\max Gr} \times (PSH)_{month} \tag{7.26}$$

and the energy consumed during one month is given by:

$$(E_{cons})_{month} = n_i (E_{cons})_{day} \tag{7.27}$$

where n_i is the number of days of each month.
The monthly balance is given by:

$$E_{balance} = (E_{gen})_{month} - (E_{cons})_{month} = N_{sG} \times N_{pG} \times P_{\max Gr} \times (PSH)_{month} - n_i (E_{cons})_{day} \tag{7.28}$$

This is easily implemented in a PSpice file 'seasonal.cir' listed in Annex 7. An example of the results obtained for $N_{pG} = 3$ and $N_{pG} = 4$ corresponding to the Sacramento example are given in Figure 7.8.

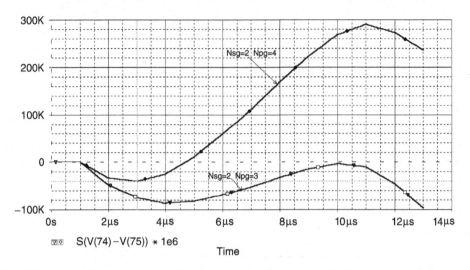

Figure 7.8 Seasonal energy balance. *Warning*: x-axis units are internal PSpice units (microseconds) corresponding to real time units of hours (1 µs corresponds to 1 month)

As can be seen the maximum seasonal deficit during the year is 40 294 Wh for the configuration 2 × 4, whereas the configuration 2 × 3 has a much larger deficit of energy. Of course, this result will also depend on the orientation of the surface of the PV generator, because, as described in Chapter 1, the yearly total energy available and the monthly variation depend largely on the orientation. Looking at the radiation data in Figures 1.10 and 1.11 an angle of inclination close to the latitude will optimize the total energy collected in the year and also reduce the differences between the maximum and the minimum values.

7.6 Simplified Sizing Procedure for the Battery in a Standalone PV System

We have seen how the energy flows in and out of an ideal battery in two time frames: day and year, and considering the functionality that has to be fulfilled by the battery in a standalone PV system, we can now introduce a simplified method to compute the battery size required by:

$$E_{battery} = \left\{ MAX \left[(E_{balance})_{max} + E_{backup}, E_{cycle} \left(\frac{1}{x} \right) \right] \right\} \frac{1}{y \, \eta_{cd}} \qquad (7.29)$$

where

(a) E_{backup} is the energy stored to account for a certain number of days of system operation without solar power generation (due to system failure, maintenance or a number of consecutive days with little radiation from the sun);

(b) $E_{balance\ max}$ is the maximum seasonal deficit during one full year of operation;

(c) E_{cycle} is the energy deficit originated by power mismatch or night load, and x is the battery daily cycling factor;

(d) y is the maximum depth of discharge factor; and

(e) η_{cd} is the battery charge–discharge efficiency.

are the main parameters of the battery size. Equation (7.29) is written in Wh. Most of the batteries are specified in Ah instead. So,

$$C = \frac{E_{battery}}{V_{cc}} \qquad (7.30)$$

In the PV nomenclature it is also common to find the battery size normalized to the daily load L, as:

$$C_s = \frac{E_{battery}}{L} \qquad (7.31)$$

such that the units of C_s are days.

Example 7.5

Calculate the battery size required for a PV system having to supply a load of 3000 Wh/day in Sacramento. The PV array is composed of a 2 × 4 PV module array. The characteristics of each PV module are the same as in Example 7.1. Consider that the daily battery cycling has to be smaller than 15%, the maximum depth of discharge 80% and the charge–discharge efficiency of 95%. Imagine that two other similar arrays are configured, one 3 × 2 and the other 5 × 2 PV modules. Draw the C_a–C_s graph.

Using the radiation values in Section 7.5 for Sacramento, we will first note that seasonal deficit, as calculated in Section 7.5 is

$$E_{seasonal} = 40\,294\,\text{Wh} \qquad (7.32)$$

Considering a back-up storage of five days:

$$E_{backup} = 5 \times 3000 = 15\,000\,\text{Wh} \qquad (7.33)$$

Taking $x = 0.15$ and as the worst case for daily cycling is when little radiation from the sun is converted we will take the maximum daily cycling possible: 3000 Wh-day

$$E_{battery} = \left\{ MAX\left[40\,296 + 15\,000,\ 3000\left(\frac{1}{0.15}\right)\right] \right\} \frac{1}{0.8\,0.95} = 81\,317\,\text{Wh} \qquad (7.34)$$

and

$$C_s = \frac{E_{battery}}{L} = \frac{81\,317}{3000} = 27.1 \tag{7.35}$$

$$C_a = \frac{N_s \times N_p \times P_{mp} \times (\overline{PSH})}{L} = \frac{2 \times 4 \times 85 \times 5.35}{3000} = 1.21 \tag{7.36}$$

If we run PSpice for the two other sizes, using the same netlist with N_p values from 3 to 6 we get the results shown in Table 7.4, which are plotted in Figure 7.9. As can be seen there is a trade-off between array size and battery size for the same system.

Table 7.4 Values of seasonal energy deficit for several system sizes

$N_p \times N_s$	C_a	Seasonal energy deficit (Wh)	C_s from seasonal deficit (includes a backup storage of 5 days)	C_s from equation (7.36)
3 × 2	0.909	1 45 647	48.54	48.54
4 × 2	1.21	40 296	27.1	27.1
5 × 2	1.51	18 162	14.54	14.54
6 × 2	1.81	3 255*	7.11	8.71

*Seasonal deficit plus autonomy smaller than daily cycling.

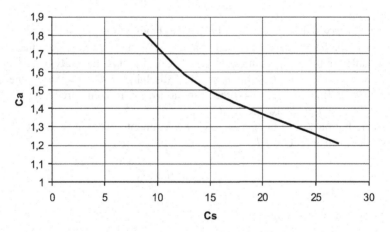

Figure 7.9 Graph of C_a–C_s for Example 7.5

7.7 Stochastic Radiation Time Series

Photovoltaic modules and arrays are placed in suitable surfaces, which in the more general case, are not horizontal but have an inclination angle. Solar energy system design most often requires the values of monthly average radiation values and in some cases design tools require hourly radiation data. This information for a particular place in the world can be generated from measurements of radiation and meteorological data available from atmospheric observatories around the world. The data from these observatories basically concern the global and diffuse radiation at a horizontal surface and a methodology and models have to be applied to produce daily or hourly values at an inclined surface of global, direct and diffuse radiation. The objective of these tools is to generate a time series of data, with the same statistical properties as the data available from measurements. For example, the tool METEONORM employs one of these methods and in particular the daily values are obtained using the Collares–Pereira [7.5] model, which uses Markov chains, and hourly values are then generated from these daily values.

Photovoltaic system designers are interested in the availability of such representative time series of radiation to more accurately size systems and estimate reliability. We will run long-term simulations using synthetic radiation time series to evaluate the loss of load probability in Section 7.8 below, and in this section Example 7.6 illustrates the nature of these radiation time series.

Example 7.6

Write a PSpice file of a subcircuit describing a stochastically produced, one-year radiation time series for a PV system located in Madrid (Spain), 40.39° of latitude North at a 40° inclined surface facing south.

The results of the data series are taken from METEONORM and a stimulus file is written with the name 'madrid.stl', available at the book's website. The .cir file to plot the resulting time series is the following:

```
*madrid.cir
.include madrid.stl
v1 1 0 stimulus vmadrid
r1 1 0 1
.tran 0.1u 365u 0 0.1
.probe
.end
```

The result is shown in Figure 7.10 for the 365 days of the year. Remember that in order to speed up simulations, the internal PSpice time unit is set to one microsecond corresponding to a real time unit of one day in this file.

This time series has the same average value, variance and autocorrelation as the measured time series.

Moreover, hourly values of radiation are also available from commercial software and the output files can be converted into PSpice stimulus files by adapting the format, as with the daily radiation values in Example 7.6.

Figure 7.11 shows the hourly radiation data plotted with PSpice from the data generated using METEONORM [7.4] for Madrid (Spain) at an inclined surface of 40°. Figure 7.11 shows only the first 10 days for sake of clarity. As can be seen the time axis has internal PSpice units of microseconds corresponding to real time units of hours (1 microsecond corresponds to 1 hour).

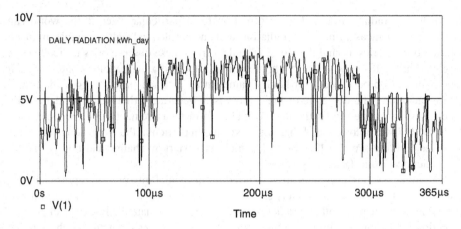

Figure 7.10 PSpice plot of daily radiation data stochastically generated by METEONORM for Madrid (Spain) at an inclined surface of 45°. *Warning*: *x*-axis is time with internal units of microseconds and real units are days (1 µs corresponds to 1 day)

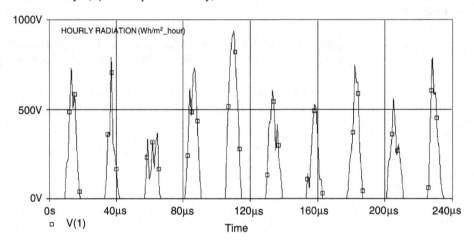

Figure 7.11 PSpice plot of stochastically generated hourly radiation values for the first 10 days of the year for Madrid (Spain) using METEONORM software, inclination angle of 40° facing south. *Warning*: *x*-axis units are internal PSpice units (microseconds) corresponding to real time units of hours (1 µs corresponds to 1 hour)

Using this procedure '.stimulus' files for long time series of radiation data can be generated to help the simulation of PV systems, as will be illustrated in the sections below.

7.8 Loss of Load Probability (LLP)

Advanced tools for PV system sizing and design often use a parameter called loss of load probability (LLP). This parameter is defined by making a simplifying assumption about the system operation. This is illustrated in Figure 7.12 where the PV generator is connected to the load as well as an auxiliary generator (e.g., a diesel generator). The LLP definition

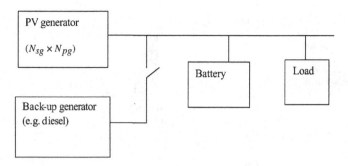

Figure 7.12   PV system with back-up diesel generator

generally assumes that the load consumes power only at night, and the auxiliary generator supplies the required energy to the system in order to fill the storage up to full charge.

Then the LLP is given by:

$$LLP = \frac{\sum E_{aux}}{N \times L} \tag{7.37}$$

where N is the number of days of analysis. In order to compute the value of the LLP in a given system, the procedure generally followed uses a stochastically generated time series of radiation data during a number of days generally more than one year. A simulation is then performed computing the energy stored at the battery at the end of every day and prior to the load connection. If the energy stored is larger than the daily load L, the auxiliary generator is not connected. On the contrary, if the energy stored is lower than L, then the auxiliary generator is connected and supplies the difference up to full charge of the battery. The value of LLP is now computed by adding all the energy delivered by the auxiliary generator during the period of N days.

The way we have implemented the procedure in PSpice requires some data manipulation in order to reproduce the sequence of events predicted in the LLP definition. This is shown in Figure 7.13 where it can be seen that a pulse of energy generated equivalent to a real day is implemented, and the same is done for the load which is also concentrated in a pulse at the end of the day. In between, the auxiliary generator is required to provide extra energy only if the energy stored at the battery at the end of the generation pulse is smaller than the typical load for one day. The action taken is to run the diesel generator or any other kind of auxiliary generator to fill up the battery. The whole process and results are illustrated in Example 7.7.

Example 7.7

Using a time series of stochastically generated daily radiation values in Barcelona for one year, simulate a PV system with a constant load of 2000 Wh-day and $N_{sG} = 2$ and $N_{pG} = 5$. The PV module maximum power under one sun standard conditions is 85 W. The initial value of the energy stored is 20 000 Wh.

(a) Generate the generation pulses, 20% wide of the day length and five times the generated energy in the day.

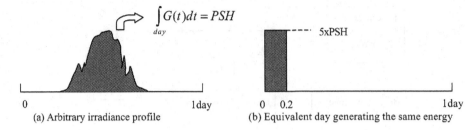

$$\int_{day} G(t)dt = PSH$$

5xPSH

0   1day   0   0.2   1day
(a) Arbitrary irradiance profile   (b) Equivalent day generating the same energy

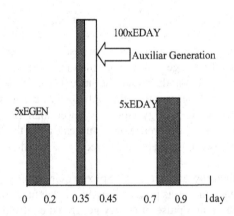

$$\int_{day} P_L(t)dt = EDAY$$

5xEDAY

0   1day   0   0.7   0.9   1day

(c) Arbitrary consumption profile   (d) Equivalent day consuming the same energy EDAY

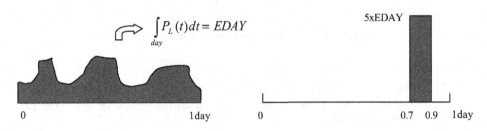

100xEDAY

Auxiliar Generation

5xEGEN

5xEDAY

0   0.2   0.35 0.45   0.7   0.9   1day

Figure 7.13   Model of an equivalent day to compute the LLP

This is performed by the following PSpice file:

```
.subckt generation 700 730 731 740 params: pmaxr = 1, nsg = 1, npg = 1
* Sample and hold circuit
sw1 731 720 711 700 switch
.model switch vswitch roff = 1e8 ron = 0.0001 voff = 0 von = 5
vctrl1 711 700 0 pulse (0,5,0,0,0,0.01u,1u)
csh 720 700 10u
sw2 720 700 721 700 switch
vctrl2 721 700 0 pulse (0,5,0.2u,0,0,0.01u,1u)
* end of Sample and hold
ggen 700 730 value = {5*v(720)/1000*pmaxr/1000*nsg*npg}
egen_out 740 700 value = {sdt(1e6*5*v(720)/1000*pmaxr/1000*nsg*npg)}
.ends generation
```

which basically performs two functions:

- generate a daily radiation generation pulse using a sample and hold circuit shown in Figure 7.14. It can be seen that the sampling phase tracks the value of the radiation signal during a short time while the sample switch is closed, and then holds the value until the discharge switch closes; and

- generate an equivalent current pulse with an amplitude of five times the daily load (ggen) and compute the time integral value (egen).

(b) Create the load pulse and the auxiliary generator pulse files

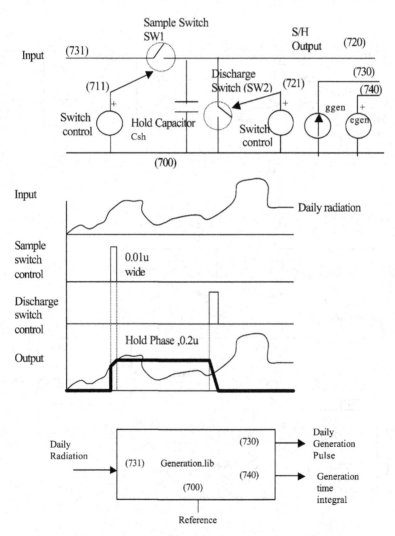

Figure 7.14 Generation subcircuit including sample and hold circuit generation current source (ggen) and time integral generation source (egen)

The load pulse is achieved by means of the circuits shown in Figure 7.15 where the circuit shown in (a) is implemented by the PSpice file 'cons.lib' listed in Annex 7 and the auxiliary generator required to produce the auxiliary energy pulse is implemented by means of the circuit shown in Figure 7.15(b) with the contents of the following PSpice file 'aux_gen.lib' also listed in Annex 7.

The sample and hold circuit included in this file has the objective to check when the value of the energy stored is smaller than the daily load and then holds this condition until the reservoir has been fully filled up.

(c) Calculate the LLP

Finally the LLP value is calculated with the help of the model shown in Figure 7.16, which includes the subcircuits of the daily radiation for one year (radiation_1year-lib), the

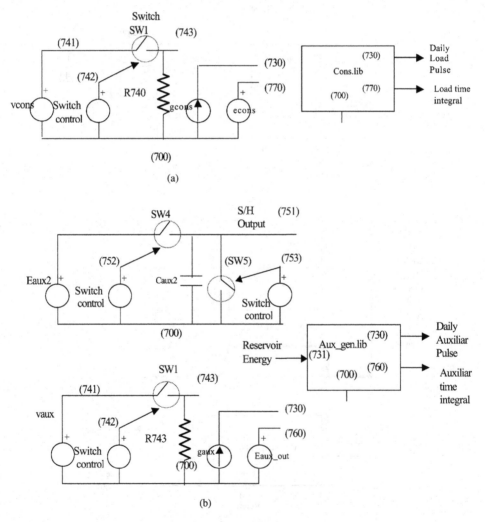

Figure 7.15 Subcircuits generating a load pulse and time integral of load (a) and the auxiliary generator pulse (b)

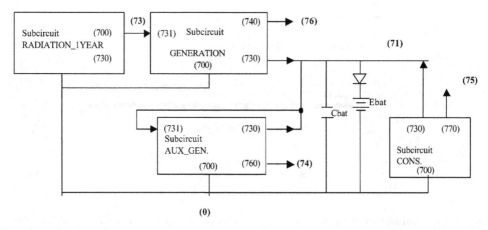

Figure 7.16 PSpice block diagram used to compute the LLP of a PV system

generation subcircuit (generation.lib), the load subcircuit (cons.lib), the auxiliary generator subcircuit (aux_gen.lib) as shown in the file 'llp.cir' listed in Annex 7.

It can be seen that the battery is modelled here by a simple capacitor where the generation, load and auxiliary energy pulses are fed (it should be noticed that the value of the capacitor is 1 μF because the internal PSpice time is set to 1 microsecond (one day real time corresponds to 1 μs internal PSpice time), this means that the time integral function implemented by the capacitor reverts the units:

$$v(71) = \frac{1}{C_{bat}} \int [i(ggen) + i(gaux) - i(gcons)] \mathrm{d}t \qquad (7.38)$$

where the internal PSpice units of $v(71)$ are volts, returns the value in real units of kWh. A detailed graph of the 12th and 13th day of the year is given in Figure 7.17, where the

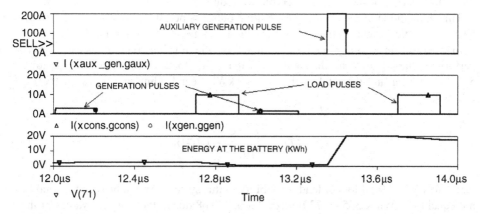

Figure 7.17 Detail of the 12th day of the year where the stored energy at the end of the generation pulse is smaller than the daily energy load and then the auxiliary generator fills up the battery storage. *Warning*: x-axis units are internal PSpice units (microseconds) corresponding to real time units of hours (1 μs corresponds to 1 hour)

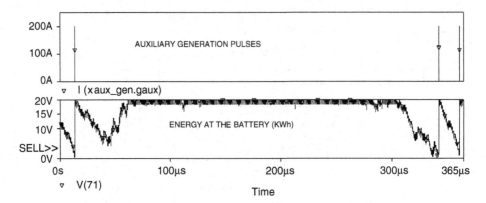

Figure 7.18 Calculation of the LLP in one year of stochastically generated daily radiation values. *Warning*: internal time units are microseconds corresponding to real time units of days (1 μs = 1 day)

generation, load and auxiliary generation of energy are shown and the evolution of the energy stored at the battery is plotted as the bottom graph. As can be seen, the generation of energy is concentrated in the first two hours of the day (which means that the amplitude of the generation pulse has to be multiplied by five to preserve the total radiation received in the full day). Once the generation pulse has come to an end, and before the load pulse arrives, the system checks the value of the stored energy. If the value is smaller than the expected daily load, then an auxiliary generation pulse fills the energy reservoir. We have arbitrarily selected the generation pulse amplitude to be 100 times the daily load and the width variable according to Figure 7.13. If no extra generation is required, this generator is not connected.

In the example shown in Figure 7.17 the auxiliary generation is only required at the 13th day due to the low radiation received and to the low value of the remaining energy that day.

A full view of the 365 days of the year is shown in Figure 7.18, where it can be seen that three full charges of the energy reservoir are required.

Time integration of the auxiliary generation pulses is performed in Figure 7.19, where the PSpice variable 'eeaux' is plotted as a function of time, in the top graph.

In order to gain accuracy in the LLP calculation, the time integral of the load pulses is also performed and shown in the bottom graph. The total auxiliary energy required is 55.38 kWh_year and the total load is 770.69 kWh_year leading to a value of the LLP as follows,

$$LLP = \frac{53.38}{770.69} = 0.0718 \tag{7.39}$$

that is to say 7.18% of loss of load probability in this system. As can be seen the total load is not equal to 2 kWh × 365 (= 730) instead it is 770.68 due to the non-zero values of the rise and fall times of the load pulses.

It becomes clear, from the LLP definition, that by selecting a higher value of the battery size and/or a higher value of the array size, the LLP can be reduced.

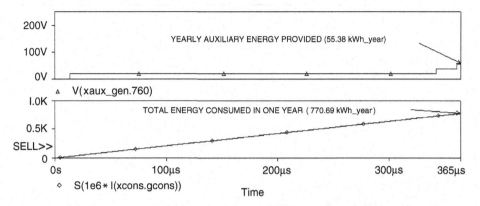

Figure 7.19 Auxiliary energy provided by the external generator (top) and energy consumed from day one to day *n* with an indication of the total energy consumed at the end of the year. *Warning*: *x*-axis units are internal PSpice units (microseconds) corresponding to real time units of days (1 µs corresponds to 1 day)

7.9 Comparison of PSpice Simulation and Monitoring Results

It may be useful to assess modelling accuracy and validity by means of comparison of modelling results with monitoring data. It may be impossible to have access to monitoring results for long periods of time, simply because similar installations data may not be available. It may be, however, enough to have a few days of monitoring data of installations using similar types of PV modules and batteries and adjust the parameter values of the models involved. A comparison of this kind is shown in this section.

As an example consider a conventional standalone PV system including a PV array, power-conditioning electronics, a battery and a load.

Example 7.8

Consider a standalone PV system composed of eight PV modules (two series and four parallel) with $I_{scM} = 2.45$ A, $V_{ocM} = 19.25$ V under standard AM1.5 conditions connected to 12 ATERSA 7TSE 70 battery elements each of 2 V nominal. The load can be modelled by a current source. The objective of this example is to simulate the PV response of such a system when subject to an irradiance profile described by a stimulus file 'irrad.stl' and compare the results to monitoring results described by PSpice files 'Imodapril.stl' (PV array output current) and 'vbatapril.stl' (monitored battery voltage values). These files are available at the book's website.

Solution

We write the PSpice netlist (stand-alone.cir), that allows the simulation of the system behaviour is as follows:

```
***** stand_alone.cir

xmodule 0 1 2 3 moduleRs params: iscmr = 9.8, tempr = 28, vocmr = 38.5, ns = 66, np = 4,
+nd = 1, pmaxr = 380
.inc moduleRs.cir

xbat1 3 0 7 batstd params: ns = 12, SOCm = 13200, k = .8, D = 1e-5, SOC1 = 0.95
.inc batstd.cir

.inc irrad.stl
vmesur 1 0 stimulus Virrad

.inc iload.stl

*** included real data files from monitoring process in order to compare results
.inc Imodapril.stl
im 10 0 stimulus modulecurrent
rmo 10 0 10 000
.inc vbatapril.stl
vcompara 8 0 stimulus vbattery
rc 8 0 10 00 000

if 3 0 stimulus load
.tran 1s 3110 00s
.probe
.end
```

The load current source 'if' is also implemented by a stimulus file and comes from real
monitoring data (file 'iload.stl') implemented in the load current source i_f and shown in
Figure 7.20.

The above standalone PV system has been simulated and monitored over a period of three
and a half consecutive days in Barcelona (Spain) for which irradiance (Figure 7.21) and
temperature were also measured every two minutes.

Figure 7.22 shows a comparison between simulation and measurements of the PV array
output current.

Moreover a comparison between battery voltage measured and simulated is shown in
Figure 7.23 and the battery state of charge evolution is shown in Figure 7.24.

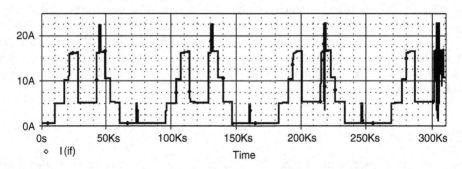

Figure 7.20   Load current

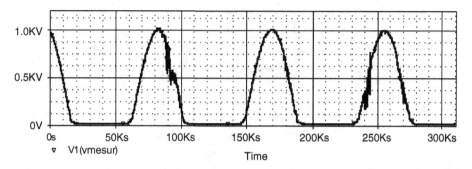

Figure 7.21   Irradiance data

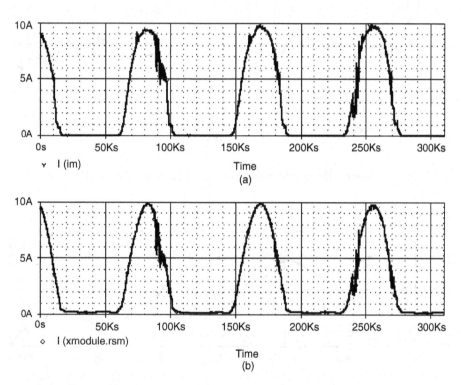

Figure 7.22   PV array output current (a) measured and (b) simulated

The results shown in Example 7.8 validate the simulation results and opens the way to perform detailed simulations of the system under average or limited working conditions.

7.10   Long-Term PSpice Simulation of Standalone PV Systems: A Case Study

We have seen so far the use of PSpice models to evaluate the LPP probability, to simulate the short-term response of a PV system and to compare simulation with measurements. It is

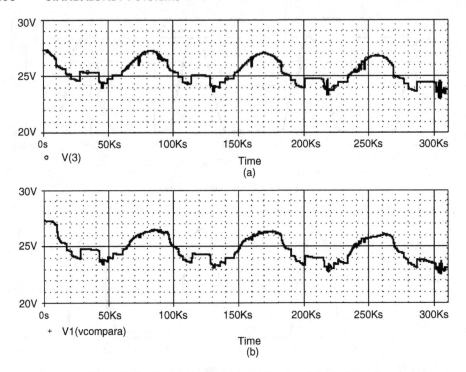

o V(3)

Time

(a)

+ V1(vcompara)

Time

(b)

Figure 7.23 Battery voltage (a) simulated and (b) measured

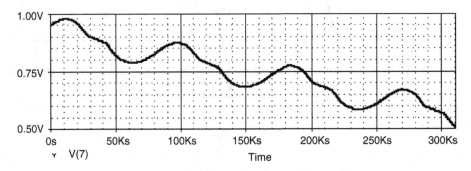

Y V(7)   Time

Figure 7.24 Battery SOC behaviour from PSpice simulation

often required to estimate the long-term response of a given PV system, because they are designed to serve electricity needs in long periods of time. This is a difficult task because it requires manipulation of a large quantity of data and, to be realistic, the use of accurate models for the PV system components, mainly the PV generator and the battery. These two models are highly non-linear and the numerical solution of the dynamical equations involved generally are time consuming and subject to numerical convergence problems.

Although PSpice is a tool meant to deal with non-linear devices such as diodes, transistors or digital circuits, convergence problems sometimes arise, although a number of parameters are made available to the user in order to achieve numerical convergence. The reader must be warned that a first try of a transient simulation of complex circuits with non-linear devices usually requires some adjustment of the values of these parameters.

Apart from these problems, long-term simulation of PV system response is possible using PSpice resulting in short computing times. This is because of the use, as we have already done in this and in previous chapters, of different internal PSpice time units and real time units, as will be shown below.

In Section 7.7 it was shown that long-term, statistically significant time series of monthly, daily or hourly radiation values are broadly available (if this is not the case we refer the reader to Chapter 9 where it is shown how to generate random time series of radiation values using PSpice). This information along with corresponding time series of ambient temperature can be fed to the PV generator model and then a full PV system can be built, as for example the one shown in Figure 7.25.

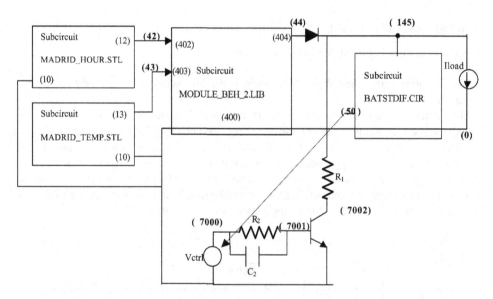

Figure 7.25   PSpice circuit to simulate long-term response of a PV system

As can be seen the stimulus files 'madrid_hour.stl' and 'madrid_temp.stl' are used along with a PV generator model connected through a diode to a battery model and a current source acting as a load. In order to illustrate the behaviour of the system over a full year, avoiding full power conditioning and control circuitry, only an overflow control is used, which is implemented by a bipolar transistor, switching ON when the state of charge of the battery reaches a certain level. This overflow circuit includes an extra load current.

We consider a standalone PV system able to serve an application consuming 3000 Wh_day at 24 V in Madrid (Spain), facing south and inclined 40°.

From equation (7.10) the load equivalent current can be calculated to be 24.761 A, and from data of average radiation daily load for Madrid (5.04 kWh/m²_day) taken from METEONORM results for this location and surface inclination, the values of the number of series and parallel PV modules can be calculated using equations (7.13) and (7.14) giving the closest integer values $N_{sG} = 2$ and $N_{pG} = 6$.

If we now use the file 'seasonal.cir' to estimate the value of the seasonal deficit of energy from the monthly values of radiation, we find that the deficit is zero, so the size of the battery is now evaluated considering five days of energy and battery efficiency and daily cycling from equation (7.29),

$$E_{battery} = \frac{3000}{0.15 \times 0.95 \times 0.8} = 26\,315\,\text{Wh}$$

where a daily cycling of 15%, a battery efficiency of 95% and a maximum depth of discharge of 80%, have been assumed.

With this estimate, we can now proceed to simulate a full year considering random irradiance and temperature hourly values generated from METEONORM software. These are included in the files:

- 'madrid_hour.stl', for the hourly radiation values; and

- 'madrid_temp.stl' for the ambient temperature.

The complete netlist describing this system is the file 'stand_alone_Madrid.cir' listed in Annex 7.

The model used here for the PV generator is a level 2 model of generator_beh.lib which includes some precautions to avoid numerical convergence problems when the irradiance is close to zero and has the values of the numerical convergence parameters adjusted in the .options statement. The PSpice netlist 'generator_beh_2.lib' is also listed in Annex 7. Let us see first the time evolution of the daily radiation (top), ambient temperature (middle) and battery state of charge (bottom) in Figure 7.26 for the first 31 days of the time series data. We are assuming that the initial value of the battery state of charge is 90%.

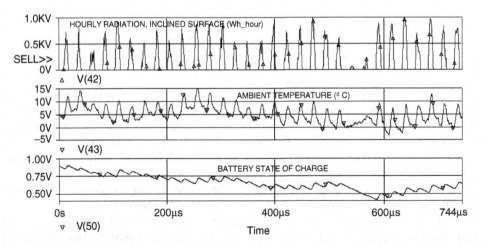

Figure 7.26 January simulation of a standalone PV system using the values of hourly irradiance and temperature stochastically generated by METEONORM. *Warning*: x-axis units are internal PSpice units (microseconds) corresponding to real time units of hours (1 μs corresponds to 1 hour)

As can be seen the state of charge degrades in the period of time observed and reaches a level below 50%. If we now look at the full year picture, some of the details are of course lost, but we see in Figure 7.27, where the state of charge evolution and the overflow transistor current are plotted, that the system is most of the time producing more energy than demanded by the load, with the exception of the beginning and the end of the year, which are the months of lowest sun radiation; the system evolves and a net flow of energy from the battery to the load is clearly observed.

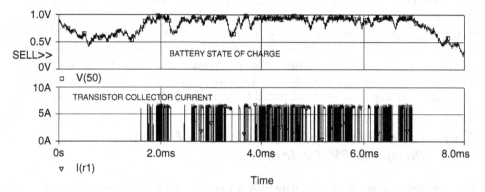

Figure 7.27 State of charge and overflow current evolution for Stand_alone_madrid. *Warning*: x-axis units are internal PSpice units (microseconds) corresponding to real time units of hours (1 μs corresponds to 1 hour)

Moreover, the overflow transistor is forced to draw additional current to avoid over-charging the battery. A closer look at the overflow current is shown in Figure 7.28 where a four-day picture of the time dynamics of the SOC and the transistor current can be seen. When the SOC reaches 95% the overflow transistor is switched ON drawing current from the system.

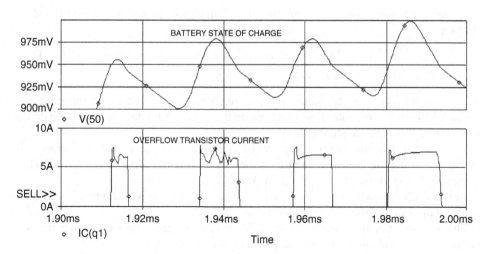

Figure 7.28 Detail of the evolution of the state of charge and overflow current through the transistor. *Warning*: x-axis units are internal PSpice units (microseconds) corresponding to real time units of hours (1 μs corresponds to 1 hour)

If we look again at Figure 7.27 we see that at the end of the year the value of the SOC is 0.287. Taking into account that the next month will again be a low radiation month such as the first one shown in the graph, we realize that this level will not be enough to guarantee full supply over the year because of the additional discharge after the twelfth month. If the size of the battery is increased by 30% the system does not suffer from this accumulated deficit of charge in a run exceeding one year.

It can be concluded that this procedure allows us to refine the size of a PV system in a simple way. The simulation shown in this section requires a CPU simulation time of the order of one minute depending on PC performance. It has to be added that changes of the data used in this system may require adjustment of the numerical convergence parameters as well as of the transistor and diode model parameters. A final remark is related to the transistor circuit used. As the internal time scale is microseconds, the transistor has to switch in this time scale although in the real system it would switch at a much smaller frequency. This means that a virtual speed-up capacitor has to be added to the base circuit as shown in Figure 7.25.

7.11 Long-Term PSpice Simulation of a Water Pumping PV System

Water pumping is one of the most frequently used PV applications in areas where totally autonomous pumping systems are required. For long-term simulations, what is really required is a quasi-steady-state model for the pump because the irradiance and temperature transients are slow compared to the electrical and mechanical time constants of the motor-pump association. In this section a one-year simulation of a directly connected PV generator–water pump system is described.

Example 7.9

Consider a PV array composed of 10×4 PV modules, with the following AM1.5 G characteristics:

$$I_{sc} = 5\,\text{A}, \quad V_{oc} = 23\,\text{V}, \quad coef\_vocm = -0.0828, \quad P_{max} = 85\,\text{W}, \, NOCT = 47,$$

$$I_m = 4.726\,\text{A}, \, V_m = 17.89\,\text{V}, \qquad n_s = 36$$

The characteristics of the motor and the pump are considered to be the same as in Chapter 5. Simulate the water flow produced by the system using the one-year radiation and temperature files used in Section 7.10.

We first simplify the motor–pump model applying quasi-steady-state conditions, this means neglecting the terms depending on the time (terms including L_a, L_f and J). This is written as 'pump_quasi_steady.lib' file in Annex 7.

Next a .cir file 'water_pump.cir' shown in Annex 7, is written including the radiation and temperature data and the PV generator model.

The resulting flow is shown in Figure 7.29 for 12 consecutive days of the random series included in 'madrid_hour.lib' and 'madrid_temp.lib' files.

The plot has been selected to show how the water pumping works under random irradiance and temperature conditions. As can be seen one of the 12 days shown has a very poor radiation and the motor does not have enough power at the input to get started. As

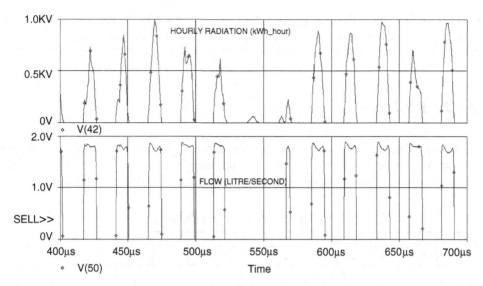

Figure 7.29 Detail of the value of the flow (litre/second). *Warning*: *x*-axis units are internal PSpice units (microseconds) corresponding to real time units of hours (1 μs corresponds to 1 hour)

can be seen at the next day, although also poor in radiation, the power is enough to overcome the starting torque and then some water can be pumped.

The total water pumped from the start of the year can be obtained in node (60) which produces the time integral of the flow. In that particular case the total volume pumped is 20.41×10^6 litre/year as shown in Figure 7.30. Although it may appear that these are very

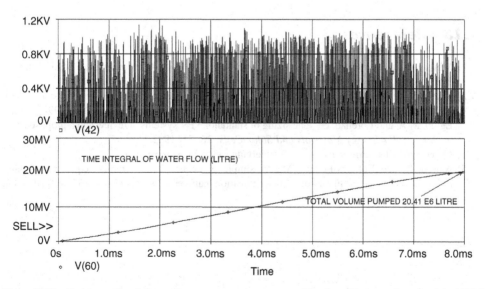

Figure 7.30 One-year simulation of a water pumping system. Hourly radiation values (top) in kWh/m²_day and time integral of the water flow (bottom). *Warning*: *x*-axis units are internal PSpice units (microseconds) corresponding to real time units of hours (1 μs corresponds to 1 hour)

time-consuming simulations, in practice the CPU time for the simulation in Figure 7.30, covering a full year of hourly radiation data, is typically less than one minute.

7.12 Problems

7.1 Consider the irradiance and temperature profiles in Table 7.5 for June and run the file psh.cir with the new values and calculate the energy generated by the same PV module described in the example and compare with the value of PSH of that month.

Table 7.5 Hourly radiation kWh/m^2 hour

Hour	6	7	8	9	10	11	12	13	14	15	16	17	18
Radiation	0.038	0.122	0.274	0.433	0.562	0.654	0.725	0.725	0.654	0.562	0.433	0.274	0.122

7.2 Calculate the response of a PV module on a typical day of 16th June neglecting in equation (7.11) the term depending on the ambient temperature. Calculate and plot the evolution of the maximum power point voltage as a function of time.

7.3 Using the data values of radiation in Sacramento in Table 7.1 calculate the values of PSH for the 12 months of the year and calculate the size of the PV array for the same load as in Example 6.1 but using the worst-case sizing (worst month in the year). Calculate the oversizing factors in that case.

7.13 References

[7.1] Ministerio de Industria y energia, *Radiación Solar sobre superficies inclinadas*, Madrid, Spain, 1981.

[7.2] Markvart, T., Castañer, L. and Egido, M.A., '*Sizing and reliability of stand-alone PV systems*', 12th European Photovoltaic Solar Energy Conference, pp. 1722–24, Amsterdam April 1994.

[7.3] Egido, M.A. and Lorenzo, E., 'The sizing of standalone PV systems: a review and a proposed new method', *Solar Energy Materials and Solar Cells*, **26**, 51, 1992.

[7.4] Meteotest Meteonorm version 4.0, Switzerland, www.meteonorm.com.

[7.5] Aguiar, R. and Collares-Pereira, M., 'A simple procedure for generating sequences of daily radiation values using a library of Markov transition matrices', *Solar Energy*, **40**(3), 269–79, 1988.

8

Grid-connected PV Systems

Summary

Grid-connected PV systems are starting to play a very important role in photovoltaic applications. This chapter describes some PSpice models for inverters and AC modules which aid the PSpice simulation of grid-connected PV systems.

Examples of sizing and energy balance for this kind of PV system are also included showing that PSpice can help engineers to obtain a good approximation of system behaviour using the proposed inverter models. Short- and long-term simulations validate the sizing procedures.

8.1 Introduction

Photovoltaic systems have a wide range of applications, from very small units for low power, as will be described in Chapter 9, to large PV power plants in the MW range. Despite this wide range the most important application fields of PV systems have historically been outer-space PV systems and standalone systems in areas with poor or absent electricity supply from the public grid.

In the last few years an important growth of grid-connected PV systems has been observed, specially in industrialized countries. Several reasons lie behind this fact apart from the traditional advantages of photovoltaic electricity:

- Utility grid-interactive PV systems are becoming more economically viable as the cost of PV components has been significantly decreasing in recent years, in particular the average cost of PV modules and inverters.

- Technical issues associated with inverters and interconnection of PV systems to the grid have been addressed by manufacturers and today's generation of inverters have enhanced reliability and reduced size.

- Utility benefits. The fact that solar electricity is produced in central hours of the day can add value to the electricity. This power peak demand can be partially supplied by dispersed grid-connected PV systems that are able to generate power at the same place where this power is used, reducing the heavy load supported by the transmission systems and achieving benefits in distribution and line support.

- National or international programmes promoting the implantation of grid-connected PV systems. Most industrialized countries have launched programmes offering different incentives to small-scale renewable energy producers.

Even with the aforementioned benefits and the significant cost reductions achieved, these systems still cannot compete with other energy resources on a pure financial analysis without reasonable funding and promotion from public bodies. The lack of standardization of interconnection requirements in different countries, for PV system components, especially inverters is also an important barrier to the market growth of grid-connected PV systems. Despite these obstacles, this market is becoming important in photovoltaic applications. Photovoltaic power generation systems are likely to become, although small compared with other power generation sources, important sources of distributed generation, interconnected with utility grids.

8.2   General System Description

Grid-connected PV systems, also called utility interactive PV systems, feed solar electricity directly to a utility power grid. These systems consist of:

- A PV generator, an array of PV modules converting solar energy to DC electricity.

- An inverter, also known as a power conditioning unit or PCU, that converts DC into AC electricity.

- System balance, including wiring and mounting structure.

- Surge and ground fault protection and metering or other components than may be required for interconnection to the AC grid.

- AC loads, electrical appliances.

When the sun is shining, the DC power generated by the PV modules is converted to AC electricity by the inverter. This AC electrical power can either supply the system's AC loads, outputting any excess to the utility grid, or may completely output all the energy produced. During dark hours, the power demanded by the loads is supplied entirely by the utility grid. The role played by batteries in standalone PV applications is replaced, in grid-connected systems, by the utility grid itself acting as energy reservoir for the system, resulting in cost and maintenance reduction. If necessary, batteries can also be included in the system in the same way as the case of standalone systems with AC output.

Figure 8.1 shows a schematic diagram of a three-phase grid-connected PV system. This schematic diagram may vary from country to country due to different national regulations, especially the circuitry involving safety and protection devices.

At can be seen in Figure 8.1, a first protection level is formed by fuses and blocking diodes between the PV array output and the main DC conductor. Surge protection elements must be

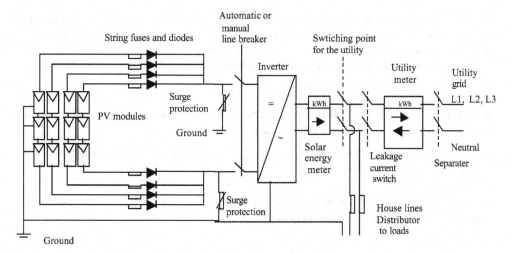

Figure 8.1 Grid-connected photovoltaic system schematic

included at the inverter input and output, the figure shows these elements only at the inverter input.

Isolation transformers can be optionally placed at the inverter output depending on the type of inverter used; some inverters include transformers of this type inside and others do not. Some compulsory national regulations require an isolation transformer, in which one of the windings has a delta connection, for connection to grids of medium, high or very high voltage. In most countries this is not required for low voltage connection.

If possible, utility safety devices and guard relays must be placed between the inverter output and the switching point with the utility grid; these elements are not shown in the figure. Finally the metering can be made as shown on the figure or at the utility grid side.

A first classification of grid-connected PV systems can be made according to size as follows:

1. *Small* – Power from 1 to 10 kWp. Typical applications are: rooftops of private houses, school buildings, car parks etc.

2. *Medium size* – Power from 10 kWp to some hundred of kWp. These kind of systems can be found in what are called building integrated PV(BIPV) systems, in roofs or facades. They may operate at higher voltages than smaller systems.

3. *Large size* – Power from 500 kWp to MWp range, centralized systems. These systems are normally operated by electric companies.

8.3 Technical Considerations

Protection systems are required to prevent damage to the PV system and also to avoid downgrading the quality of the grid electricity. Other important topics are the electrical installation procedures, electrical interference between utility grid and PV systems, EMI (electromagnetic interference) and harmonics. A wide set of standards and recommendations exist in most of the EU, the USA, Japan and Australia, which have been developed by different national bodies and cover most of these topics.

As an example of these standards and recommendations, two US national standards can be cited here:

- ANSI/IEEE std 929-1988: IEEE recommended practice for utility interface of residential and intermediate photovoltaic (PV) systems.

- IEEE std 929-2000: IEEE recommended practice for utility interface of photovoltaic (PV) systems.

A small overview of the most important problems associated with the connection of PV systems to the utility grid is described below.

8.3.1   Islanding protection

The continued operation of a grid-connected inverter when the utility grid, or a portion of the utility system, has been switched off or no electric energy can be delivered from the utility system, is known as the islanding mode of operation. Islanding can strongly affect the equipment and loads connected to the network, and can cause electric shocks to users or utility grid workers. For these reasons, inverters must identify a grid fault or disconnection and must immediately disconnect its output itself.

Much research has been done about islanding prevention on PV systems in the last few years [8.1–8.2], and nowadays most commercial inverters include acceptable islanding prevention capabilities obtained by a combination of different control algorithms. Detection of islanding can be achieved using active or passive methods [8.1].

8.3.2   Voltage disturbances

Appropriate voltage levels must be maintained at the customer's input connection to the grid. Different limits for the voltage levels have been established in different countries. The fact that grid-connected PV systems can contribute to distribution line voltage variation has to be taken into account. Inverters must sense the voltage variations at the connecting point to the grid and most inverters also include the capability of disconnection and automatic reconnection after confirmation of utility stability recovery.

8.3.3   Frequency disturbances

Inverters must generally provide internal over/under line voltage frequency shutdown. Internal shutdown should be produced if, within a few cycles, the frequency falls outside predetermined boundaries, usually ranging from ± 0.2 Hz to ± 5 Hz, around the nominal grid frequency.

8.3.4   Disconnection

As shown in Figure 8.1 a switch for utility interface disconnect or separator, is normally included in grid-connected PV systems. This switch provides safety especially for personnel involved in maintenance or for utility workers. Taking into account the above-described protections that inverters must implement, especially islanding detection and prevention, this main disconnection switch can be considered redundant, however, personal safety is the most important issue concerning grid-connected PV systems [8.3], and regional regulations in different countries must be satisfied.

8.3.5   Reconnection after grid failure

It is important to ensure correct operation of the grid for a prudent interval of time before reconnecting the inverter. This task is implemented in most commercial inverters by means of grid sensing and auto-reconnection.

Most inverters are able to reconnect themselves to the grid after observation of a number of grid cycles with correct values of voltage, amplitude and frequency.

8.3.6   DC injection into the grid

As has already been said, some inverter designs include transformers inside the inverter, suppressing any DC injection into the grid. Advantages of keeping DC out of the grid are personal safety improvement, protection of disturbances on the utility grid and saturation effects in local distribution transformers, and finally the prevention of saturation on inductive loads.

However, the number of transformerless inverters in the marketplace today, has been increasing because of technical and cost advantages. In most countries isolation transformers are not required for small PV grid-connected systems, but they are compulsory for medium or large size PV systems much of the time.

8.3.7   Grounding

The components of a PV system, including the inverter, must be grounded in accordance with the applicable national regulations in each case. Grounding conductors are necessary to conduct current when a ground fault occurs, this will minimize electrical shock hazards, fire hazards and damage to the loads and system equipment.

8.3.8   EMI

The components of a PV system, especially the inverters, are subject to varying high frequency noise emission/immunity requirements that limit the permissible radiation spectrum for a range of frequencies, usually between 150 kHz and 30 MHz.

Table 8.1 Commercial inverter characteristics

Inverter	Manufacturer	Nominal Output Power (VA)	THD (%)	Ouput freq. (Hz)	Vout (V AC)	Max. Efficiency (%)
Tauro	Atersa www.atersa.com	700–3000	<4	$50 \pm 5\%$	$220 \pm 7\%$	93
MM 5000/3000	Advanced Energy Inc. http://www.advancedenergy.com	5000	<3	$60 \pm 5\%$	$120 \pm 5\%$	92
Fronius IG	Fronius http://www.fronius.com/	1300–4600	<3.5	50	230	94.5
T series	Futronics http://www.futronics.co.uk/	1200–1500	<3	$50/60 \pm 0.02\%$	$230 \pm 3\%$	90
Sunmaster QS series	Mastervolt http://www.mastervoltsolar.com/	1500–5000	<3	50	230	94
Sunny Boy series	SMA http://www.sma.de/	700–2600	<4	50	180–265	93.6
Prosine series	Xantrex http://www.xantrex.com/	1000–1800	<3	$50/60 \pm 0.05\%$	$120/230 \pm 3\%$	90

8.3.9   Power factor

A value of unity is desired for the power factor, both at the utility grid connection as well as at the inverter output. The regulation depends again on the country, but in general the inverter operation at power factors greater than 0.85, whenever its output exceeds 10% of its rated value, is a widely established minimum requirement.

Some of the most important inverter characteristics such as input voltage range, nominal and maximum output power, total harmonic distortion (THD) and inverter efficiency have been defined previously in Chapter 6. A wide range of inverters are now available from a number of worldwide manufacturers, with different sizes and characteristics. Table 8.1 lists some of these inverters for use in PV systems and shows some interesting characteristics. For detailed information about these inverters, the manufacturer's web sites are also listed in Table 8.1.

8.4   PSpice Modelling of Inverters for Grid-connected PV Systems

In Chapter 6, two inverter models were described for PSpice simulation. A behavioural inverter model and a second inverter model developed for direct connection of the inverter input to a battery. These inverters are modelled as controlled voltage sources. In grid-connected PV systems, commercial inverters modelled as dependent current sources is more adequate. Moreover some of the design considerations in standalone inverter models can be neglected here, in particular the power control implemented in these models (which takes as inputs the load power demand and the battery SOC), has to be changed for inverters connected to the grid.

A new inverter model is proposed for grid-connected PV systems. The following netlist shows this inverter PSpice model, which implements the connectivity shown in Figure 8.2. Figure 8.3 shows the schematics of the inverter model equivalent circuit.

inverter7.cir

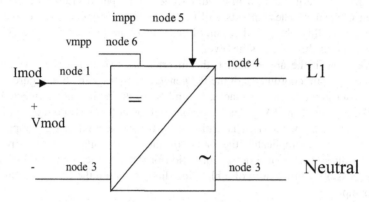

Figure 8.2. Schematic representation of the inverter model

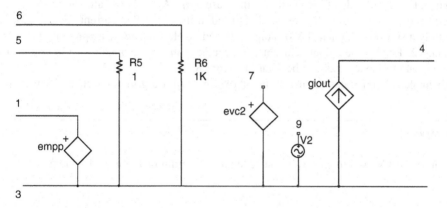

Figure 8.3. Schematic of the equivalent circuit of the inverter model

```
***** inverter7.cir / Inverter model for grid connected pv systems
.subckt inv 1 3 4 5 6 Params: nf=1
empp 1 3 value={v(6)-0.8}
r6 6 3 1000
r5 5 3 1
v2 9 3 sin(0 1 50Hz)
eVc2 7 3 value={(v(6)*v(5)*nf*1.41)/220}
giout 3 4 value={v(7)*v(9)}
.ends inv
```

Maximum power point tracking, MPPT, capability is normally included in PV inverters to obtain the maximum power from a PV array. The model shown in Figures 8.2 and 8.3 includes such capability. In order to keep the inverter model as simple as possible, no restrictions have been considered on the maximum power tracking range and minimum inverter input DC voltage. Commercial inverters, however, implement an MPP tracking function for a given range of the DC input voltage values, and it is also necessary for the DC input voltage to be larger than a minimum value for proper operation of the inverter, as described in Chapter 6. These effects will be taken into account in a more realistic model described later in this chapter. The simplified model in Figures 8.2 and 8.3 is useful to describe some of the concepts involved.

As can be seen in the above netlist, the inverter has two inputs, i.e. the values of the maximum power point coordinates of the PV generator, nodes 5 and 6 in Figures 8.2 and 8.3. At these inverter inputs, two resistances, R_5 and R_6 in the netlist, are connected.

The working point of the PV generator is forced to be at the maximum power point, MPP, by the inverter input, node 1. This is achieved by the connection of the 'empp' controlled voltage source, which replicates the value of the MPP voltage at the irradiance and temperature values at a given time. A correction of 0.8 V is included in order to take into account the possible voltage drop of a blocking diode between the PV generator output and the inverter input.

The controlled voltage source 'eVc2' calculates the necessary amplitude of the sinusoidal output current, taking into account the inverter efficiency, nf, the DC power at the inverter input and the amplitude of the output (in this case a rms AC voltage value of 220 V has been considered). The value of the amplitude is used at the controlled current source 'giout', to multiply a sinusoidal signal, 1 V of amplitude and 50 Hz frequency, supplied by the voltage source v2. Finally the 'giout' current source places the corresponding AC current at the inverter output, node 4, as can be seen in Figure 8.3.

Example 8.1 illustrates the use of the above model for a grid-connected PV system.

Example 8.1

Simulate the PV system shown in the schematic diagram of Figure 8.4, where:

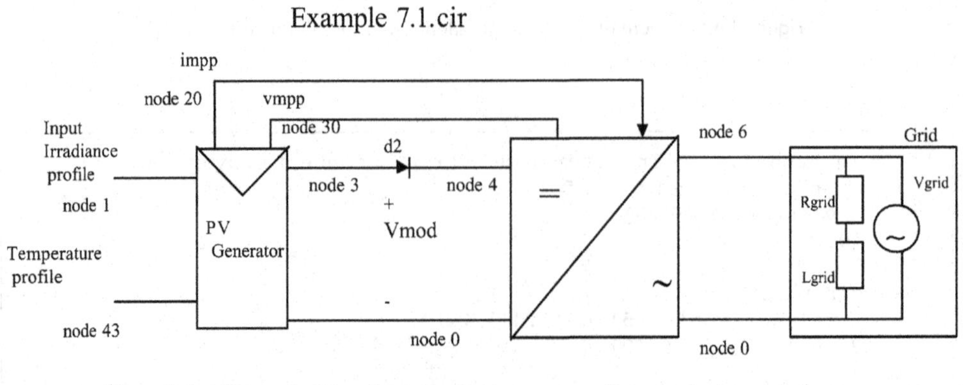

Figure 8.4. Example 8.1, schematic diagram, grounding conductors not shown

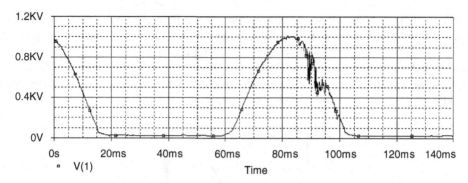

Figure 8.5 Irradiance profile for Example 8.1

- The PV generator is composed of 2×3 PV modules each having 36 solar cells series connected. The modules have a peak power of 85 W, the MPP coordinates are: immr $= 4.9$ A and Vmmr $= 17.3$ V, and the open-circuit voltage is vocmr $= 21.2$ V.

- Consider the irradiance profile given at file 'aprilmicro.stl', plotted in Figure 8.5, and a constant temperature of 12 °C. The time units of the irradiance profile have been scaled down from real measured data, by a factor 1e-6, and the x-axis ends at 140 ms corresponding to approximately 40 hours of real time.

- The inverter efficiency is considered the unity, $nf = 1$.

- Consider that the grid can be modelled as an ideal sine wave voltage source, having a V rms voltage value of 220 V and a frequency of 50 Hz, connected to a parallel load formed by the series connection of a resistor and an inductor: Lgrid $= 0.5$ mH Rgrid $= 14.4 \, \Omega$.

Write the corresponding netlist and simulate the evolution of the PV system over two days.

The netlist for the simulation of the above described PV system is included in annex 8 as file 'Example 8.1.cir', where the PV generator has been implemented using the generator_beh.lib described in Chapter 4. A little change, consisting of the elimination of the rigm resistor in the netlist of generator_beh.lib, has been made because this resistance was also included in the inverter model. This is necessary to obtain the current value of the MMP as a current to be transferred to the inverter input node 5,where this information is a voltage value.

The inverter model, inverter7.cir, presented in this section has been used for the inverter implementation and the grid has been included as a subcircuit 'grid.cir'. We first look at the inverter input where DC voltages and currents are involved. A comparison of the inverter input current, $I(d_2)$, and of the PV generator output current at the MPP, I(xinvert.r5) allows us to validate the maximum power point tracking performed by the inverter, as shown in Figure 8.6 where small differences are seen between the two waveforms.

The effect of the MPP tracking in the inverter input voltage can be seen in Figure 8.7 where the values of the inverter input voltage, V(xinvert.emmp), and of the PV generator MPP voltage, V1(xinvert.r6), are shown. The difference observed between these two voltages is 0.8 V, corresponding to the blocking diode voltage drop.

If we now look at the output AC signal, the internal PSpice time scale of the irradiance waveform in Figures 8.6 and 8.7 does not agree with the time scale required to show AC signals of 50 Hz. Two possible ways of harmonizing the two time scales are (1) enter the irradiance data in real time units, or (2) scale the frequency of the AC signals by the reverse of the same scaling factor as time. Both solutions are not practical in a PSpice environment because they require very long simulation times.

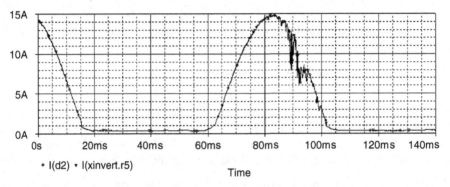

Figure 8.6 Comparison of inverter input current, I(d2) and PV generator output current at its MPP, I(xinvert.r5)

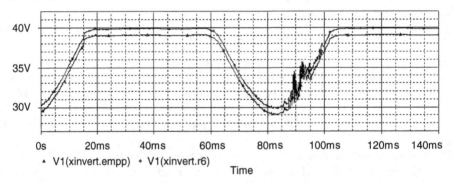

Figure 8.7 MPPT confirmation of voltage point of work at the inverter input

Instead a, good way to observe the time evolution of the AC output signals is to set a few discrete values of the irradiance and look at a length of time covering several periods of the AC signal. Only a small modification of the PSpice file is required replacing the sentences:

```
.inc april stl
vmesur 1 0 stimulus Virrad
```

by

```
.param irrad=1000
vmesur 1 0 dc {irrad}
```

and reducing the transient analysis time to one second, this will allow the observation of 50 periods of the output signal.

A second possibility is to introduce an irradiance profile with the time scale in seconds and considering different irradiance levels for a short time simulation. Therefore the inverter output behaviour can be observed without spending long simulation times.

Following the above guidelines, a new simulation has been done with a transient analysis time end of 1.3 s and four different constant irradiance levels ranging from 10 to 1000 W/m^2. Figure 8.8 shows the considered irradiance profile for this new simulation, which substitutes the previous irradiance profile considered and is included as a stimulus file in the netlist.

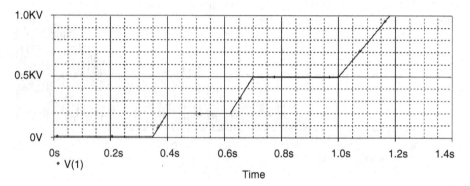

Figure 8.8   Irradiance profile

To reduce the simulation time a second modification has been included in the netlist modifying the transient analysis parameters as:

```
.tran 2ms 1.3s 0 90u
```

Figure 8.9 shows the results for the inverter output current, I(xinvert.giout), and for the AC inverter output power, (v(6)*I(xinvert.giout)/2). The effect of the irradiance value on the amplitude of the sinusoidal output current is as expected, as well as the trend of the maximum inverter output power. As the PV has a 510 W peak for an irradiance value of 1000 W/m^2 the AC inverter output power is around to 400 W, as the inverter efficiency of 90% must be taken into account. Temperature effects in the generator model also explain that the output power is below the maximum available, the simulation has been run at a constant ambient temperature of 12 °C.

Finally Figure 8.10 shows a plot of the voltage at the inverter connection to the grid. The time axis has been reduced to observe the correct frequency of 50 Hz.

8.5 AC Modules PSpice Model

Some manufacturers commercialize AC-PV modules for grid-connected applications consisting of a standard PV module and a small DC–AC inverter built into the module packaging. Sometimes more than one PV module is connected to the small inverter, and in that case the term PV-AC generator can be used instead of a PV-AC module.

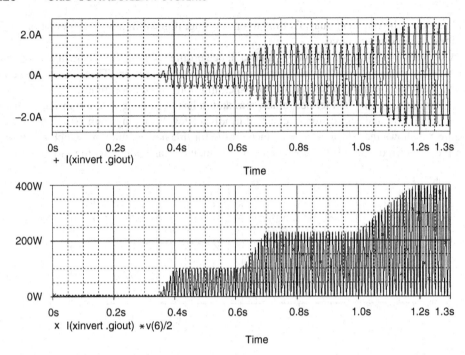

Figure 8.9 Current evolution at the inverter connection to the grid point, top, and inverter output power, bottom

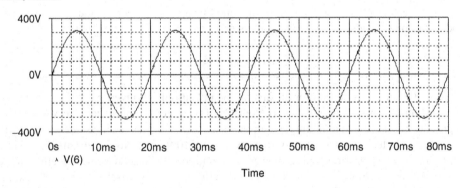

Figure 8.10 Voltage evolution at the inverter connection to the grid point, inverter output

AC modules have been considered as a possible option for wide market dissemination of grid-connected PV systems [8.4]. The main advantage of the use of AC modules is the elimination of the DC wiring and of some of the other components of a conventional grid-connected PV system, resulting in a reduction of power losses and system complexity. One of the main disadvantages is that each of the module inverters must incorporate the necessary electronic control, for example MPPT, and must also satisfy safety standards [8.5], resulting in possible increased cost.

Taking into account the inverter model described in the previous section, and the model presented in earlier chapters for a PV generator, generator_beh.lib, these two models can be easily grouped to obtain the corresponding model for an AC generator. The following netlist can be writen for this AC generator model where only changes to the subcircuit generator_beh.lib have been shown.

```
***** ACGENERATOR_BEH.LIB
.subckt acgenerator_beh 400 402 403 404 405 406 407 408 409 419 params:
+ iscmr=1, coef_iscm=1, vocmr=1, coef_vocm=1,pmaxmr=1,
+ noct=1,immr=1, vmmr=1, tr=1, ns=1, nsg=1 npg=1, nf=1

empp 404 400 value={v(409)}
vc1 9 400 sin(0 1 50Hz)
eVc2 7 400 value={(v(409)*v(408)*nf*1.4142)/220}
giout 400 419 value={v(7)*v(9)}
.ends acgenerator_beh
```

Basically the AC generator model is formed by the PV generator being directly connected to the inverter. A new node has been added to generator_beh.lib, node (419), which is the AC output current generated by the internal inverter. The AC generator includes the MPPT function, in the same way as the inverter model presented in the previous section.

Example 8.2

Simulate an AC module connected to a grid line, where:

- The PV module has 36 solar cells series connected. The module has a peak power of 85 W, MPP coordinates are: $i_{mmr} = 4.9$ A and $V_{mmr} = 17.3$ V, and its open circuit voltage is $v_{ocmr} = 21.2$ V.

- The AC module includes the described PV module and a small inverter with an efficiency of 100%.

Write the corresponding netlist and obtain the output current of the AC module for irradiances of 400 and 1000 W/m^2. Consider an ambient temperature of 16 °C. The netlist, where the irradiance profile is introduced in file irrad2E.stl, is as follows:

```
******* Example 8.2 Acgenerator
*ACgen.cir
.inc irrad2E.stl
.include acgenerator_beh.lib
xacgenerator 0 1 43 3 45 46 47 20 30 50 acgenerator_beh params:
+ iscmr=5.2, coef_iscm=0.13e-3, vocmr=21.2, coef_vocm=-0.1,pmaxmr=85,
+ noct=47,immr=4.9, vmmr=17.3, tr=25, ns=36, nsg=1 npg=1, nf=1

vm 1 0 stimulus virrad
Vtemp 43 0 dc 16
.inc grid.cir
xgr1 50 0 grid
.tran 0.002s 1s
.probe
.end
```

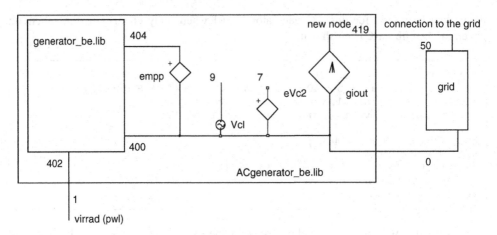

Figure 8.11 Schematic diagram for Example 8.2

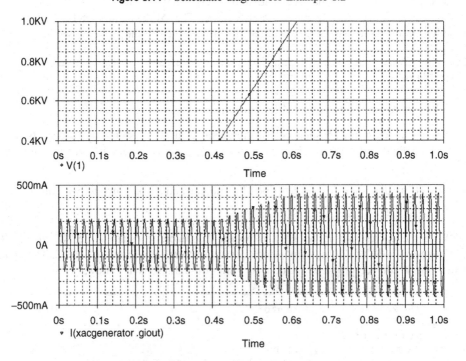

Figure 8.12 Irradiance profile, top, and AC module output current, bottom

Figure 8.11 shows the schematic diagram of the above netlist.

Figure 8.12 shows the AC module output current and the irradiance profile.

The values of the maximum AC power can be calculated from the amplitudes at the inverter output for the two considered irradiance levels:

$$Power = V_{rms}\, I_{rms} = \frac{1}{2} V_{peak}\, I_{peak}$$

which in this example is 34.34 W for 400 W/m^2 of irradiance and 63.9 W for 1000 W/m^2, as Figure 8.13 shows.

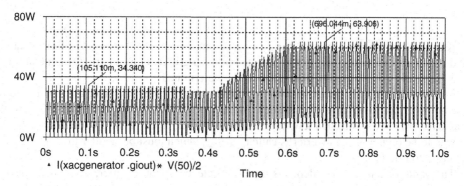

Figure 8.13   AC module output power

8.6   Sizing and Energy Balance of Grid-connected PV Systems

Several considerations must be taken into account in order to design a grid-connected PV system: PV generator size, inverter size, system wiring, grounding, surge protection, islanding prevention, connecting point between the utility grid and the inverter output etc. These considerations can also vary depending on the system size: small, medium or large. Moreover, if the electricity generated from solar photovoltaic systems is economically subsidized, then the amount of the subsidy and the regulatory issues limiting the system size may influence the final size of the grid-connected system because not only energy criteria are involved.

In the context of this book, where electrical modelling is the main focus, and despite the wide range of different design scenarios, the most logical design criteria used are based on energy considerations.

The energy balance in PV systems was introduced in Chapter 7, section 7.3. This sizing procedure can also be applied to grid-connected PV systems.

Consider that we are interested in an average design, based on the average value of peak solar hours $(\overline{PSH})$.

Equation (7.8), can be rewritten substituting L, energy consumed by loads, by E_{DC}

$$N_{sG}\, V_{mMr}\, N_{pG}\, I_{mMr}\, (\overline{PSH}) \cdot D = E_{DC} \tag{8.1}$$

where V_{mMr} and I_{mMr} are the voltage and current coordinates of the maximum power point of a single PV module under standard conditions, N_{sg} is the number of modules in series and N_{pg} the number of parallel rows forming the PV generator, and E_{DC} is the DC output energy generated by the PV generator along a time period of D days with an average value of peak solar hours equal to $(\overline{PSH})$.

The peak power, P_{DCpeak}, of the PV system can be written as

$$N_{sG} V_{mMr} N_{pG} I_{mMr} = P_{DCpeak} \tag{8.2}$$

Knowing the $(\overline{PSH})$ value of a particular location, where the PV system will be installed, and the desired value of energy to be injected into the grid in a given period of time: months, years etc., the size of the PV generator can be determined.

The first step is the correct selection of the inverter or inverters' size; this must be done according to the inverter characteristics, especially its nominal output power P_{nom}. In the standard AM1.5G spectrum $1 \, kW/m^2$ conditions of irradiance of the PV generator, the AC power at the output of the inverter is called P_{ACpeak}, which must be smaller than the nominal inverter power, P_{nom}.

The value of P_{ACpeak} also depends on the efficiency of the inverter, η, as mentioned in Chapter 6.

$$\eta = \frac{P_{ACpeak}}{P_{DCpeak}} \tag{8.3}$$

According to equation (8.3), equation (8.2) can be rewritten as follows:

$$N_{sG} V_{mMr} N_{pG} I_{mMr} = \frac{P_{ACpeak}}{\eta} \tag{8.4}$$

Considering the input requirements of the selected inverter, especially the maximum input current, I_{max}, the number of parallel PV modules, N_{pg} can be calculated as follows:

$$N_{pg} = \frac{I_{max}}{I_{mMr}} \tag{8.5}$$

Finally, the number of PV modules in series and the total necessary area for the PV generator can be estimated rewriting equation (8.4) as equation (8.6) and using equation (8.7).

$$N_{sG} = \frac{P_{ACpeak}}{\eta \, V_{mMr} \, N_{pG} \, I_{mMr}} \tag{8.6}$$

$$A = N_{pg} \, N_{sg} \, A_m \tag{8.7}$$

where A_m is the area of one of the PV modules and A is the total area required.

It must also be taken into account that a part of the P_{DC} generated by the PV modules will not be converted for the inverter, because when the power at the inverter input is lower than the minimum specified value, the inverter output is left open.

If we now think about the energy performance of the system, it will be sufficient to evaluate the rms value of the AC output energy, allowing long-term simulations to be performed using little CPU time. This idea has been successfully implemented in rather complex inverter models [8.6], and it is timely to implement a new inverter PSpice model here. The netlist is as follows:

```
************ Inverter 9.cir
.subckt inv 1 3 4 5 6 Params: nf=1, Vmin=1
empp 1 3 value={v(6)-0.8}
r6 6 3 1000
r5 5 3 1
giout 3 4 value={(v(88)*v(5)*nf)/220}
ef 88 3 value=IF {(v(6)>Vmin, V(6), v(3))}
.ends inv
```

The block diagram of this model is the same as in Figure 8.2 for 'inverter7.cir': the two inverter models have the same input/output nodes. The MPPT strategy is also the same for the two inverter models, and only two minor changes have been introduced:

- Minimum value of the inverter input voltage. A new input parameter, V_{min}, is included in the model. For input voltages below V_{min} the inverter output current is forced to be zero, while for input voltages greater than V_{min} the inverter works under normal operating conditions. This control has been implemented by means of the voltage controlled source 'ef' and an IF sentence in the above netlist:

$$\text{ef } 88\ 3\ \text{value} = \{\text{IF } (v(6) > \text{Vmin}, V(6), v(3))\}$$

- DC output current equivalent to the rms value. The inverter output remains as a controlled current source, as in the previous model 'inverter7.cir', but now the output current is a DC magnitude. The controlled current source 'giout' gives a DC current value equal to the rms current value of the AC inverter output current and takes into account the inverter efficiency, nf, the DC power at the inverter input and the rms value of the inverter output voltage. To calculate this rms output current, an rms grid voltage value of 220 V has been considered as in previous sections. This calculation has been introduced in the above netlist into the 'giout' controlled current source as:

$$\text{giout } 3\ 4\ \text{value} = \{(v(88)*v(5)*nf)/220\}$$

To illustrate the behaviour of this inverter model, consider the following netlist for the simulation of a small grid-connected PV system.

```
****** Grid connected PV system using inverter9.cir
*GCPV1.cir
.include generator_beh.lib
xgenerator 0 1 43 3 45 46 47 20 30 generator_beh params:
+ iscmr=5.2, coef_iscm=0.13e-3, vocmr=21.2, coef_vocm=-0.1,pmaxmr=85,
+ noct=47, immr=4.9, vmmr=17.3, tr=25, ns=36, nsg=4 npg=6

.inc madrid1.stl
vmesur 1 0 stimulus vmadrid_hour
```

```
.inc MadridT1.stl
vtemp 43 0 stimulus vtemp_madrid
dblock 3 4 diode
.model diode d(n=1)
.inc inverter9.cir
xinvert 4 0 16 20 30 inv params: nf=0.9,Vmin=67
Rm 16 0 1
eE 7 0 value={(sdt(1e6*220*v(16)))}
.tran 0.1us 24u
.probe
.end
```

Figure 8.14 illustrates the schematic block diagram of the above netlist. As can be seen, the PV generator is composed of 24 PV modules, $n_{sg} = 4$ $n_{pg} = 6$, each one having 85 W peak, resulting in a PV system of a 2 kW peak. A blocking 'dblock' diode has been placed between the PV modules and the inverter input.

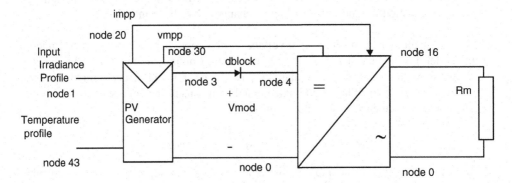

Figure 8.14 Schematic diagram of the simulated PV system

The inverter has an efficiency of 90% and the minimum inverter input voltage required for correct operation is 67 V.

Irradiance and temperature profiles have been included in the file as 'Madrid1.stl' and 'MadridT1.stl' files, respectively, where the internal PSpice time unit is the microsecond to improve simulation speed. To simulate a complete day a correspondence of 1 hour of real time to a microsecond as internal time unit is considered. These files are listed below.

```
*** Madrid1.stl                      ***** MADRIDT1.stl
.stimulus vmadrid_hour pwl           .stimulus vtemp_madrid pwl

+1u  0                               +  1u  4.7
+2u  0                               +  2u  4.8
+3u  0                               +  3u  4.4
+4u  0                               +  4u  4.3
+5u  0                               +  5u  4.2
+6u  0                               +  6u  4.4
+7u  0                               +  7u  4.1
+8u  0                               +  8u  3.8
+9u  10                              +  9u  4.1
+10u 193                             +  10u 5.2
```

+11u	218		+ 11u	5.7
+12u	520		+ 12u	7.4
+13u	729		+ 13u	9
+14u	511		+ 14u	10.2
+15u	587		+ 15u	11.4
+16u	299		+ 16u	12.1
+17u	132		+ 17u	11.9
+18u	2		+ 18u	11.1
+19u	0		+ 19u	9.6
+20u	0		+ 20u	8.8
+21u	0		+ 21u	8.1
+22u	0		+ 22u	7.3
+23u	0		+ 23u	6.1
+24u	0		+ 24u	6.5

Figure 8.15 and 8.16 show the corresponding irradiance and temperature profiles listed above.

A simple resistor, Rm, has been placed at the inverter output to simulate the load. Figure 8.17 shows the inverter rms output current, v(16), and inverter DC input voltage, v(4). As can be seen the inverter output current goes to zero when the inverter input voltage is lower that the minimum inverter input voltage, fixed to 67 V in this case.

Finally, Figure 8.18 shows a plot of the rms energy generated by the PV system in Wh along the simulated day, in this case 5719.4 Wh/day. To evaluate the generated energy, a

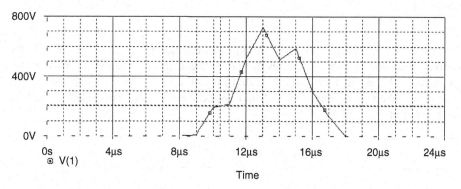

Figure 8.15 Irradiance profile corresponding to Madrid1.stl file

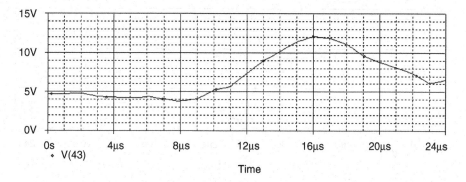

Figure 8.16 Temperature profile corresponding to MadridT1.stl file

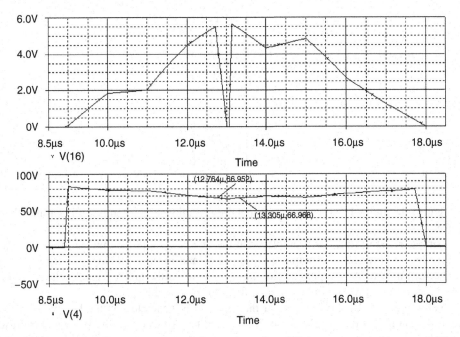

Figure 8.17 Inverter DC output current v(16), warning units of V represents A and inverter DC input voltage v(4)

controlled voltage source has been introduced into the file 'eE' using the sdt PSpice function; this voltage source evaluates the total AC generated power considering an AC rms voltage of 220 V at the utility grid. A scale factor of 1×10^6 has also been included in the equation to take into account the time unit correction. The 'eE' controlled voltage source was defined in the netlist as follows:

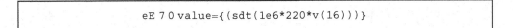

```
eE 7 0 value={(sdt(1e6*220*v(16)))}
```

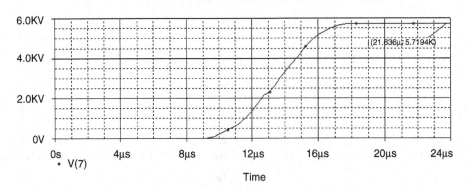

Figure 8.18 Plot of the generated energy in Wh/day, v(7)

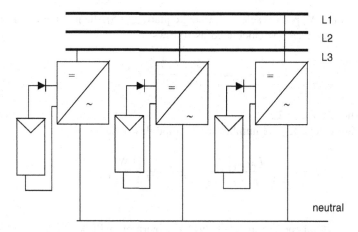

L1
L2
L3

neutral

Figure 8.19   Schematic diagram of PV system of Example 8.3

Example 8.3 shows the design of a medium-size grid-connected PV system and the simulation results of one-year operation using the inverter model presented in this chapter.

Example 8.3

Design and simulate a medium-size grid-connected PV system for a commercial rooftop located at Madrid, Spain. The schematic diagram of the system is as shown in Figure 8.19, where three monofasic inverters are connected to three PV generators to obtain a three-phase grid-connected PV system. In this case:

- Each one of the PV generators is formed by modules of 85 Wpic, the coordinates of the maximum power point are: $i_{mmr} = 4.9$ A, $V_{mmr} = 17.3$ V. The module area is 0.64 m^2.

- Inverter parameters are: $nf = 0.91$, $V_{min} = 1$ V, AC output voltage 220 V rms.

Consider a value of $(\overline{PSH}) = 4.5$ h/day for Madrid, a total energy of 18 MWh injected into the grid for a year of operation and design the inverter and PV generator sizes. Finally estimate the necessary total area for the PV modules.

The study can be restricted to one inverter connected to a PV generator, because of the symmetry of the system, in other words, just one output line of the three output phases can be considered. The energy injected by a line will be the third part of the total energy: 6 MWh/year. Using equations (8.3) and (8.1), the required inverter nominal power can be estimated.

$$P_{nom} > P_{ACpeak} = \eta \, \frac{E_{AC}}{D \, \overline{PSH}} = \frac{6 \times 10^6 \, \text{Wh}}{365 \, \text{day} \, 4.5 \, \dfrac{\text{h}}{\text{day}}} = 3324.2 \, \text{W} \tag{8.8}$$

An inverter of 4 kW peak is selected. Using equation (8.5), the number of parallel modules of each PV generator can be calculated as follows:

$$N_{pg} = \frac{I_{max}}{I_{mMr}} = \frac{115\,\text{A}}{5.2\,\text{A}} = 22.11$$

We will choose $N_{pg} = 22$.

Equation 8.6 calculates the number of PV modules in series:

$$N_{sG} = \frac{P_{ACpeak}}{\eta\,V_{mMr}\,N_{pG}\,I_{mMr}} = \frac{4\,\text{kW}}{0.91\;17.3\,\text{V}\;22\;5.2\,\text{A}} = 2.22$$

We choose $N_{sg} = 2$.

The necessary total rooftop area for the PV generators will be:

$$A = 3 \times N_{sg} \times N_{pg} \times A_m = 84.48\,\text{m}^2$$

Simulate a period of one year of operation considering the irradiance and temperature profiles given in files 'irradmadrid.stl' and 'TempM.stl', where data have been introduced in microsecond time units, one microsecond represents one hour of real time. These irradiance and temperature files are the ones used in Chapter 7.

The corresponding netlist for the simulation is the following:

```
******* Example 8.3. cir
.include generator_beh.lib
xgenerator 0 1 43 3 45 46 47 20 30 generator_beh params:
+ iscmr=5.2, coef_iscm=0.13e-3, vocmr=21.2, coef_vocm=-0.1,pmaxmr=85,
+ noct=47,immr=4.9, vmmr=17.3, tr=25, ns=36, nsg=2 npg=22
.inc irradmadrid.stl
vmesur 1 0 stimulus vmadrid_hour
.inc TempM.stl
vtemp 43 0 stimulus vtemp_Madrid
d2 3 4 diode
.model diode d(n=1)
.inc inverter9.cir
xinvert 4 0 16 20 30 inv params: nf=0.91,Vmin=1
Rm 16 0 1
eE 7 0 value={(sdt(1e6*220*v(16))))}
.tran 0.5lus 8760u
.probe
.end
```

As can be seen the simulation is restricted again to one inverter connected to a PV generator, done the symmetry of the system, in other words, just one output line of the three output phases has been simulated.

Figure 8.20 shows the plot of the total generated energy over the year, note that the total energy corresponds to three times the energy supplied to a line.

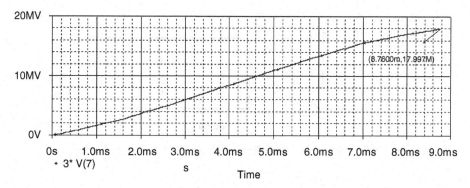

Figure 8.20   Total generated energy in MWh

As can be observed in Figure 8.20 the total generated energy at the end of the year by the whole PV system is 17.997 MWh.

Finally Figure 8.21 shows a simulation result detail corresponding to January. The evolution of the current (rms Amps), and the energy (kWh), injected into L1 along this first month are plotted.

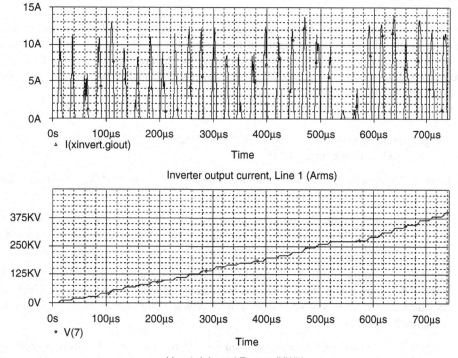

Inverter output current, Line 1 (Arms)

Line 1, Injected Energy (KWh)

Figure 8.21

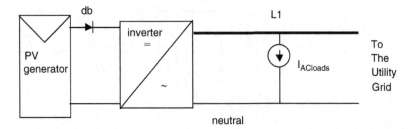

Figure 8.22   Schematic block diagram of the PV system of Example 8.4

The inverter models introduced in this section are also useful for small PV grid-connected PV systems when AC loads are present, such as small residential installations. Example 8.4 shows that this kind of simulation can achieve, in a very simple manner and without spending long simulation times, important information about PV system size design and energy balance.

Example 8.4

Design and simulate a small grid-connected PV system for a residential rooftop located at Madrid, Spain. The schematic diagram of the system is as shown in Figure 8.22, where a monofasic inverter is connected to a PV generator. Finally, AC loads present in the system are modelled, as shown in Figure 8.22, by a current source that reproduces the daily load demand profile, AC rms current.

Detailed description of the system:

- PV generator: composed of modules of 85 Wpeak, MPP coordinates are $i_{mmr} = 4.9$ A, $v_{mmr} = 17.3$ V, module area$=0.7\text{m}^2$.

- Inverter: efficiency, $nf = 0.85$, minimum input DC voltage $= 40$ V, Maximum input current $= 15$ A, voltage at Line 1 $= 220$ V rms.

- Temperature and irradiance profiles: given at files 'MadridT2.stl' and 'madrid2.stl', respectively. These files correspond to two days' data and time units are microseconds, one microsecond represents one hour of real time, as can be seen in Figure 8.23.

- The AC rms current profile demanded by the loads is available at file 'Madrid_load.stl', where data corresponding to two days for a typical residential location has been included, considering an average consumed energy of 6 kWh/day. This current load profile is shown in Figure 8.24.

- A blocking diode has been placed between the PV generator and the inverter input.

Design the inverter and PV generator sizes considering $(\overline{PSH}) = 2.8\,\text{h/day}$ at Madrid in January and a total energy of 700 Wh consumed after two days of operation from the grid-connected PV system and loads.

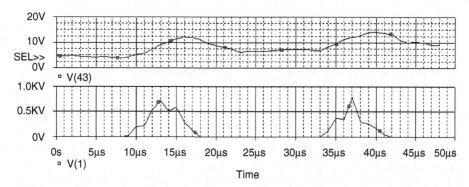

Figure 8.23 Temperature (°C), v(43), and irradiance (W/m²), v(1), profiles

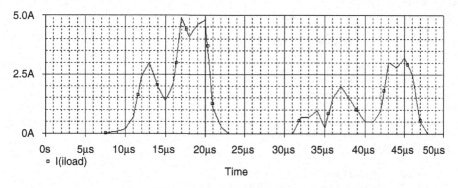

Figure 8.24 AC loads current profile (A rms)

Using again equation (8.8), and taking into account the total energy supplied by the PV system, that is energy consumed by the load minus energy supplied by the grid, the inverter size can be obtained as follows:

$$P_{nom} > P_{ACpeak} = \frac{E_{AC}}{D\,\overline{PHS}} = \frac{11\,300\,\text{Wh}}{2\,\text{days}\,2.8\,\dfrac{\text{h}}{\text{day}}} = 2017.14\,\text{W}$$

An inverter of 2.5 kW$_{peak}$ will be sufficient.

$$N_{pg} = \frac{I_{max}}{I_{mMr}} = \frac{15\,\text{A}}{4.9\,\text{A}} = 3.06$$

$N_{pg} = 3$ is selected.

$$N_{sg} = \frac{P_{ACpeak}}{\eta\,V_{mMr}N_{pg}I_{mMr}} = \frac{2107.14\,\text{W}}{0.91\,\,17.3\,\text{V}\,3\,4.9\,\text{A}} = 8.7$$

$N_{sg} = 9$ is selected.

The necessary total area for the PV generator is:

$$A = 0.7\,\text{m}^2 \times 3 \times 9 = 18.9\,\text{m}^2$$

Write the corresponding netlist for the PV system simulation.
The netlist is as follows:

```
******* Example 8.4 Small grid connected PV system with AC loads
*Example8.4.cir
.include generator_beh.lib
xgenerator 0 1 43 3 45 46 47 20 30 generator_beh params:
+ iscmr=5.2, coef_iscm=0.13e-3, vocmr=21.2, coef_vocm=-0.1,pmaxmr=85,
+ noct=47,immr=4.9,vmmr=17.3,tr=25,ns=36,nsg=9 npg=3
.inc madrid2.stl
vmesur 1 0 stimulus vmadrid_hour
.inc MadridT2.stl
vtemp 43 0 stimulus vtemp_madrid
db 3 4 diode
.model diode d(n=1)
.inc inversor9.cir
xinvert 4 0 16 20 30 inv params: nf=0.85,Vmin=40
Rm 16 0 1
.include Madrid_load.stl
iload 16 200 stimulus iload_madrid
Vref 200 0 dc 0

eE 7 0 value={(sdt(1e6*220*v(16))))}
eEgrid 101 0 value={(sdt(1e6*220*i(Vref))))}

.tran 0.51u 48u
.probe
.end
```

As shown in the netlist, the grid has been replaced by a simple resistor, Rm, as in Example
8.3. This is sufficient to obtain the current evolution at L1 as the current evolution at Rm, as
can be seen in Figure 8.25, where positive values of current indicate that the AC rms current,
injected to the L1 line by the inverter, is larger than the AC rms current demanded by the
load at this time.

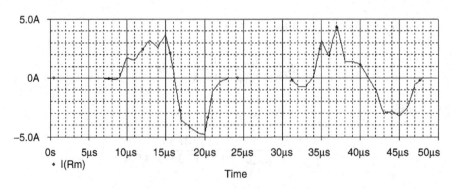

Figure 8.25   Current evolution at L1

As a result of the current load demand the corresponding energy consumed, associated to the AC loads, can be evaluated by including the following sentences:

$$eE \ 7 \ 0 \ value = \{(sdt(1e6*220*v(16)))\}$$
$$Vref \ 200 \ 0 \ dc \ 0$$

where V_{ref} has been introduced just as a current sensor, and the voltage-controlled source 'eE' evaluates the energy consumed by the loads as explained in Example 8.3. The plot in Figure 8.26 shows the voltage waveform at 'eE' source terminals, v(7), which is the consumed energy profile.

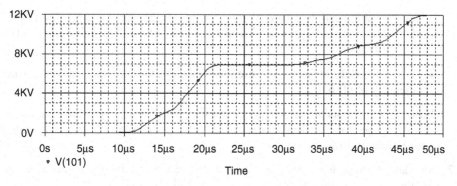

Figure 8.26 Energy demanded by the AC loads (kWh)

This should be the consumed energy profile from the utility grid at the building in the absence of the PV system, resulting in a total energy of 12 kWh.

A second controlled voltage source has been included in the netlist to evaluate, in a similar manner, the real energy profile at L1 when the PV system is present, by means of the voltage-controlled source eEgrid:

$$eEgrid \ 101 \ 0 \ value = \{(sdt(1e6*220*i(Vref)))\}$$

A plot of this real energy profile, v(101), is shown in Figure 8.27.

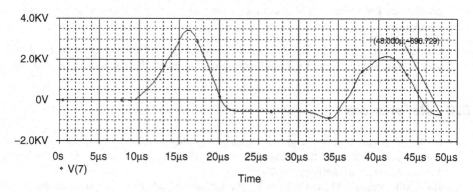

Figure 8.27 Evolution of the energy (kWh) profile at line L1

As can be seen the PV system does not supply enough energy and at the end of the two simulated days a deficit of 696.729 Wh must be supplied by the utility grid.

8.7 Problems

8.1 Simulate an AC generator connected to a grid line. The AC generator is formed by $2nsg \times 2npg$ modules and a small inverter with the following characteristics.

- The PV module has 36 solar cells series connected. Module has a p_{ic} power of 85 W, MPP coordinates are $i_{mmr} = 4.9$ A and $V_{mmr} = 17.3$ V, and its open circuit voltage is $v_{ocmr} = 21.2$ V.

- The AC module includes the described PV module and a small inverter with an efficiency of 90%.

Write the corresponding netlist and obtain the current and the power at the output of the AC generator, for irradiances of 250 and 1000 W/m². Consider an ambient temperature of 19 °C.

8.2 Simulate a small PV system designed for a residential rooftop, with a power of 5 kWpic. The details of the PV system are as follows:

- PV generator: formed by 90 modules of 55 Wpic, $n_{sg} = 10$, $n_{pg} = 9$. The PV modules are formed by 36 6″ solar cells, and the module characteristics are the following: MPP: $I_{mmr} = 3.4$ A, $V_{mmr} = 16.2$ V. Short circuit current $I_{scmr} = 3.7$ A and open circuit voltage $V_{ocmr} = 20.5$ V.

- Inverter parameters: nominal power $= 4.00$ kW, one output line: AC voltage 220 V rms, efficiency: $nf = 0.85$. Minimum input DC voltage $= 30$ V.

- AC loads are also connected to the inverter output. Select any daily power load demand but maintain a total energy demand of 4 MWh/year.

Evaluate the simulation of this PV system over a year. Temperature and irradiance profiles are known at the PV system location and must be introduced as .stl files in the simulation.

Estimate the energy balance of the system, obtaining the total energy demanded to the grid or injected into it and the total energy generated by the PV system over one year.

8.9 References

[8.1] He, W., Markvart, T., and Arnold, R., 'Islanding of grid-connected PV generators: experimental results', *Proc. of the 2nd World Conference and Exhibition on PV Solar Energy Conversion*, Vienna, Austria, July 1998, pp. 2772–5.

[8.2] Roop, M.E., Begovic, M., and Rohatgi A., 'Prevention of islanding in grid-connected photovoltaic systems', *Progress in Photovoltaics: Research and Applications*, **7**, 39–59, 1999.

[8.3] Hernández, J.C., Vidal, P.G., and Almonacid, G., 'Analysis of personal electric risk in photovoltaic systems using standard electronic circuit simulator', *Proc. of the 14th European Photovoltaic Solar Energy Conference*, Barcelona, Spain, June 1997, pp. 1062–5.

[8.4] Wills, R.H., Hall, F.E., Strong, S.J., and Wohlgemuth, J.H., 'The AC photovoltaic module', *Proc. of the 25th IEEE PVSC*, Washington DC, May 1996, pp. 1231–4.

[8.5] Woyte, A., Belmans, R., Mercierlaan, K., and Nijs, J., 'Islanding of grid connected AC module inverters', *Proc. of the 28th IEEE PVSEC*, Anchorage, Alaska, September 2000, pp. 1683–6.

[8.6] Moreno, A., Julve, J., Silvestre, S., and Castañer, L., 'SPICE macromodelling of photovoltaic systems', *Progress in Photovoltaics: Research and Applications*, 8, 293–306, 2000.

9

Small Photovoltaics

Summary

This chapter addresses common PV practical applications composed of small numbers of PV cells or modules operating under natural or artificial light. An effective irradiance value is calculated to take into account different spectra of artificial light. Correspondence between radiometric and photometric quantities is used. The random generation of $I(V)$ characteristics of small PV modules is performed and Monte Carlo randomly generated numbers are used to generate PSpice time series of radiation, which can be used to see the PV system operation in detail. This method is used to show how a solar pocket calculator works, as well as a flash light and a street light system.

9.1 Introduction

Photovoltaic systems are known by the average person, not only by the media coverage of large PV plants or autonomous standalone systems, but also, by the PV power supply of small systems in a wrist watch, a pocket calculator, a car sunroof and many other applications where small PV arrays are a cheap solution for consumer applications or professional systems. This chapter covers the specific issues raised by these small PV systems.

9.2 Small Photovoltaic System Constraints

In general, small photovoltaic systems are used either as an entirely autonomous system or as an auxiliary power supply to extend the life of the system beyond the lifetime of a battery pack. This means that in general these systems include a battery and hence most of the concepts developed in Chapter 7 for standalone systems are still valid. There are, however, some differences. Because the solar (or more generally light) energy availability is often unpredictable (think for instance in a wrist watch), the nature and spectrum of the 'light' energy is very different from the spectrum and radiation data used in conventional standalone or grid connected systems (think for instance in a desk light). Moreover the load can also be of a different nature than that of standard home appliances, for example LEDs can be a common load used in lighting applications. And finally the overall system requirements may be of a different nature because the application may only require to enforce the battery supply but not fully balanced autonomy from solar energy. These specific and different matters are the subject of this chapter.

9.3   Radiometric and Photometric Quantities

Some small photovoltaic systems operate under artificial light in offices or homes and this raises the question of how to design such systems, when the electrical properties of solar cells or modules are generally known under standard illumination conditions (either AM1.5 G or AM0). Artificial light has a different spectral irradiance than the sunlight and, moreover, the measure of artificial light magnitudes is given in photometric units rather than in radiometric units. The main difference is that photometric quantities are a measure of visible light and are weighted by the responsivity of the human eye by means of the function $V(\lambda)$, which is the CIE sensitivity curve. In fact there are two CIE sensitivity curves, one for the photopic response of the eye and one for the scotopic (eye adapted to the dark) response. For the purpose of this book we will refer to photopic response. Figure 9.1 shows the CIE photopic responsivity curve, which is given in relative values normalized to unity at a wavelength of 550 nm. The numerical data are available in the file 'cie.stl' in Annex 9.

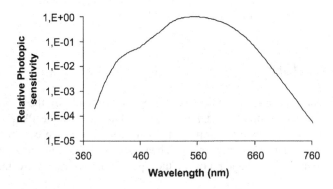

Figure 9.1   Photopic CIE responsivity curve $V(\lambda)$

Table 9.1 summarizes some important magnitudes and units for comparison.

In general a small PV system will be receiving light measured in units of illuminance, which is what really will tell us how much energy is available to the solar cells. On the other hand, artificial light sources are rated by two magnitudes:

- electrical power consumed (given in watts); and
- luminous radiant flux produced (given in lumen).

Table 9.1   Radiometric and photometric magnitudes

Radiometric magnitude	Units	Photometric magnitude	Units	Conversion factor at 555 nm of wavelength (photopic flux)
Radiant flux	W	Luminous flux Φ_v	lm (lumen)	1 lumen $= 1.464 \times 10^{-3}$ W 1 W $= 683$ lm
Irradiance	W/m^2 G	Illuminance G_v	Lux	1 lux $= 1$ lm/m^2 1 W/m^2 $= 683$ lux
Radiant intensity	W/sr	Luminous intensity	lm/sr	1 lm/sr $= 1$ candel

9.4   Luminous Flux and Illuminance

The transformation from the luminous radiant flux produced by a light source to the illuminance received by the PV system depends on the relative geometry of the source–receiver path. The following concepts are of interest.

9.4.1   Distance square law

The steradian (sr) is the unit of a solid angle. The solid angle subtended by the surface of its sphere at its centre is equal to 4π steradians.

Consistent with this definition, if we are interested in the relative value of the illuminance at two distances of the radiant source, d_1 and d_2, a given luminous flux of one lumen will produce illuminance values related by:

$$\frac{G_{v1}}{G_{v2}} = \frac{d_2^2}{d_1^2}$$
(9.1)

which is the distance square law.

9.4.2   Relationship between luminous flux and illuminance

The illuminance G_v in lux at a given distance of a light source relates to the luminous flux Φ_v in lumen, considering the geometry in Figure 9.2, as

$$\Phi_v = G_v d^2 \Omega$$
(9.2)

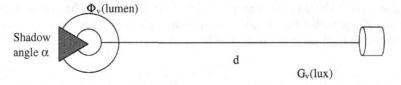

Figure 9.2   Geometry used to define photometric magnitudes

where d is the distance between source and target and Ω is the solid angle of the lamp. If the lamp can be considered as a point isotropic source, then the solid angle is 4π. If the radiation is restricted to a smaller solid angle due to shadowing, then the solid angle in equation (9.2) is given by,

$$\Omega = 4\pi - 2\pi\left(1 - \cos\frac{\alpha}{2}\right) \tag{9.3}$$

where α is the shadowed angle.

Example 9.1

Consider that we measure 20 lux at a distance of 5 m from a street light. Calculate the value of the luminous flux of the lamp which has a shadow angle of 60°.

$$\Omega = 4\pi - 2\pi(1 - \cos 30°) = 11.619\,\text{sr} \tag{9.4}$$

Then the luminous flux is calculated:

$$\Phi_v = 20(5)^2 11.619 = 5809.5\,\text{lm} \tag{9.5}$$

9.5 Solar Cell Short Circuit Current Density Produced by an Artificial Light

The short circuit current produced by an illuminated solar cell was calculated in Chapter 2. Remember that the short circuit current density is given by,

$$J_{sc} = \int_0^\infty 0.808QE(\lambda)\lambda I_\lambda \, d\lambda \tag{9.6}$$

which is the integral over all wavelengths of the spectral short circuit current density when the solar cell is illuminated by a standard spectrum such as AM1.5; the irradiance is given by the integral of the spectral irradiance I_λ.

$$G = \int_0^\infty I_\lambda \, d\lambda \tag{9.7}$$

When an artificial light is concerned, the information we generally have is the value of the illuminance G_v provided by the artificial light at a given distance from the source. The illuminance of a light spectrum is defined by:

$$G_v = \int_{360\,\text{nm}}^{760\,\text{nm}} I_{\lambda(art)} K_m V(\lambda) \, d\lambda \tag{9.8}$$

where K_m is the luminous efficacy and equals 683 lux/W/m^2 for a photopic CIE curve, $I_{\lambda(art)}$ is the spectral irradiance of the artificial light, and $V(\lambda)$ is the photopic CIE eye responsivity (for daylight conditions).

The spectrum of artificial light sources is normally available in normalized form to a maximum value of unity, $I_{\lambda(art)norm}$, and an example is shown in Figure 9.3 [9.1].

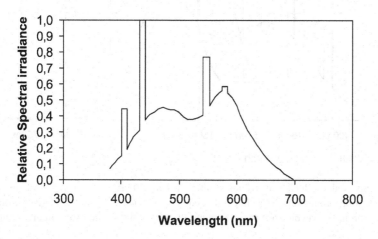

Figure 9.3 Typical spectrum of a fluorescent light (after normalization, from Alex Ryer, in http://www.intl-light.com/handbook *Light Measurement Handbook*, 1988)

It is convenient to select a value for a scale factor, K, such that the value of the illuminance of the normalized artificial light spectrum is 1 lux as follows:

$$K \int_{360\,nm}^{760\,nm} I_{\lambda(art)norm} K_m V(\lambda)\, d\lambda = 1 \qquad (9.9)$$

Now, the integral in equation (9.9) can be calculated using a PSpice file as follows:

```
*normalization.cir
.include fluorescent_rel.stl
.include cie.stl
.param k=0.0292
vfluor 90 0 stimulus vfluorescent_rel
vcie 91 0 stimulus vcie
elux 92 0 value={sdt(v(90)*v(91)*k*683*1e6)};computes the illuminance
.tran 0.01u 0.770u 0.38u 0.01u
.probe
.end
```

where the fluorescent light normalized spectrum and the CIE curve are both included by .stl files as sources in nodes (90) and (91), respectively. The computation of the integral of equation (9.9) is performed by the e-device 'elux2' for given values of the normalization

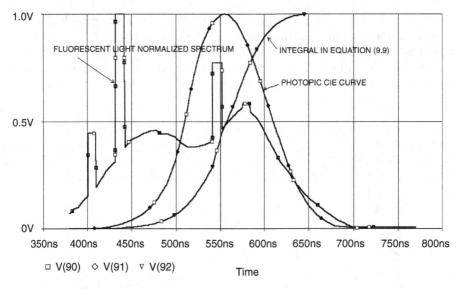

Figure 9.4 Spectral irradiance of a fluorescent light and CIE responsivity of the human eye, both normalized to a maximum value of 1. Integral in equation (9.9) using $K = 0.0292$ gives a value of 1 lux after integration in the visible spectrum. *Warning*: the *x*-axis shows the wavelength in units of nm

constant K and of the illuminance value G_v as illustrated in Figure 9.4, where the wavelength integral of the product of the spectral irradiance and the CIE curve is shown using a value for $K = 0.0292$ W/m²µm. With this value the integral in equation (9.9) is unity.

Once the value of K is known, the spectral irradiance corresponding to an illuminance of 1 lux is given by:

$$I_{\lambda\_1\_(art)} = KI_{\lambda(art)norm} = 0.0292I_{\lambda(art)norm} \tag{9.10}$$

The value of the short circuit current produced by a solar cell when illuminated by this artificial light of 1 lux illuminance can be calculated provided the quantum efficiency of the solar cell is known:

$$J_{sc(art)} = \int_0^\infty 0.808QE(\lambda)\lambda I_{\lambda(art)} \, d\lambda \tag{9.11}$$

Therefore, in the general case of an illuminance G_v which is different from unity, the short circuit current collected will be:

$$J_{sc(art)} = G_v 0.0292 \int_0^\infty 0.808QE(\lambda)\lambda I_{\lambda\_1\_(art)} \, d\lambda \tag{9.12}$$

Figure 9.5 shows the spectral current density calculated for the silicon solar cell described and simulated in Chapter 2, but in this case comparing the effect of the light source spectrum, namely AM1.5 G at 100 W/m^2, with a fluorescent light producing an illuminance of 5000 lux. For the artificial light computation the 'jsc_silicon_art.cir' listed in Annex 9 has been used, calling a new 'jsc_art.lib' subcircuit, which is also listed in the Annex.

The result of the simulation is shown in Figure 9.5.

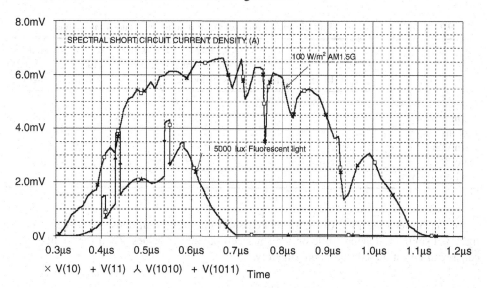

Figure 9.5 Comparison between the spectral current density under AM1.5 G and 100 W/m^2 (upper graph) and under 5000 lux artificial fluorescent light. *Warning*: x-axis is the wavelength in microns and the y-axis is the spectral current density in mA/cm^{2}µm

As can be seen the longer wavelengths of the spectrum present in the AM1.5 light spectrum produce higher density current than in the fluorescent light, which has poor long wavelength content.

9.5.1 Effect of the illuminance

As shown above, neglecting temperature effects on the short circuit current, the effect of the value of the illuminance on the short circuit current is linear:

$$J_{sc(art)} = J_{sc\_1(art)} G_v \qquad (9.13)$$

where $J_{sc\_1(art)}$ is the value of the short circuit current density for a value of the illuminance of 1 lux.

9.5.2 Effect of the quantum efficiency

As described in Chapter 2, the value of the quantum efficiency at a given wavelength depends on the values of the reflection coefficient, absorption coefficient, geometrical parameters such as the thickness of the cell and emitter depth, semiconductor parameters such as lifetimes and mobilities, and on technology parameters such as the surface recombination velocities.

If we run the simulation of a solar cell with several values of these parameters, the results can be compared as shown in Table 9.2.

Table 9.2 Comparison of the short circuit currents of several solar cells and light sources

	Baseline cell#1	Cell#2 Baseline with surface recombination velocities 1 cm/s	Cell#3 is equal to Cell#2 with enhanced diffusion lengths
J_{sc} (AM1.5 @1 kW/m^2) (mA/cm^2)	31.811	32.672	36.67
J_{sc}(1 lux) (nA/cm^2)	121	123	128.9
Ratio (1 lux/AM1.5)	3.8×10^{-6}	3.79×10^{-6}	3.51×10^{-6}

As can be seen the ratio of the generated short circuit current density between artificial light and sunlight, depends little on the solar cell parameter values. Then for a silicon solar cell it can be estimated that approximately,

$$\frac{J_{sc(art)}(1\,\text{lux})}{J_{scr(\text{AM1.5@1 kW/m}^2)}} \approx 3.8 \times 10^{-6} \tag{9.14}$$

Of course, different solar cells and different light sources will require a new and specific computation of this ratio.

If the solar cell is illuminated by an arbitrary combination of natural and artificial light, then the resulting short circuit current considering equation (9.14) is given by:

$$J_{sceff} \approx J_{scr(\text{AM1.5})}\left(\frac{G}{G_r} + 3.8 \times 10^{-6}G_v\right) = J_{scr(\text{AM1.5})}\frac{G_{eff}}{G_r} \tag{9.15}$$

where G_{eff} is the effective irradiance, which is given (taking into account that $G_r = 1000$), by:

$$G_{eff} = G + 3.8 \times 10^{-3}G_v \tag{9.16}$$

Example 9.2

Consider a light source of 8000 lumen in a room and we have a pocket calculator with a small solar array of four devices in series with a total surface of 2 cm^2. The light has a reflector which aproximately produces a hemispherical radiation of the light. The distance from the light source to the top of the table where the pocket calculator is located is 1.6 m. The solar array has been rated at standard AM1.5 conditions to produce 20 mA/cm^2 short circuit current density. Calculate the short circuit current produced under the artificial light assuming that the operating temperature is 25 °C. The first thing to calculate is the illuminance received by the solar array.

$$G_v = \frac{\Phi_v}{d^2\Omega} = \frac{8000}{1.6^2 2\pi} = 497.6 \, \text{lux} \tag{9.17}$$

where a solid angle of 2π has been used to account for the hemispherical radiation. Next the short circuit current density is calculated,

$$J_{sc(art)} \approx 3.8 \times 10^{-6} J_{sc(AM1.5)} G_v = 3.8 \times 10^{-6} \times 20 \times 10^{-3} \times 497.6 = 37.8 \, \mu\text{A/cm}^2 \tag{9.18}$$

As the array is made of four cells in series and the area of one solar cell is 0.5 cm^2, the total short circuit current is

$$I_{sc(art)} \approx 37.8 \times 10^{-6} \times 0.5 = 18.9 \, \mu\text{A} \tag{9.19}$$

9.6 I(V) Characteristics Under Artificial Light

As has been shown in Section 9.5 the short circuit current under artificial light can be calculated provided the quantum efficiency of the solar cell, light spectrum and illuminance are known. The open circuit voltage and the full $I(V)$ characterisitics can also be known because the dark saturation density current is independent of the light spectrum as described in Chapter 2, and hence the value calculated from the rated values of short circuit current and open circuit voltage under AM1.5 standard conditions is valid for artificial light calculations as well.

9.7 Illuminance Equivalent of AM1.5 G Spectrum

Indoor photovoltaic devices receive a random mixture of natural and artificial light. The mixture is random in time for the same user and in location or activity for different users. Natural light in office spaces or homes will have a different spectral distribution than the standard AM1.5 we are using in this book, and depends on the environment. Again, as the data available from solar cell manufacturers have been measured under AM1.5 G standard spectra a conversion factor or equivalence between the irradiance and illuminance of the AM1.5 spectrum will be useful. The illuminance of the AM1.5 G spectrum (1000 W/m^2) is given by:

$$G_v = \frac{1000}{962.5}\int_{360\,nm}^{760\,nm} I_\lambda K_m V(\lambda)\,d\lambda \tag{9.20}$$

where again I_λ is the spectral AM1.5 irradiance. The reference irradiance of AM1.5 G (1000 W/m^2) is given by

$$G_r = \frac{1000}{962.5}\int_0^\infty I_\lambda\,d\lambda \tag{9.21}$$

Both equations (9.20) and (9.21) can be easily evaluated using PSpice (file illuminance_am15g.cir) and the resulting equivalence is 109.87 lux per W/m^2.

```
*illuminance_am15g.cir
.include am15g.lib
.include cie.stl
.param g=1000
xam15g 90 0 am15g
vcie 91 0 stimulus vcie
eilluminance 92 0 value={sdt(v(90)*683*v(91)*g/1000*1e6)};irradiance
.tran 0.01u 0.770u 0.38u 0.01u
.probe
.end
```

Example 9.3

Consider two light sources, one having an AM1.5 G (file am15g.lib) spectrum and the other having a fluorescent light spectrum (file fluorescent_rel.stl) and a solar cell with the baseline parameters used in Example 2.2 in Chapter 2.

(a)   Calculate the value of the AM1.5 G irradiance required for the solar cell to generate a short circuit density current of 60.5 μA/cm^2, which is the same value generated under a fluorescent spectrum of 500 lux illuminance.

The value of the short circuit current under AM1.5 G spectrum for this solar cell is 31.81 mA/cm^2 so,

$$J_{sc} = \frac{G}{G_r}J_{sc}(1\text{ sun}) = 60.5\,\mu\text{A/cm}^2 \tag{9.22}$$

and then,

$$G = \frac{G_r 60.5 \, \mu\text{A/cm}^2}{31.81 \, \text{mA/cm}^2} = 1.90 \, \text{W/m}^2 \qquad (9.23)$$

(b) Using the file corresponding to the fluorescent light calculate the value of the irradiance of this source at 500 lux.

Writing a file:

```
*irradiance_art.cir
.include fluorescent_rel.stl
.param gv = 500
.param k = 0.0292
vfluor 90 0 stimulus vfluorescent_rel
eIRRAD 92 0 value = {sdt(v(90)*k*gv*1e6)};irradiance
.tran 0.01u 0.770u 0.38u 0.01u
.probe
.end
```

v(92) returns the value of the irradiance corresponding to 500 lux of this fluorescent light, which is equal to $1.694 \, \text{W/m}^2$.

9.8 Random Monte Carlo Analysis

One of the interesting potentials of PSpice is the capability to carry out Monte Carlo and worst case analysis. One way to represent conditions under which portable equipment works in an office or home environment, is the use of irradiance and illuminance values randomly selected and mixed in a given manner.

This is easily incorporated in PSpice due to its built-in Monte Carlo analysis and considering that the irradiance at the plane of the PV converter is an effective irradiance, resulting from the mixture of natural light, represented by an AM1.5 spectrum of a given irradiance, and artificial light represented by a fluorescent light of a given illuminance.

Taking into account that the recommended illuminance levels in housing or office space depend on the type of activity and the visual effort required, it is generally recommended that an activity requiring an average visual effort, or high visual effort for a short time, is supported by typically 500 lux. This illuminance can be raised to 1000 lux for high precision activities or can be reduced to 250 lux for low visual effort activities.

One way to generate random values of the irradiance and illuminance, is to connect resistors across the current generator used for the irradiance and for the illuminance, with values randomly varied within a given tolerance and probability distribution function. This is allowed in the Monte Carlo analysis of PSpice. We start by describing the Monte Carlo analysis command:

Syntax for Monte Carlo analysis

The general form is:

.mc <nr of runs> <analysis> <output variable> <function> <seed> <options>

where,

<nr of runs> is the number of analyses to be performed;
<analysis> is the analysis type, DC, tran or AC;
<output variable> set to YMAX computes the absolute maximum difference
between the nominal run and each run;
<seed> value has to be an odd integer from 1 to 32 767;
<options> list or print the values of the random variables used in each
run.

A random generation of mixtures of natural and artificial light has to be generated
according to equation (9.16). In order to do so we give the value for the nominal irradiance
and illuminace to two g-devices, namely, 'gnat' and 'gart', respectively and we connect one
resistor across the dependent current sources. We then give a tolerance value to the resistor
model as:

$$\text{.model resist res } (r = \{xnat\} \text{ dev/uniform } 40\%)$$

where the nominal value is 'xnat' and the tolerance is 40% with a uniform probability
distribution. Different probability distributions may be used and we refer the reader to the
PSpice manual.

Finally equation (9.16) can be implemented and the effective irradiance calculated for any
value inside the tolerance. This is shown in the PSpice file 'irradiance_eff.lib' as follows:

```
*irradiance_eff.lib
.subckt irradiance_eff 900 920 params:irrad=1, xnat=1, gv=1, xart=1
gnat 900 910 value={irrad}
rnat 910 900 resist {xnat}
.model resist res (r=xnat dev/uniform 40%)
gart 900 930 value={gv}
rart 930 900 resist_art {xart}
.model resist_art res (r={xart} dev/uniform 40%)
eirradeff 920 900 value={v(910)+3.8e-3*v(930)}
.ends irradiance_eff
```

The above circuit implements equation (9.16), which has been derived for silicon solar cells
and hence the accuracy of the results may suffer if solar cells made from different material
are considered. This is due to the different quantum efficiencies although this effect may be
small because equation (9.16) is in fact a relative result for the same solar cell under two
different sources of light. For the sake of simplicity we will use equation (9.16) throughout
the book.

This subcircuit can now be used to produce a random series of $I(V)$ characteristics of a PV module including a Monte Carlo analysis and a PV module as shown in the file 'montecarlo.cir' as follows:

```
*montecarlo.cir
.include irradiance_eff.lib
.include module_beh.lib
xirrad_eff 0 92 irradiance_eff params: irrad = 10, gv = 500, xnat = 1, xart = 1
xmodule 0 92 93 94 95 96 97 98 99 module_beh params: iscmr = 8.5e-3 coef_iscm = 0 vocmr = 2.8
+ coef_vocm = 0 pmaxmr = 14.2e-3 tr = 25 noct = 47 ns = 4 immr = 7.06e-3 vmmr = 2.01
vbias 94 0 dc 0
vtemp 93 0 dc 25
.dc vbias 0 3.5 0.01
.mc 10 dc i(vbias) ymax seed = 9321 list output all
.probe
.end
```

As can be seen, in this file we model a circuit including a PV module represented by the subcircuit described in Chapter 4, 'module_beh', and we particularize for a series array of four solar cells with the parameters shown. These parameter values may be representative of an amorphous silicon solar cell array used in pocket calculators. As can also be seen, we have considered as zero the temperature coefficient of both current and voltage due to the fact that these applications in general work at room temperature, which we have considered constant and equal to 25 °C. A DC analysis allows the Monte Carlo runs to be performed.

In order to verify that the random number generator produces uncorrelated couples of values for the natural and artificial light, a plot is shown in Figure 9.6 and the $I(V)$ plots of 10 Monte Carlo runs are shown in Figure 9.7.

The resulting values of the coordinates of the maximum power point are shown in Figure 9.8.

As can be seen, the average value of the voltage at the maximum power point is reasonably constant at 2.494 V and a standard deviation of 3 mV, whereas the current at the maximum power point has an average value of 88.45 µA with a standard deviation of 12.29 µA.

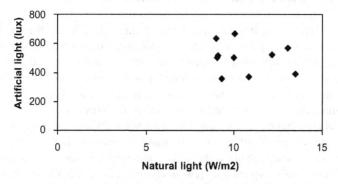

Figure 9.6 Plot of randomly generated values of natural and artificial light

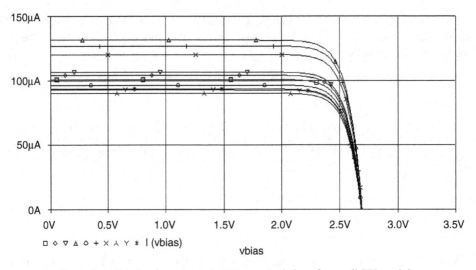

Figure 9.7 Randomly generated $I(V)$ characteristics of a small PV module

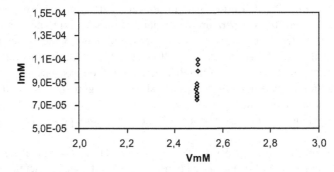

Figure 9.8 Values of the coordinates of the maximum power point for a randomly generated couple of values of natural and artificial light

9.9 Case Study: Solar Pocket Calculator

One consumer product of widespread use is the pocket calculator with a small array of PV cells supplying some of the power required for operation. Today CMOS electronics work at a low voltage supply, typically under 2 V, and consume very little power, therefore a small PV array can provide full operation of the calculator or extend the life of a battery. Normally these devices are used in offices, home, schools, markets and shops where the light is a mixture of natural and artificial light as described in the previous section.

Pocket calculators can be directly powered by a small PV array or they can have a mixed battery–PV generator combined system. The main requirement is that the power produced by the PV system is enough to allow the system to work. The power consumed by a pocket calculator is very small, typically a few microamperes at 1.5 V. This can be provided typically by a small array of four amorphous silicon solar cells in series of roughly 0.5 cm^2

area. Amorphous silicon solar cells are available from a number of manufacturers and use technologies from single to triple junctions in order to improve efficiency. The operation of a solar pocket calculator is illustrated in the following example.

Example 9.4

Consider a pocket calculator with a solar cell series array of four cells $2\,cm^2$, with the following characteristics at AM1.5 G $(1000\,W/m^2)$ $V_{ocr} = 2.8\,V$, $I_{scr} = 8.5\,mA$, $I_m = 7.06\,mA$, $V_m = 2.01\,V$.

Consider that the solar array is typically illuminated with 500 lux of artificial fluorescent light and by a natural light having an irradiance of $10\,W/m^2$. Both the natural and artificial light intensity varies randomly with a tolerance of 40% and with uniform probability distribution.

The circuit we are considering is shown in Figure 9.9. As can be seen the battery has a diode in series to avoid reverse current, because a non-rechargeable battery is considered in this case. We have considered a constant load given by $3.5\,\mu A$, which is typical of the low power CMOS circuitry used in these products. The result of the simulation using 'pocket_calculator.cir' file listed in Annex 9 is shown in Figure 9.10.

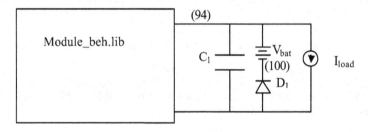

Figure 9.9   Circuit for a pocket calculator

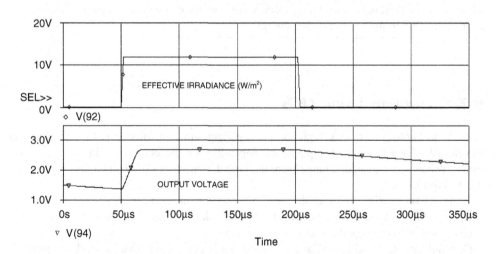

Figure 9.10   (top) Effective irradiance; (bottom) Output voltage

For the operation of the circuit, we consider that the light source is of pulsed nature for illustrative purposes. As can be seen at the beginning with no light the output voltage is that of the battery minus the voltage drop (1.3 V approximately). When the light pulse occurs, the capacitor current and the output array current sharply increase, meaning that the load supply is provided by the PV array which also charges the capacitance. Simultaneously the output voltage rises and sets at approximately 2.6 V indicating that the diode is reverse biased and the battery does not deliver current to the load. In order to see more exactly the operation of the circuit imagine we use only the artificial light source of 500 lux, the results are shown in Figure 9.11.

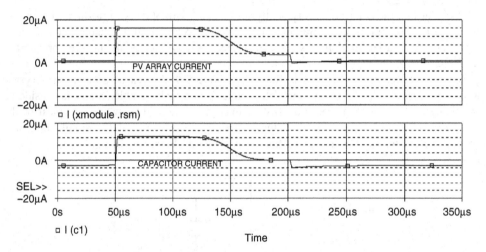

Figure 9.11 PV array in a pocket calculator circuit when powered by 500 lux of artificial light (top) and capacitor current (bottom)

As can be seen the output current of the array charges the capacitor and provides the load, while when the light pulse stops, the capacitor delivers the charge to the load until the battery is on again and continues to provide energy.

9.10 Lighting Using LEDs

Today's technology provides high luminosity light emitting diodes (LED) at different wavelengths for a number of applications depending on the colour and intensity requirements. Wavelengths available range from the blue 430 nm of a GaN LED to the red 660 nm for AlGaAs LED.

Table 9.3 summarizes some of the available components and typical rating voltages and currents [9.2]. The increasing luminosity of these devices makes them suitable to many applications in traffic signals or in general lighting.

Generally the LEDs are rated at a given electrical current input and provide a luminous intensity (given in candels). LEDs do not emit light isotropically but at angle which varies

Table 9.3 Summary of some characteristics of LEDs

Wavelength (nm)	Material	Voltage (V)	Current (mA) typical	Intensity (mcd) at 20 mA	Output angle (°)
430 (blue)	GaN	4.8	20	12	20
473 (blue)	AlInGaP	3.5	20	1100	15
565 (green)	GaP	2.2	20	40–60	60
595 (yelow)	AlInGaP	1.8	20	100–400	60
605 (orange)	AlInGaP	1.9	20	1300–8000	30–8
626 (red)	AlInGaP-AlGaAs	1.85–1.9	20	20–1200	130–50

Adapted from RS catalogue 2001–2002.

from one to another as can be seen in Table 9.3. We are interested in the luminous flux resulting from this luminous intensity, taking into account the solid angle of viewing. Noting that the luminous flux is the integral of the luminous intensity over the solid angle of emission and considering a uniform luminous intensity, it follows that:

$$\text{Luminous flux} = \text{Luminous intensity} \times \text{solid angle}$$

And finally considering that the solid angle relates to the viewing angle γ as

$$\Omega = 2\pi\left(1 - \cos\frac{\gamma}{2}\right) \tag{9.24}$$

the value of the luminous flux can be calculated.

Example 9.5

Calculate, from the data in Table 9.3, the values of the luminous flux corresponding to a LED with a viewing angle of 30° and a luminous intensity of 1300 mcd at 2 V and 20 mA. Calculate the luminous efficacy defined as the luminous flux divided by the electrical power. The solid angle corresponding to 30° is calculated as

$$\Omega = 2\pi\left(1 - \cos\frac{30}{2}\right) = 0.213 \, \text{sr} \tag{9.25}$$

and then the luminous flux is

$$\text{flux} = 1300 \, \text{mcd} \times 0.213 \, \text{sr} = 0.276 \, \text{lm}$$

The luminous efficacy is then given by

$$\text{Efficacy} = \frac{\text{Flux}}{\text{Electrical} \cdot \text{power}} = \frac{0.276}{2 \times 20 \times 10^{-3}} = 6.9\frac{\text{lm}}{\text{W}} \tag{9.26}$$

9.11 Case Study: Light Alarm

In this section we will analyse the case of a flashing light signal which may be useful for example in building autonomous alarms, or for caution or danger signals.

The use of high luminosity LEDs for this kind of application is possible due, in general, to the relative low power required and the effectiveness of the LED light.

As a typical example we will consider an application of a flashing light composed of a number NL_{ED} of high luminosity orange LEDs with a total autonomy life of at least the battery life, which may be of five years plus.

We first start by calculating the predicted power consumed by this system by assuming that the light will flash once every second at the rated 20 mA as shown in Figure 9.12. The flash lasts 85 ms. Consider that a current of 0.5 mA is continuously consumed by a timer (type 555) circuitry which controls the flash.

(a) Calculation of the equivalent average current

The equivalent value is the addition of the average of the waveform shown in Figure 9.12 to the rest of the consumption by the electronics.

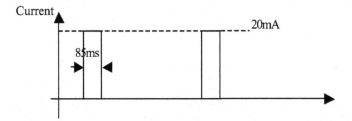

Figure 9.12 Waveform of the pulsed current consumed by a LED

The result is:

$$I_{eq} = 0.5 \times 10^{-3} + 20 \times 10^{-3} \frac{85\text{ms}}{1\text{s}} \text{x} N_{LEDP} = 0.5 + 1.7 N_{LEDP} (\text{mA}) \qquad (9.27)$$

where N_{LEDP} is the number of parallel rows of LEDs.

(b) Calculation of the PV array size

Taking into account the results in Chapter 7,

$$I_{PV} = I_{eq} \frac{24}{PSH_{\min}} = 6 + 20.4 \, N_{LEDP} (\text{mA}) \qquad (9.28)$$

where we considered a current safety factor and the minimum value of PSH. In these applications, as the random nature of the sun radiation is accompanied by the random nature of the location where the system is operating, it is advisable to take as a value for PSH the lowest value at the horizontal surface. In this example we will consider $PSH = PSH_{min} = 2 \, kwh/m^2$-day.

(c) Calculate the capacity of a battery considering that the autonomy is to be a number of days, N_{days}, without sun radiation.

$$C_{bat} = I_{eq} \times 24 \times N_{days} \, Ah = (0.012 + 0.0408 \, N_{LEDP})N_{days}(Ah) \qquad (9.29)$$

Figure 9.13 shows a plot of the required battery capacity for a variable number of parallel rows of LEDs and three sets of autonomy, 10, 20 and 30.

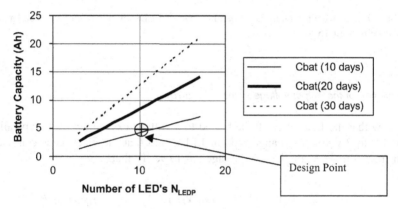

Figure 9.13 Required battery capacity

(d) Consider that the solar cells available for the application have a rated AM1.5 (1 kW/m²) of 37.5 mA/cm². Calculate the area required of N_{sG} series connected solar cells for the array.

Setting the equality between the current required for the PV array with the area times the density short circuit current, we have

$$I_{PV} = 6 + 20.4 \times 10^{-3} N_{LEDP} = A \times 37.5 \times 10^{-3} \qquad (9.30)$$

and hence the area A is given in cm^2 by,

$$A = N_{sG}[0.16 + 0.544 N_{LEDP}] \qquad (9.31)$$

which will be the total PV array area. A plot of equation (9.31) is shown in Figure 9.14.

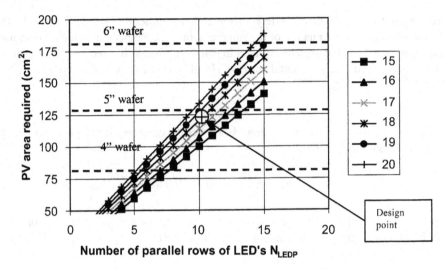

Figure 9.14 Design space for a number of parallel rows of LEDs ranging from 3 to 15 and a number of series solar cells from 15 to 20

(e) Calculation of the output voltage of the solar array

As we know that the LEDs have to be biased to a voltage exceeding 1.5 V, typically in the range of 1.8 to 2 V for the orange type of LEDs, and that the electronics can usually be powered from 3 to 18 V, the battery voltage has to verify that

$$V_{PV} > V_{bat} + V_d = V_d + N_{LEDS}V_{LED} + V_{driver} + N_{LEDP} I_{LED} R \qquad (9.32)$$

If we go for a standard battery of $V_{bat} = 6$ V nominal which is standard, then we will have to limit the number of LEDs in series to two and set the output voltage of the PV generator larger than approximately 8 V to take into account diode losses and temperature effects.

(f) Calculation of the number of solar cells in series, N_{sG}

The number N_{sG} depends on the ratings of the solar cell selected. Assuming that a medium rated silicon solar cell is selected, we can consider that at 1 kW/m² the maximum power point voltage has to exceed the V_{PV} calculated in (c) above.

$$N_{sG} > \frac{V_{PV}}{V_{mM}} = \frac{8}{0.55} = 14.54 \qquad (9.33)$$

This means that at least 15 devices have to be connected in series taking into account that the rated maximum power point voltage in equation (9.33) is at 1000 W/m², which will only happen during a few days in the summer. This result shows additional restrictions to be added to the design space in Figure 9.14.

(g) Final adjustment

Taking into account the values of the safety factors and also the values of some important boundary conditions, the final values have to be adjusted. These boundary conditions are the availability of solar cell arrays and batteries.

The marketplace offers a large variety of rechargeable lead–acid sealed batteries at 6 V with capacities ranging from (in the range of interest considered) 1 to 12 Ah.

Imagine that the specifications require a number of LEDs, $N_{LEDP} = 10$ (we consider a number of LEDs in series in each row of $N_{LEDS} = 2$). A possible solution from Figure 9.14 is to take a full 5″ wafer and cut it into 16 pieces of equivalent area.

The design point corresponds to a required PV area of 100.8 cm^2. If we use a full 5″ wafer we allow for a current safety factor of the ratio of the full wafer divided by the area required. This means a safety factor of a roughly 25%. This, of course, will be useful to account for the cutting losses of the full wafer. Looking at the graph of the battery capacity and assuming a number of days of totally autonomous operation without any charging from the sun, a battery of 4 Ah will approximately cover 10 days of total lack of sun.

9.11.1   PSpice generated random time series of radiation

Simulation of applications such as described in Section 9.11 often calls for a 'worst month' design where the designer looks for a reliable design especially for winter months when little sun radiation is available. The approach we follow in this case is to assume the January data for sun radiation which in many cases is the worst case.

We have seen in Chapters 7 and 8 how to use stochastically generated time series of radiation using stimulus files in Pspice. The availability of such time series is linked to the availability of specifically developed commercial or propietary software. In order to extend the capabilities of designers who only have access to PSpice, we have developed in this section a method to generate random time series of daily radiation only using PSpice.

Among several methods used, the one in Reference [9.3] can be easily implemented. It assumes that the clearness index (k_h) can be modelled by a first-order autoregressive stochastic process. k_h is the ratio between the solar radiation at a horizontal surface to the extra-atmospherical radiation.

It can then be written that

$$u(i) = \rho u(i-1) + \varepsilon \qquad (9.34)$$

where $u(i)$ is the difference between the clearness index at day i and the monthly average value of this same parameter, ϵ is a gaussian random number comprised between -1 and $+1$ and ρ is the autocorrelation coefficient, which is assumed to be 0.25 and practically independent of the locality [9.3].

We will apply these concepts for illustrative purposes to the case study of Section 9.11 where a high reliability system is pursued, mainly in winter time. A random series of values of radiation for the 31 days of January can be generated (it may also be of course extended to longer periods of time).

We start by running a PSpice file (random_gen.cir) in order to generate a series of 31 random numbers. We use the Monte Carlo analysis with a nominal value for the average value of the horizontal daily radiation and then allow the random generator a tolerance of 15% with Gaussian distribution. The file used is the following:

```
*random_gen.cir
g1 0 1 value = {1}
r1 1 0 resist 1
.model resist res (r = 1 dev/gauss 15%)
vaux 2 0 dc 1
.dc vaux 0 1 0.1
.mc 31 dc v(1) ymax list seed = 9321 output all
.probe
.print dc v(1)
.end
```

where a g-device of value of unity is placed in parallel with a resistor of nominal value 1 Ω with a tolerance of 15%. The dummy DC analysis is performed to allow the Monte Carlo routine to be executed and the values of the random numbers are produced in the .out file. The unity is subtracted from the random numbers so that they are distributed between −1 and +1, as shown in Table 9.4 and then used in equation (9.34) to generate a list of the 31 values of daily radiation.

Table 9.4 Gaussian random numbers generated by Monte Carlo analysis with 15% tolerance

0	−0,181	0,06	0,12	0,0002	0,038	0,091	0,163
−0,063	0,05	0,225	−0,157	−0,039	0,025	−0,356	0,099
−0,344	−0,19	−0,055	0,12	0,3	0,1	0,1	−0,161
0,12	−0,19	−0,192	−0,191	−0,151	−0,105	−0,099	

The procedure does not ensure that the monthly average value of the radiation is the nominal value used, so the random values are multiplied by a constant.

We start by writing a subcircuit 'hour_rad_bartoli.lib' containing the random daily radiation values arranged in a voltage source (v_random). The number of the day is also written in another voltage source named v_n, as can be seen in Annex 9.

These random daily radiation values, arranged in a PWL voltage source last 24 hours of real time corresponding to 24 μs of internal time. Following the procedure described in Annex 11, the randomly generated values of the daily radiation are converted into hourly values of radiation. This procedure does not create a synthetic hourly radiation time series as desribed in Chapter 7, but instead a daily theoretical irradiance curve with the same time integral value as the daily radiation. More elaborated methods used to obtain hourly radiation time series, such as the ones described in references [9.4] and [9.5] for example, would be preferable, but in the absence of a suitable code to compute these values the procedure described here may help in the long-term simulation of PV systems. The illustration of this procedure is shown in the next section.

9.11.2   Long-term simulation of a flash light system

According to the previous discussion a system like that shown in Figure 9.15 can now be simulated.

The design parameters used for the PV module are as follows:

$$I_{scM} = 126\,\text{mA}, \qquad V_{ocmr} = 10.4\,\text{V}, \quad P_{maxr} = 0.9828\,\text{W},$$

$$I_{mMr} = 122.4\,\text{mA}, \qquad V_{mMr} = 8.32\,\text{V}, \qquad N_s = 16$$

For the battery:

$$\text{Initial state of charge} = 50\%, \quad \text{size of battery} = 24\,\text{Wh},$$

$$\text{nominal battery voltage} = 6\,\text{V}$$

Load
$$I_{equivalent\ load} = 17.5\,\text{mA}$$

We have relaxed the value of the short circuit current to take into account that generally for these type of small applications, rated values for modules are perhaps optimistic.

Using these values we produced the PSpice file 'flash_light.cir' corresponding to the circuit shown in Figure 9.15 and listed in Annex 9, where it can be seen that the subcircuit 'hour_rad_31_bartoli.lib' is used to generate the hourly radiation values following the equations described in Annex 3. The resulting data values are filtered with an RC circuit to smooth the data profile and help the numerical convergence of the simulation. These filtered data are fed into the model of a PV module 'module_beh.lib' and the output is connected to a node where the battery and the load are also connected. We have considered an equivalent DC load current to model the power consumption by the LEDs and because in these applications the alarm light may be necessary during daytime or during night-time. Finally

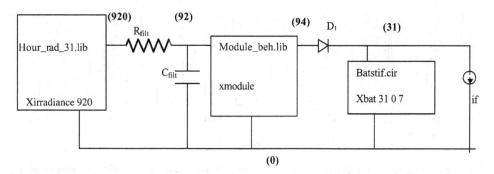

Figure 9.15   Circuit of the flash light including random series of hourly radiation values, module and battery

in this case we have considered the ambient temperature constant and equal to 5 °C for this winter month.

The first result of the simulation is shown in Figure 9.16 where the random daily radiation values generated for January at a site located 41.2° N of latitude are plotted as a function of time. Every step in the figure corresponds to one day.

Figure 9.17 shows the translation of the daily radiation values in Figure 9.16 into hourly radiation values for the first four days of January at that particular location.

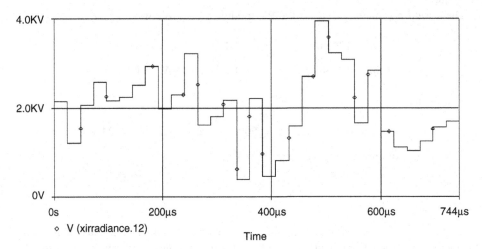

Figure 9.16 Plot of the daily radiation values for the 31 days of January at a location with an average radiation of 2000 Wh/m²-day. *Warning*: the y-axis is given in kWh/m²-day and the x-axis is the time in hours of that month starting with time 0 and ending at time 744 hours of real time corresponding to the 31 days

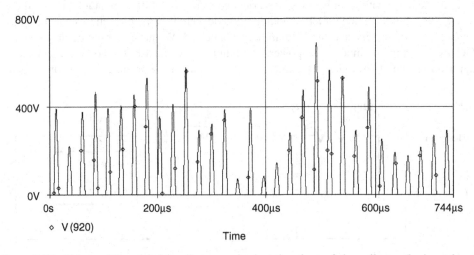

Figure 9.17 Values of the global irradiance received at the plane of the collector (horizontal and 41.2° N latitude) of the 31 days of January. *Warning*: x-axis internal time unit is in microseconds and real time unit is hours with an average daily horizontal irradiation of 2000 W/m²

The average value of the daily radiation is 2000 Wh/m²_day according to the specification.

Figure 9.18 shows the time evolution of two of the magnitudes of the circuit, the upper graph plots the PV output current as a function of time and the bottom plot is the SOC waveform.

It can be seen that even starting the worst month with 90% of the nominal capacity, the system is not able to hold the state of charge inside the limits. This result suggests that probably the system reliability will be increased if a battery with more initial capacity is

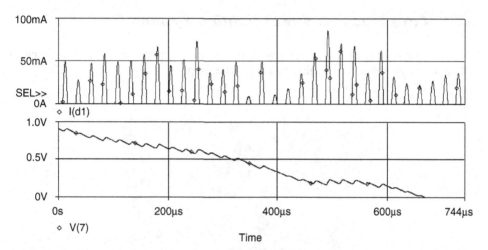

Figure 9.18 Evolution of SOC (bottom graph) for a sequence of 31 days of January with an average horizontal irradiation of 2000 Wh_day, starting with an initial value of SOC = 0.9. The upper graph shows the PV generator output current. *Warning*: internal time units are microseconds (1 microsecond corresponds to one hour). SOC has no units

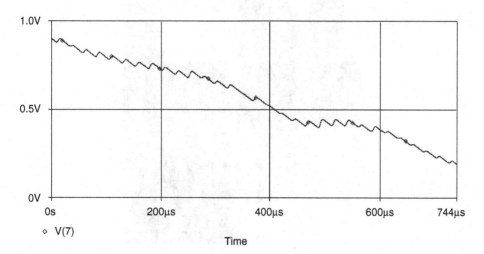

Figure 9.19 The same example as in Figure 9.18 but with a battery of 36 Wh instead of 24 Wh. *Warning*: x-axis internal time units are microseconds and real time units are hours

used. For commercially available sizes, the next could possibly be a 6 V battery, 6 Ah capacity, resulting in 36 Wh. Applying this in the simulation, we find the result shown in Figure 9.19 where the SOC remains inside the limits.

The procedure described in this example allows a refinement of the design values obtained and may also consider different random numbers or different tolerance values for the parameters involved. It is remarkable that the total CPU time needed to simulate the graph in Figure 9.18 is of the order of 40 seconds depending on the computer used.

9.12 Case Study: A Street Lighting System

Street lighting is becoming an important issue for many communities and local councils willing to use photovoltaics for this purpose. An example is shown in Figure 9.20.

Figure 9.20 Example of a street light. Reprinted with permission by Solar Ingenieria 2000, Castelldefels, Spain

The lamps are generally conceived to meet safety and lighting requirements. The specifications are as follows:

- Light flux from 3000 to 5000 lm.

- Enough energy to provide light during 8 hours in winter time at night. This is equivalent to a current load consumption of 3.08 A at 12 V.

One possible solution frequently found is 165 Wpeak of PV module and 200 Ah battery capacity.

The time series we will use are the same as in the previous section but with an average daily radiation value of 1734 W/m^2_day, for January and an inclination angle of 30°, which are realistic figures for an average Mediterranean location. The file 'street_light_bartoli.cir' is shown in Annex 9.

This file corresponds very much to a similar circuit as that in Figure 9.15 with the difference that the temperature profile is provided by the circuit 'temp_profile.lib' in Annex 9. The result of the simulation is shown in Figure 9.21 where the hourly radiation values, cell temperature profile and battery SOC are plotted as a function of time.

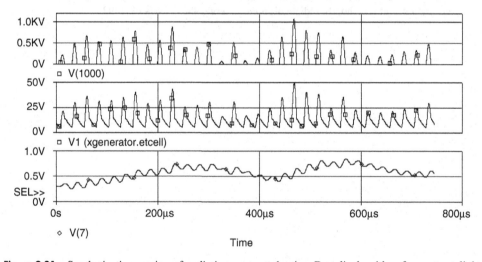

Figure 9.21 Synthetic time series of radiation generated using Bartoli algorithm for a street light system. Hourly radiation (W/m^2) (top), cell temperature (°C) (middle) and battery state of charge SOC (bottom). *Warning*: x-axis internal time units are microseconds and real time units are hours

As can be seen if the battery nominal capacity is set to 2400 Wh even by starting the month with a low value of the SOC, this parameter remains inside the limits.

9.13 Problems

9.1 Following the approach and files used in Example 9.5, simulate a system of the same size but having to supply 7 μA of constant load current. Increase the size of the individual solar cell to 0.7 cm^2 and simulate again.

9.2 Size a simplified light alarm where two arrays of three LEDs each have to be supplied working in the same conditions as in Section 9.11. Simulate the resulting system adjusting the values of the battery parameters in order to have the SOC inside limits.

9.3 Generate a time series of daily radiation values for the month of July in a latitude of 40° N with a daily average of 5 kWh/m^2-day, and a Gaussian random number, tolerance 10%.

9.14   References

[9.1] Ryer, A., *Light Measurement Handbook*, http://www.intl-light.com/handbook.

[9.2] *RS Electronic Components Catalogue, Edition* 2000.

[9.3] Bartoli, B., Coluzzi, B., Cuomo, V., Francesca, M. and Serio, C., 'Autocorelation of daily global solar radiation', *Nuovo Cimento*, **40**(2), 113–21, 1983.

[9.4] Graham, V.A. and Hollands, K.G.T., 'A method to generate synthetic hourly solar radiation globally', *Solar Energy*, **44**(6), 333–41, 1990.

[9.5] Aguiar, R. and Collares-Pereira, M., 'TAG: a time dependent autoregresive gaussian model for generating synthetic hourly radiation', *Solar Energy*, **49**(3), 167–74, 1992.

Annex 1

PSpice Files Used in Chapter 1

File name	Description	Section	Includes	Figures
learning.cir	First RC circuit in PSpice	1.3		1.4
Rc.lib	Subcircuit for an RC circuit	1.4		
Learning_subckt.cir	Using two RC subcircuits	1.4	rc.lib	
am15g.lib	AM1.5G subcircuit	1.6		
am15g.cir	Circuit equivalent to the AM1.5G spectrum	1.6	am15g.lib	1.7
wavelength.lib	Subcircuit with the wavelength in microns	1.7		
am0.lib	AM0 subcircuit	1.7		
am0.cir		1.7	Am0.lib	1.9
black_body.lib	Subcircuit with a black body spectrum	1.7		
black_body.cir	PV module conversion to arbitrary values of irradiance and temperature	1.8	black_body.lib	1.9

```
**********************************************
*                                            *
*              LEARNING.CIR                   *
*                                            *
**********************************************

* CIRCUIT TO LEARN ABOUT PSPICE
* COMPRISES A RESISTOR AND A CAPACITOR

* NODES
*    (0) GROUND
*    (1) INPUT
*    (2) OUTPUT

* COMPONENTS
* RESISTOR SYNTAX: rxx node_a node_b value
r1 1 2 1K; resistor between node (1) and node (2) value 1KOhm
```

```
* CAPACITOR SYNTAX : cxx node_a node_b value
c1 2 0 1n; capacitor btween node (2) and node (0) value 1nF

*INPUT

* SYNTAX FOR A PULSE VOLTAGE SOURCE:
*    vxx node+ node- pulse ( initial_value pulse_value delay risetime
*       falltime pulse_length, period)
v1 1 0 pulse (0 5 0 1u 1u 10u 20u)

*ANALYSIS TYPE
* SYNTAX FOR TRANSIENT ANALYSIS: .tran tstep tstop tstart tmax
* tstep: printing increment
* tstop: final simulation time
* tstart: (optional) start of printing time
* tmax: ( optional) maximum step size of the internal time step

.tran 0 40u

*OUTPUT
.probe; starts the post processing of the data
.plot tran v(1) v(2); plots the transients
.end; ends the file


**********************************************
*                                            *
*              AM15G.LIB                      *
*                                            *
**********************************************


* AM1.5G 37°tilt angle, global spectral irradiance
* (modified trapezoidal integration)
* Total irradiance 962.5W/m2
* from Hulstrom, R., Bird, R., Riordan,C., Solar Cells, vol. 15, pp. 365-391, (1985)

* NODES
*    (11) spectral irradiance in (W/m2micron)
*    (12) reference

.subckt am15g 11 10
v_am15g 11 10 pwl
+    0.295u   0
+    0.305u   9.2
+    0.315u   103.9
+    0.325u   237.9
+    0.335u   376
+    0.345u   423
+    0.350u   466.2
+    0.360u   501.4
+    0.370u   642.1
+    0.380u   686.7
+    0.390u   694.6
+    0.400u   976.4
+    0.410u   1116.2
+    0.420u   1141.1
+    0.430u   1033
+    0.440u   1254.8
+    0.450u   1470.7
+    0.460u   1541.6
+    0.470u   1523.7
+    0.480u   1569.3
```

```
+    0.490u    1483.4
+    0.500u    1492.6
+    0.510u    1529
+    0.520u    1431
+    0.530u    1515.4
+    0.540u    1494.5
+    0.550u    1504.9
+    0.570u    1447.1
+    0.590u    1344.9
+    0.610u    1431.5
+    0.630u    1382.1
+    0.650u    1368.4
+    0.670u    1341.8
+    0.690u    1089
+    0.710u    1269.0
+    0.718u    973.7
+    0.724u    1005.4
+    0.740u    1167.3
+    0.753u    1150.6
+    0.758u    1132.9
+    0.763u    619.8
+    0.768u    993.3
+    0.780u    1090.1
+    0.8u      1042.4
+    0.816u    818.4
+    0.824u    765.5
+    0.832u    883.2
+    0.84u     925.1
+    0.86u     943.4
+    0.88u     899.4
+    0.905u    721.4
+    0.915u    643.3
+    0.925u    665.3
+    0.93u     389
+    0.937u    248.9
+    0.948u    302.2
+    0.965u    507.7
+    0.980u    623
+    0.994u    719.7
+    1.040u    665.5
+    1.070u    614.4
+    1.1u      397.6
+    1.12u     98.1
+    1.13u     182.2
+    1.137u    127.4
+    1.161u    326.7
+    1.18u     443.3
+    1.2u      408.2
+    1.235u    463.1
+    1.290u    398.1
+    1.32u     241.1
+    1.35u     31.3
+    1.395u    1.5
+    1.443u    53.7
+    1.463u    101.3
+    1.477u    101.7
+    1.497u    167.1
+    1.52u     253.1
+    1.539u    264.3
+    1.558u    265
+    1.578u    235.7
+    1.592u    238.4
```

```
+    1.61u      220.4
+    1.63u      2.35.6
+    1.646u     226.3
+    1.678u     212.5
+    1.740u     165.3
+    1.8u       29.2
+    1.86u      1.9
+    1.92u      1.2
+    1.96u      20.4
+    1.985u     87.8
+    2.005u     25.8
+    2.035u     95.9
+    2.065u     58.2
+    2.1u       85.9
+    2.148u     79.2
+    2.198u     68.9
+    2.27u      67.7
+    2.36u      59.8
+    2.45u      20.4
+    2.494u     17.8
+    2.537u     3.1
+    2.941u     4.2
+    2.973u     7.3
+    3.005u     6.3
+    3.056u     3.1
+    3.132u     5.2
+    3.156u     18.7
+    3.204u     1.3
+    3.245u     3.1
+    3.317u     12.6
+    3.344u     3.1
+    3.45u      12.8
+    3.573u     11.5
+    3.765u     9.4
+    4.045u     7.2

.ends am15g


*********************************************
*                                           *
*            WAVELENGTH.LIB                  *
*                                           *
*********************************************

* NODES
*    (11) wavelength (micron) ( output)

*    (10) reference

.subckt wavelength 11 10
Vlambda 11 0 pwl 0.0001u          0.0001
+    0.1u       0.1
+    0.295u     0.295
+    0.305u     0.305
+    0.315u     0.315
+    0.325u     0.325
+    0.335u     0.335
+    0.345u     0.345
+    0.355u     0.355
+    0.365u     0.365
+    0.375u     0.375
```

```
+    0.385u    0.385
+    0.395u    0.395
+    0.405u    0.405
+    0.415u    0.415
+    0.425u    0.425
+    0.435u    0.435
+    0.445u    0.445
+    0.455u    0.455
+    0.465u    0.465
+    0.475u    0.475
+    0.485u    0.485
+    0.495u    0.495
+    0.505u    0.505
+    0.515u    0.515
+    0.525u    0.525
+    0.535u    0.535
+    0.545u    0.545
+    0.555u    0.555
+    0.565u    0.565
+    0.575u    0.575
+    0.585u    0.585
+    0.595u    0.595
+    0.605u    0.605
+    0.615u    0.615
+    0.625u    0.625
+    0.635u    0.635
+    0.645u    0.645
+    0.655u    0.655
+    0.665u    0.665
+    0.675u    0.675
+    0.685u    0.685
+    0.695u    0.695
+    0.6983u   0.6983
+    0.7u      0.7
+    0.71u     0.71
+    0.72u     0.72
+    0.7277u   0.7277
+    0.73u     0.73
+    0.74u     0.74
+    0.75u     0.75
+    0.7621u   0.7621
+    0.77u     0.77
+    0.78u     0.78
+    0.79u     0.79
+    0.8u      0.8
+    0.8059u   0.8059
+    0.825u    0.825
+    0.83u     0.83
+    0.835u    0.835
+    0.8465u   0.8465
+    0.86u     0.86
+    0.87u     0.87
+    0.875u    0.875
+    0.8875u   0.8875
+    0.9u      0.9
+    0.9075u   0.9075
+    0.915u    0.915
+    0.925u    0.925
+    0.93u     0.93
+    0.94u     0.94
+    0.95u     0.95
+    0.955u    0.955
```

```
+    0.966u   0.966
+    0.975u   0.975
+    0.985u   0.985
+    1.018u   1.018
+    1.082u   1.082
+    1.094u   1.094
+    1.098u   1.098
+    1.101u   1.101
+    1.128u   1.128
+    1.131u   1.131
+    1.144u   1.144
+    1.178u   1.178
+    1.264u   1.264
+    2u       2
+    4u       4

.ends wavelength


*********************************************
*                                           *
*              AM0.LIB                       *
*                                           *
*********************************************

* AM0 spectral irradiance
*
* Total irradiance 1353 W/m2
* from M.P.Thekaekara, A.J. Drummond, D.G. Murcray, P.R.Gast,
* E.G.Laue and R.C Wilson, 'Solar Electromagnetic radiation'
* NASA SP 8005, (1971).

* NODES

*    (11) spectral irradiance in (W/m2micron)
*    (12) reference


.subckt am0 11 10
v_am0 11 10 pwl 0.120u    .1
+    0.14u    0.03
+    0.15u    0.07
+    0.16u    0.23
+    0.17u    0.63
+    0.18u    1.25
+    0.19u    2.71
+    0.2u     10.7
+    0.21u    22.9
+    0.22u    57.5
+    0.23u    66.7
+    0.24u    68
+    0.25u    70.4
+    0.26u    180
+    0.27u    232
+    0.28u    222
+    0.29u    482
+    0.295u   584
+    0.3u     514
+    0.305u   603
+    0.310u   689
+    0.315u   764
+    0.32u    830
+    0.325u   975
```

```
+    0.33u     1059
+    0.335u    1081
+    0.34u     1074
+    0.345u    1069
+    0.35u     1093
+    0.355u    1083
+    0.360u    1068
+    0.365u    1132
+    0.370u    1181
+    0.375u    1157
+    0.38u     1120
+    0.385u    1098
+    0.390u    1098
+    0.395u    1089
+    0.4u      1429
+    0.405u    1644
+    0.415u    1774
+    0.425u    1693
+    0.435u    1663
+    0.445u    1922
+    0.455u    2057
+    0.465u    2048
+    0.475u    2044
+    0.485u    1976
+    0.495u    1960
+    0.505u    1920
+    0.515u    1833
+    0.525u    1852
+    0.535u    1818
+    0.545u    1754
+    0.555u    1720
+    0.565u    1705
+    0.575u    1719
+    0.585u    1712
+    0.595u    1682
+    0.605u    1647
+    0.61u     1635
+    0.62u     1602
+    0.63u     1570
+    0.64u     1544
+    0.65u     1511
+    0.66u     1486
+    0.67u     1456
+    0.68u     1427
+    0.69u     1402
+    0.7u      1369
+    0.71u     1344
+    0.72u     1314
+    0.73u     1290
+    0.74u     1260
+    0.75u     1235
+    0.76u     1211
+    0.77u     1185
+    0.78u     1159
+    0.79u     1134
+    0.8u      1109
+    0.81u     1013
+    0.82u     1060
+    0.83u     1036
+    0.84u     1013
+    0.85u     990
+    0.86u     968
```

```
+    0.87u    947
+    0.88u    925
+    0.89u    908
+    0.9u     891
+    0.91u    880
+    0.92u    869
+    0.93u    858
+    0.94u    847
+    0.95u    837
+    0.96u    820
+    0.97u    803
+    0.98u    785
+    0.99u    767
+    1u       748
+    1.05u    668
+    1.1u     593
+    1.15u    535
+    1.2u     485
+    1.25u    438
+    1.3u     397
+    1.35u    358
+    1.4u     337
+    1.45u    312
+    1.5u     288
+    1.6u     245
+    1.65u    223
+    1.7u     202
+    1.75u    180
+    1.8u     159
+    1.85u    142
+    1.9u     126
+    1.95u    114
+    2u       103
+    2.1u     90
+    2.2u     79
+    2.3u     69
+    2.4u     62
+    2.5u     55
+    2.6u     48
+    2.7u     43
+    2.8u     39
+    2.9u     35
+    3u       31
+    3.1u     26
+    3.1u     22
+    3.3u     19.2
+    3.4u     16.6
+    3.5u     14.6
+    3.6u     13.5
+    3.7u     12.3
+    3.8u     11.1
+    3.9u     10.3
+    4u       9.5
+    4.1u     8.7
+    4.2u     7.8
+    4.3u     7.1
+    4.4u     6.5
+    4.5u     5.9
+    4.6u     5.3
+    4.7u     4.8
+    4.8u     4.5
+    4.9u     4.1
```

```
+    5u         3.83
+    6u         1.75
+    7u         0.99
+    8u         0.60
+    9u         0.380
+    10u        0.250
+    11u        0.170
+    12u        0.120
+    13u        0.087
+    14u        0.055
+    15u        0.049
+    16u        0.038
+    17u        0.031
+    18u        0.024
+    19u        0.02
+    20u        0.016
+    25u        0.0061
+    30u        0.003
+    35u        0.0016
+    40u        0.00094
+    50u        0.00038
+    60u        0.00019
+    80u        0.00007
+    100u       0.00003

.ends am0


*********************************************
*                                           *
*              AM0.CIR                       *
*                                           *
*********************************************

* NODES

*    (11) AM0 irradiance in (W/m2micron)
*    (0) reference

xam0 11 0 am0
r1 11 0 1
.include am0.lib

.tran 0.1u 4u
.probe
.plot tran v(11)
.end
```

Annex 2

PSpice Files Used in Chapter 2

Table A2.1 PSpice files used in Chapter 2

File name	Description	Section	Includes	Figures
Silicon_abs.lib	Absorption coefficient for silicon	2.3.1		2.4
Jsc.lib	Subcircuit computing the spectral short circuit current, total short circuit current, quantum efficiency and spectral response	2.3.2		2.5
Jsc_silicon.cir	File using jsc.lib to compute the same magnitudes for given parameter values	2.3.2 Example 2.1 Example 2.2 Example 2.3 Example 2.4	am15g.lib wavelength.lib silicon_abs.lib jsc.lib	2.7 2.8 2.9 2.10
Cell_1_si.cir	Plots $I(V)$ curve for a silicon solar cell	2.9 Example 2.5		2.12
Cell_1_gaas.cir	Plots $I(V)$ curve for a GaAs solar cell	2.9 Example 2.6		2.13

```
*********************************************
*                                           *
*              SILICON_ABS.LIB              *
*                                           *
*********************************************

* ABSORTION COEFFICIENT FOR SILICON
* DATA TAKEN FROM PC1D LIBRARY
*

* NODES

*    (11) ABSORPTION COEFFICIENT (1/CM)
*    (10) REFERENCE

.subckt silicon_abs 11 10
Vabs_si 11 10 pwl    0.295u    1.890E+06
```

```
+    0.305u    1.545E+06
+    0.315u    2.080E+06
+    0.325u    1.875E+06
+    0.335u    1.155E+06
+    0.345u    1.100E+06
+    0.355u    1.060E+06
+    0.365u    8.860E+05
+    0.375u    5.070E+05
+    0.385u    2.260E+05
+    0.395u    1.385E+05
+    0.405u    9.250E+04
+    0.415u    6.765E+04
+    0.425u    5.180E+04
+    0.435u    4.165E+04
+    0.445u    3.380E+04
+    0.455u    2.800E+04
+    0.465u    2.360E+04
+    0.475u    2.000E+04
+    0.485u    1.705E+04
+    0.495u    1.485E+04
+    0.505u    1.280E+04
+    0.515u    1.100E+04
+    0.525u    9.735E+03
+    0.535u    8.865E+03
+    0.545u    7.600E+03
+    0.555u    6.775E+03
+    0.565u    6.195E+03
+    0.575u    5.685E+03
+    0.585u    5.100E+03
+    0.595u    4.585E+03
+    0.605u    4.245E+03
+    0.615u    3.955E+03
+    0.625u    3.685E+03
+    0.635u    3.145E+03
+    0.645u    3.150E+03
+    0.655u    2.895E+03
+    0.665u    2.650E+03
+    0.675u    2.435E+03
+    0.685u    2.255E+03
+    0.695u    2.085E+03
+    0.6983u   1.930E+03
+    0.7u      2.000E+03
+    0.71u     1.860E+03
+    0.72u     1.710E+03
+    0.7277u   1.400E+03
+    0.73u     1.580E+03
+    0.74u     1.460E+03
+    0.75u     1.340E+03
+    0.7621u   1.234E+03
+    0.77u     1.133E+03
+    0.78u     1.039E+03
+    0.79u     9.510E+02
+    0.8u      8.690E+02
+    0.8059u   8.300E+02
+    0.825u    6.880E+02
+    0.83u     6.550E+02
+    0.835u    6.245E+02
+    0.8465u   5.535E+02
+    0.86u     4.830E+02
+    0.87u     4.340E+02
+    0.875u    4.115E+02
+    0.8875u   3.890E+02
```

```
+    0.9u       3.080E+02
+    0.9075u    2.720E+02
+    0.915u     2.555E+02
+    0.925u     2.240E+02
+    0.93u      2.090E+02
+    0.94u      1.820E+02
+    0.95u      1.570E+02
+    0.955u     1.455E+02
+    0.966u     1.240E+02
+    0.975u     1.040E+02
+    0.985u     8.705E+01
+    1.018u     3.990E+01
+    1.082u     6.200E+00
+    1.094u     4.700E+00
+    1.098u     3.500E+00
+    1.101u     3.500E+00
+    1.128u     2.000E+00
+    1.131u     1.5
+    1.144u     0.68
+    1.178u     0.065
+    1.264u     0

.ends silicon_abs


*********************************************
*                                           *
*               JSC.LIB                      *
*                                           *
*********************************************

*   SUBCIRCUIT CALCULATING THE ANALYTICAL SOLAR CELL MODEL FOR SPECTRAL
*   SHORT CIRCUIT CURRENT DENSITY, TOTAL SHORT CIRCUIT CURRENT DENSITY AND
*   QUANTUM EFFICIENCY AND SPECTRAL RESPONSE

*NODES
*(200) REFERENCE
*(201) INPUT,  WAVELENGTH IN MICRON
*(202) INPUT, ABSORPTION COEFFICIENT IN 1/CM
*(203) INPUT, SPECTRAL IRRADIANCE IN W/M2MICRON
*(204) INPUT, REFLECTION COEFFICIENT
*(205) OUTPUT, EMITTER SPECTRAL SHORT CIRCUIT CURRENT DENSITY ( A/CM2MICRON)
*(206) OUTPUT, BASE SPECTRAL SHORT CIRCUIT CURRENT DENSITY(A/CM2MICRON)
*(207) OUTPUT, TOTAL SHORT CIRCUIT CURRENT DENSITY(A/CM2)
*(208) OUTPUT, TOTAL INTERNAL QUANTUM EFFICIENCY(%)
*(209) OUTPUT, TOTAL INTERNAL SPECTRAL RESPONSE(A/W)

*NORMALIZATION FACTOR TO 1000W/M2. 1000/962.5

****** subcircuit
.subckt jsc 200 201 202 203 204 205 206 207 208 209 PARAMS:we=1,lp=1,dp=1,se=1,
+wb=1, ln=1, dn=1, sb=1
.param q=1.6e-19

**emitter component

egeom3 230 200 value={1.6e-19*v(202)*v(203)*(1000/962.5)*v(201)*(1e16/19.8)*
+ lp*(1-V(204))/(v(202)*lp+1)}
egeom0 231 200 value={cosh(we/lp)+se*(lp/dp)*sinh(we/lp)}
egeom1 232 200 value={se*(lp/dp)*cosh(we/lp)+sinh(we/lp)}
egeom2 233 200 value={se*(lp/dp)+v(202)*lp-exp(-v(202)*we)*v(232)}
ejsce 205 200 value={v(230)/(v(202)*lp-1)*(-v(202)*lp*
+ exp(-v(202)*we)+v(233)/v(231))};short circuit
```

```
eqee 234 200 value={v(205)*19.8/(q*v(203)*(1000/962.5)*
+ (1-V(204))*v(201)*1e16)};quantum efficiency

**base component

egeom30 240 200 value={cosh(wb/ln)+sb*ln/dn*sinh(wb/ln)}
egeom31 241 200 value={sb*ln/dn*(cosh(wb/ln)-exp(-v(202)*wb))}
egeom32 242 200 value={sinh(wb/ln)+v(202)*ln*exp(-v(202)*wb)}
egeom33 243 200 value={q*v(202)*v(203)*v(201)*1e16/19.8*1000/962.5*
+ (1-v(204))*ln*exp(-we*v(202))/((ln*v(202))^2-1)}
ejscb 206 200 value={v(243)*(v(202)*ln-(v(241)+v(242))/v(240))};short circuit
eqeb 244 200 value={v(206)*19.8/(q*v(203)*(1000/962.5)*(1-v(204))*v(201)*1e16)}
******* total short circuit current
ejsc 207 200 value={1e6*sdt(v(205)+v(206))}
eqe 208 200 value={(v(234)+v(244))*100}; total quantum efficiency in %
******* quantum efficiency
esr 209 200 value={(v(234)+v(244))*0.808*v(201)}

.ends jsc
```

Annex 3

PSpice Files Used in Chapter 3

Table A3.1   PSpice files used in Chapter 3

File name	Description	Section	includes	Figures
cell_1.lib	Level 1 subcircuit of a solar cell	3.2		
cell_1.cir	Plots the $I(V)$ curve of level 1	3.2	cell_1.lib	3.3
irradiance.cir	Effects of the irradiance	3.3, 3.4, 3.5	cell_1.lib	3.4, 3.5 Tables 3.1, 3.2, 3.3, 3.4
cell_2.lib	Includes two diode series and shunt resistances	3.7		
cell_2.cir	Plots $I(V)$ curves using model level 2	3.7	cell_2.lib	3.8, Table 3.5
example3_3.cir	Effects of series resistance values	3.7	cell_2.lib	3.8
shunt.cir	Effects of parallel resistance	3.10	cell_2.lib	3.9
diode_rec.cir	Effects of the recombination diode	3.11	cell_2.lib	3.10
cell_3.lib	Effects of temperature	3.12		
temp.cir	Effects of temperature	3.12	cell_3.lib	3.11, Table 3.6
cell_4.lib	Built-in diode temperature effects and short circuit temperature coefficient	3.12		
temp_2.cir	Shows the temperature effects	3.12 Example 3.5	cell_4.lib	3.12
cell_5.lib	Space radiation effects	3.13		
space.cir	Space radiation effects	Example 3.6	cell_5.lib	3.13
cell_beh.lib	Behavioural model	3.14		
cell_beh.cir	Behavioural circuit	3.14 Example 3.7	Cell_beh.lib	3.15
cell_pwl.cir	Piecewise linear sources	3.15	Cell_beh.lib	3.16

```
**********************************************
*                                            *
*              EXAMPLE3_3.CIR                 *
*                                            *
**********************************************

*   EFFECT OF THE SERIES RESISTANCE
*   NODES

*   (0) REFERENCE
*   (31) OUTPUT
*   (32) IRRADIANCE (W/M2)

.include cell_2.lib
xcell2 0 31 32 cell_2 params:area=126.6 j0=1e-11 j02=0
+ jsc=0.0343 rs={RS} rsh=1000
.param RS=1
vbias 31 0 dc 0
virrad 32 0 dc 1000
.plot dc i(vbias)
.dc vbias 0 0.6 0.01
.step param RS list 1e-4 1e-3 2e-3 5e-3 1e-2 2e-2 5e-2 1e-1
.probe
.end


**********************************************
*                                            *
*                SHUNT.CIR                    *
*                                            *
**********************************************

*   EFFECT OF THE SHUNT RESISTANCE
*   NODES

*   (0) REFERENCE
*   (31) OUTPUT
*   (32) IRRADIANCE (W/M2)

.include cell_2.lib
xcell2 0 31 32 cell_2 params:area=126.6 j0=1e-11 j02=0
+ jsc=0.0343 rs=1e-6 rsh={RSH}
.param RSH=1
vbias 31 0 dc 0
virrad 32 0 dc 1000
.plot dc i(vbias)
.dc vbias 0 0.6 0.01
.step param RSH list 10000 1000 100 10 1 0.1
.probe
.end


**********************************************
*                                            *
*              DIODE_REC.CIR                  *
*                                            *
**********************************************

*   EFFECT OF THE RECOMBINATION DIODE
*   NODES

*   (0) REFERENCE
*   (31) OUTPUT
*   (32) IRRADIANCE (W/M2)
```

```
.include cell_2.lib
xcell2 0 31 32 cell_2 params:area=126.6 j0=1e-11 j02={J02}
+ jsc=0.0343 rs=1e-6 rsh=1000
.param J02=1
vbias 31 0 dc 0
virrad 32 0 dc 1000
.plot dc i(vbias)
.dc vbias 0 0.6 0.01
.step param J02 list 1e-8 1e-7 1e-6 1e-5 1e-4
.probe
.end


*********************************************
*                                           *
*               CELL_3.LIB                  *
*                                           *
*********************************************


*    MODEL LEVEL 3
*    INCLUDES THE EXTRAPOLATED
*    TO 0K BANDGAP VALUE FOR SILICON FOR THE
*    TEMPERATURE MODEL, EG=1.17 FOR TEMPERATURE ANALYSIS
*    TWO DIODES (DIFFUSION,D1 AND RECOMBINATION,D2)
*    CURRENT SOURCE
*    SERIES RESISTANCE  RS
*    SHUNT RESISTANCE    RSH
*    BUILT-IN SPICE DIODE MODEL D
*    NODES
*    (300)REFERENCE
*    (301)INTERNAL NODE
*    (302)INPUT, IRRADIANCE
*    (303) OUTPUT
.subckt cell_3 300 303 302 params:area=1, j0=1, jsc=1, j02=1, rs=1, rsh=1

girrad 300 301 value={(jsc/1000)*v(302)*area}
d1 301 300 diode
.model diode d(is={j0*area},eg=1.17)
d2 301 300 diode2
.model diode2 d(is={j02*area}, n=2)
rs 301 303 {rs}
rsh 301 300 {rsh}
.ends cell_3


*********************************************
*                                           *
*               CELL_5.LIB                  *
*                                           *
*********************************************


*    MODEL LEVEL 5
*    SPACE RADIATION DEGRADATION MODEL
*    STANDARD AM0 IRRADIANCE 1353 W/m2
*    SINGLE DIODE (DIFFUSION,D1)
*    CURRENT SOURCE
*    SERIES RESISTANCE    RS IS INTERNALLY CALCULATED
*    BUILT-IN SPICE DIODE MODEL D
*    INPUT PARAMETERS: AREA,CELL TEMPERATURE,BOL DATA,FLUENCE
*    DEGRADATON CONSTANTS FOR CURRENT AND VOLTAGE

*    NODES
*    (300)REFERENCE
*    (301)INTERNAL NODE
```

```
*    (302)INPUT, IRRADIANCE
*    (303) OUTPUT

.subckt cell_5 300 303 302 params:area=1, temp=1, jscbol=1, pmaxbol=1, vocbol=1, f=1
+ ki=1, fi=1, kv=1, fv=1
girrad 300 301 value={(jscbol-ki*log10(1+(f/fi)))/1353*v(302)*(area)}
d1 301 300 diode
.model diode d(is={(jscbol-ki*log10(1+(f/fi)))*area*exp(-(vocbol-kv*log10
+(1+f/fv))
+ /(8.66e-5*(temp+273)))})
.func uvet() {8.66e-5*(temp+273)}
.func vocnorm() {vocbol/uvet}
.func rs() {vocbol/(jscbol*area)- pmaxbol*(1+vocnorm)/(jscbol**2*area*(vocnorm-
+ log((vocnorm)+0.72)))}
rs 301 303 {rs()}
.ends cell_5


*******************************************
*                                         *
*               SPACE.CIR                 *
*                                         *
*******************************************


*    EFFECT OF THE SPACE RADIATION
*    NODES

*    (0) REFERENCE
*    (31) OUTPUT
*    (32) IRRADIANCE (W/M2)

.include cell_5.lib
xcell5 0 31 32 cell_5 params:area=8 temp=27 jscbol=0.0436
+ pmaxbol=0.0208 vocbol=0.608 f={F}
+ ki=5.26E-3 fi=3.02e13 kv=0.042 fv=2.99e12
vbias 31 0 dc 0
.param F=1

virrad 32 0 dc 1353; One sun AM0

.plot dc i(vbias)
.probe
.dc vbias 0 0.6 0.01
.step param F list 1e10 1e11 1e12 1e13 1e14 1e15 1e16
.end


*******************************************
*                                         *
*               CELL_BEH.LIB              *
*                                         *
*******************************************


*    BEHAVIOURAL MODEL OF A SOLAR CELL

*    INPUT PARAMETERS:AREA,AM1.5 JSCR,AM1.5 VOCR,AM1.5 PMAXR
*    AM1.5 VMR,AM1.5 IMR, CURRENT TEMP COEFF.,VOLTAGE TEMP.COEFF,
*    NOCT, REFERENCE TEMPERATURE

*    NODES
*    (300)    REFERENCE
*    (301) INTERNAL NODE
*    (302) INPUT, IRRADIANCE
*    (303)    INPUT, AMBIENT TEMPERATURE
*    (304) OUTPUT
```

```
*    (305)    OUTPUT, (VOLTAGE) VALUE=SHORT CIRCUIT CURRENT(A) AT
*    IRRADIANCE AND TEMPERATURE
*    (306)    OUTPUT, OPEN CIRCUIT VOLTAGE AT IRRADIANCE ANDTEMPERATURE
*    (307) OUTPUT, (VOLTAGE) VALUE=CELL OPERATING TEMPERATURE(°C)
*    (308) OUTPUT, MPP CURRENT
*    (309) OUTPUT, MPP VOLTAGE
.subckt cell_beh 300 302 303 304 305 306 307 308 309 params: area=1,
+ jscr=1, coef_jsc=1, vocr=1, coef_voc=1,pmaxr=1,
+ noct=1,jmr=1 , vmr=1, tr=1
girrad 300 301 value={v(302)/1000*(jscr*area+coef_jsc*area*(v(307)-25))}
eisc 305 300 value={v(302)/1000*(jscr*area+coef_jsc*area*(v(307)-25))}
evoc 306 300 value={if (v(305)>1e-11, vocr+coef_voc*(v(307)-25)+8.66e-5*
+ (v(307)+273)*log(v(305)/(area*jscr)),0)}
etcell 307 300 value={v(303)+(noct-20)/800*v(302)}

gidiode 301 300 value={area*v(305)/(exp(v(306)/(8.66e-5*(v(307)+273)))-1)*
+ (exp(v(301)/(8.66e-5*(v(307)+273)))-1)}
rs 301 304 {vocr/(jscr*area)-pmaxr/(jscr**2*area*(vocr/0.0258-log
+ ((vocr/0.0258)+0.72))/(1+vocr/0.0258))}
.func frs() {vocr/(jscr*area)-pmaxr/(jscr**2*area*(vocr/0.0258-log
+ ((vocr/0.0258)+0.72))/(1+vocr/0.0258))}

gim 300 308 value={jmr*area*v(302)/1000+coef_jsc*area*(v(303)-25)}
rim 308 300 1
evm 309 300 value={8.66e-5*(v(307)+273)*log(1+(v(305)-v(308))/v(305)*(exp(v(306)/
+ (8.66e-5*(v(307)+273)))-1))-v(308)*frs}
.ends cell_beh


*********************************************
*                                           *
*              CELL_BEH.CIR                 *
*                                           *
*********************************************

*    USING BEHAVIOURAL MODEL OF THE SOLAR CELL
*    NODES
*    (0)       REFERENCE
*    (32) INPUT, IRRADIANCE
*    (33)      INPUT, AMBIENT TEMPERATURE
*    (34) OUTPUT
*    (35)      OUTPUT, (VOLTAGE) VALUE=SHORT CIRCUIT CURRENT(A) AT
*    IRRADIANCE AND TEMPERATURE
*    (36)      OUTPUT, OPEN CIRCUIT VOLTAGE AT IRRADIANCE ANDTEMPERATURE
*    (37) OUTPUT, (VOLTAGE) VALUE=CELL OPERATING TEMPERATURE(°C)
*    (38) OUTPUT, MPP CURRENT
*    (39) OUTPUT, MPP VOLTAGE
.include cell_beh.lib
xcellbeh 0 32 33 34 35 36 37 38 39 cell_beh params:area=1, tr=25, jscr=0.0375,
+ pmaxr=0.0184, vocr=0.669, jmr=35.52e-3, vmr=0.518, noct=47,
+ coef_jsc=12.5e-6, coef_voc=-3.1e-3

virrad 32 0 dc 500
vtemp 33 0 dc 25
vbias 34 0 dc 0

rim 38 0 1

.option stepgmin
.plot dc i(vbias)
.probe
.dc vbias 0 0.7 0.001

.end
```

Annex 4

PSpice Files Used in Chapter 4

Table A4.1 PSpice files used in Chapter 4

File name	Description	Section	Includes	Figures
series.cir	$I(V)$ curve of two solar cells in series	4.2.1	cell_2.lib	4.2, 4.3, 4.4
bypass.cir	$I(V)$ curve of 12 solar cells in series with bypass diode	4.2.3 Example 4.1	cell_2.lib	4.6, 4.7
Shunt2.cir	$I(V)$ curve of two solar cells in parallel	4.3	cell_2.lib	4.8
example4_2.cir	Effects of shadow in parallel solar cells (case A)	Example 4.2	cell_2.lib	4.9
example4_2b.cir	Effects of shadow in parallel solar cells (case B)	Example 4.2	cell_2.lib	4.9
module_1.lib	PV module subcircuit	4.4		4.10
module_1.cir	PV module circuit to plot $I(V)$ curve	4.4, 4.7	module_1.lib	4.11, 4.15
module_conv.cir	PV module conversion to arbitrary values of irradiance and temperature	4.5.1 Example 4.3	module_1.lib	
module_beh.lib	Behavioural PV module model	4.6		4.13
module_beh.cir	Computes the $I(V)$ plots using the behavioural model	4.6	module_beh.lib	4.14
6x3_array.cir	PV series parallel array with the same irradiance	4.8	cell_2.lib	
shadow.cir	Same as 6x3_array.cir with some cells shadowed	Example 4.5		4.16, 4.17
generator_beh.lib	Model for a generator of parallel–series association of PV modules	4.9		4.19
generator_beh.cir	Circuit file to simulate the $I(V)$ curves of a generator	4.9	Generator_beh.lib	

```
*******************************************
*                                         *
*              BYPASS.CIR                 *
*                                         *
*******************************************

*    INCLUDES 12 SOLAR CELLS IN SERIES AND A BYPASS DIODE
*    ACROSS SOLAR CELL NUMBER 6
.include cell_2.lib


xcell1 45 43 42 cell_2 params:area=126.6 j0=1e-11 j02=1E-9
+ jsc=0.0343 rs=1e-3 rsh=100000
xcell2 47 45 44 cell_2 params:area=126.6 j0=1e-11 j02=1E-9
+ jsc=0.0343 rs=1e-3 rsh=100000


xcell3 49 47 46 cell_2 params:area=126.6 j0=1e-11 j02=1E-9
+ jsc=0.0343 rs=1e-3 rsh=100000
xcell4 51 49 48 cell_2 params:area=126.6 j0=1e-11 j02=1E-9
+ jsc=0.0343 rs=1e-3 rsh=100000
xcell5 53 51 50 cell_2 params:area=126.6 j0=1e-11 j02=1E-9
+ jsc=0.0343 rs=1e-3 rsh=100000
xcell6 55 53 52 cell_2 params:area=126.6 j0=1e-11 j02=1E-9
+ jsc=0.0343 rs=1e-3 rsh=100000
xcell7 57 55 54 cell_2 params:area=126.6 j0=1e-11 j02=1E-9
+ jsc=0.0343 rs=1e-3 rsh=100000


xcell8 59 57 56 cell_2 params:area=126.6 j0=1e-11 j02=1E-9
+ jsc=0.0343 rs=1e-3 rsh=100000
xcell9 61 59 58 cell_2 params:area=126.6 j0=1e-11 j02=1E-9
+ jsc=0.0343 rs=1e-3 rsh=100000
xcell10 63 61 60 cell_2 params:area=126.6 j0=1e-11 j02=1E-9
+ jsc=0.0343 rs=1e-3 rsh=100000
xcell11 65 63 62 cell_2 params:area=126.6 j0=1e-11 j02=1E-9
+ jsc=0.0343 rs=1e-3 rsh=100000
xcell12 0 65 64 cell_2 params:area=126.6 j0=1e-11 j02=1E-9
+ jsc=0.0343 rs=1e-3 rsh=100000


vbias 43 0 dc 0
virrad1 42 45 dc 1000
virrad2 44 47 dc 1000
virrad3 46 49 dc 1000
virrad4 48 51 dc 1000
virrad5 50 53 dc 1000
virrad6 52 55 dc 0
virrad7 54 57 dc 1000
virrad8 56 59 dc 1000
virrad9 58 61 dc 1000
virrad10 60 63 dc 1000
virrad11 62 65 dc 1000
virrad12 64 0 dc 1000

*bypass diode
dbypass 55 53 diode
.model diode d

.dc vbias 0 8 0.01
.probe
.plot dc i(vbias)
.end
```

```
*******************************************
*                                         *
*              EXAMPLE4_2.CIR             *
*                                         *
*******************************************

*    TWO SOLAR CELLS IN PARALLEL SAME IRRADIANCE
*    NODES
*    (0) REFERENCE
*    (43) OUTPUT

.include cell_2.lib
xcell1 0 43 42 cell_2 params:area=8 j0=1e-11 j02=0
+ jsc=0.0343 rs=0.5 rsh=100
xcell2 0 43 44 cell_2 params:area=8 j0=1e-11 j02=0
+ jsc=0.0343 rs=0.5 rsh=100

vbias 43 0 dc 0
virrad1 42 0 dc 500
virrad2 44 0 dc 500
.plot dc i(vbias)

.dc vbias 0 0.6 0.01
.probe
.end

*******************************************
*                                         *
*              EXAMPLE4_2B.CIR            *
*                                         *
*******************************************

*    TWO SOLAR CELLS IN PARALLEL ONE SHADOWED
*    NODES

*    (0) REFERENCE
*    (43) OUTPUT

.include cell_2.lib
xcell1 0 43 42 cell_2 params:area=8 j0=1e-11 j02=0
+ jsc=0.0343 rs=0.5 rsh=100
xcell2 0 43 44 cell_2 params:area=8 j0=1e-11 j02=0
+ jsc=0.0343 rs=0.5 rsh=100

vbias 43 0 dc 0
virrad1 42 0 dc 1000
virrad2 44 0 dc 0
.plot dc i(vbias)

.dc vbias 0 0.6 0.01
.probe
.end

*******************************************
*                                         *
*              MODULE_CONV.CIR            *
*                                         *
*******************************************

*IMPLENTS THE IRRADIANCE-TEMPERATURE CONVERSION METHOD
* BASED ON Dv
*    NODES
*    (0)REFERENCE
```

```
*     (42)INPUT, IRRADIANCE
*     (43) BIAS
*     (44) CIRCUIT OUTPUT


.include module_1.lib
xmodule 0 43 42 module_1 params:ta=25,tr=25, iscmr=5, pmaxmr=85, vocmr=22.3,
+ ns=36, np=1, nd=1
vbias 43 0 dc 0
virrad 42 0 dc 800

*CONVERSION EQUATIONS
.param vocmr=22.3
.param tr=25
.param ta=25
edv 43 44 value={vocmr*(0.06*log(1000/v(42))+0.004*(ta-tr)+0.12e-3*v(42))}
.dc vbias 0 23 0.1
.probe
.end
*********************************************
*                                           *
*               MODULE_BEH.LIB              *
*                                           *
*********************************************

*BEHAVIOURAL MODEL OF A PV MODULE
*    INPUT PARAMETERS:AREA,AM1.5 JSCMR,AM1.5 VOCMR,AM1.5 PMAXMR
*    AM1.5 VMMR,AM1.5 IMMR, CURRENT TEMP COEFF.,VOLTAGE TEMP.COEFF,
*    NOCT, REFERENCE TEMPERATURE


*    NODES
*    (400)     REFERENCE
*    (401) INTERNAL NODE
*    (402) INPUT, IRRADIANCE
*    (403)     INPUT, AMBIENT TEMPERATURE
*    (404) OUTPUT
*    (405)     OUTPUT, (VOLTAGE) VALUE=SHORT CIRCUIT CURRENT(A) AT
*    IRRADIANCE AND TEMPERATURE
*    (406)     OUTPUT, OPEN CIRCUIT VOLTAGE AT IRRADIANCE ANDTEMPERATURE
*    (407) OUTPUT, (VOLTAGE) VALUE=CELL OPERATING TEMPERATURE(°C)
*    (408) OUTPUT, MPP CURRENT
*    (409) OUTPUT, MPP VOLTAGE


.subckt module_beh 400 402 403 404 405 406 407 408 409 params:
+ iscmr=1, coef_iscm=1, vocmr=1, coef_vocm=1,pmaxmr=1,
+ noct=1,immr=1 , vmmr=1, tr=1, ns=1, np=1

girrad 400 401 value={v(402)/1000*(iscmr+coef_iscm*(v(407)-25))}

eisc 405 400 value={v(402)/1000*(iscmr+coef_iscm*(v(407)-25))}
evoc 406 400 value={if (v(405)>1e-11, vocmr+coef_vocm*(v(407)-25)+8.66e-5*
+ (v(407)+273)*log(v(405)/(iscmr)),0)}

etcell 407 400 value={v(403)+(noct-20)/800*v(402)}

gidiode 401 400 value={v(405)/(exp(v(406)/(ns*8.66e-5*(v(407)+273)))-1)*
+ (exp(v(401)/ (ns*8.66e-5*(v(407)+273)))-1)}
rsm 401 404 {vocmr/(iscmr)-pmaxmr/(iscmr**2*(vocmr/(ns*0.0258)-log
+ ((vocmr/(ns*0.0258))+0.72)/(1+vocmr/(ns*0.0258)))}
.func frsm() {vocmr/(iscmr)-pmaxmr/(iscmr**2*(vocmr/(ns*0.0258)-log
+ ((vocmr/(ns*0.0258))+0.72))/(1+vocmr/(ns*0.0258)))}
```

```
gim 400 408 value={immr*v(402)/1000+coef_iscm*(v(403)-25)}
r imm 408 400 1
evmm 409 400 value={if (v(402)>0.001, ns*8.66e-5*(v(407)+273)*log(1+(v(405)-
v(408))/v(405)*(exp(v(406)/
+ (ns*8.66e-5*(v(407)+273)))-1))-v(408)*frsm,0)}
.ends module_beh


**********************************************
*                                            *
*              6X3_ARRAY.CIR                  *
*                                            *
**********************************************

*    INCLUDES 18 SOLAR CELLS ( SERIES ASSOCIATION OF 6 SETS OF 3 PARALLEL EACH)
*    AND ABYPASS DIODES
*    NODES

*    (0) REFERENCE
*    (41) OUTPUT
.include cell_2.lib

xcell1 43 41 42 cell_2 params:area=8 j0=1e-11 j02=0
+ jsc=0.0343 rs=0.1 rsh=1000
xcell2 43 41 44 cell_2 params:area=8 j0=1e-11 j02=0
+ jsc=0.0343 rs=0.1 rsh=1000

xcell3 43 41 46 cell_2 params:area=8 j0=1e-11 j02=0
+ jsc=0.0343 rs=0.1 rsh=1000
xcell4 45 43 48 cell_2 params:area=8 j0=1e-11 j02=0
+ jsc=0.0343 rs=0.1 rsh=1000
xcell5 45 43 50 cell_2 params:area=8 j0=1e-11 j02=0
+ jsc=0.0343 rs=0.1 rsh=1000
xcell6 45 43 52 cell_2 params:area=8 j0=1e-11 j02=0
+ jsc=0.0343 rs=0.1 rsh=1000
xcell7 47 45 54 cell_2 params:area=8 j0=1e-11 j02=0
+ jsc=0.0343 rs=0.1 rsh=1000

xcell8 47 45 56 cell_2 params:area=8 j0=1e-11 j02=0
+ jsc=0.0343 rs=0.1 rsh=1000
xcell9 47 45 58 cell_2 params:area=8 j0=1e-11 j02=0
+ jsc=0.0343 rs=0.1 rsh=1000
xcell10 49 47 60 cell_2 params:area=8 j0=1e-11 j02=0
+ jsc=0.0343 rs=0.1 rsh=1000
xcell11 49 47 62 cell_2 params:area=8 j0=1e-11 j02=0
+ jsc=0.0343 rs=0.1 rsh=1000
xcell12 49 47 64 cell_2 params:area=8 j0=1e-11 j02=0
+ jsc=0.0343 rs=0.1 rsh=1000
xcell13 51 49 66 cell_2 params:area=8 j0=1e-11 j02=0
+ jsc=0.0343 rs=0.1 rsh=1000
xcell14 51 49 68 cell_2 params:area=8 j0=1e-11 j02=0
+ jsc=0.0343 rs=0.1 rsh=1000
xcell15 51 49 70 cell_2 params:area=8 j0=1e-11 j02=0
+ jsc=0.0343 rs=0.1 rsh=1000
xcell16 0 51 72 cell_2 params:area=8 j0=1e-11 j02=0
+ jsc=0.0343 rs=0.1 rsh=1000
xcell17 0 51 74 cell_2 params:area=8 j0=1e-11 j02=0
+ jsc=0.0343 rs=0.1 rsh=1000
xcell18 0 51 76 cell_2 params:area=8 j0=1e-11 j02=0
+ jsc=0.0343 rs=0.1 rsh=1000

vbias 41 0 dc 0
virrad1 42 43 dc 1000
```

```
virrad2 44 43 dc 1000
virrad3 46 43 dc 1000
virrad4 48 45 dc 1000
virrad5 50 45 dc 1000
virrad6 52 45 dc 1000
virrad7 54 47 dc 1000
virrad8 56 47 dc 1000
virrad9 58 47 dc 1000
virrad10 60 49 dc 1000
virrad11 62 49 dc 1000
virrad12 64 49 dc 1000
virrad13 66 49 dc 1000
virrad14 68 49 dc 1000
virrad15 70 49 dc 1000
virrad16 72 49 dc 1000
virrad17 74 49 dc 1000
virrad18 76 49 dc 1000

*bypass diodes
d1 43 41 diode
d2 45 43 diode
d3 47 45 diode
d4 49 47 diode
d5 51 49 diode
d6 0 51 diode
.model diode d(is=1e-6 , n=1)


.dc vbias -1 4 0.01
.probe
.plot dc i(vbias)
.end


*********************************************
*                                           *
*              SHADOW.CIR                    *
*                                           *
*********************************************


*    INCLUDES 18 SOLAR CELLS ( SERIES ASSOCIATION OF 6 SETS OF 3 PARALLEL EACH)
*    AND ABYPASS DIODES
*    NODES

*    (0) REFERENCE
*    (41) OUTPUT
*    SOLAR CELLS NUMBER 1,2,3,5,6,9,17 AND 18 SHADOWED
.include cell_2.lib

xcell1 43 41 42 cell_2 params:area=8 j0=1e-11 j02=0
+ jsc=0.0343 rs=0.1 rsh=1000
xcell2 43 41 44 cell_2 params:area=8 j0=1e-11 j02=0
+ jsc=0.0343 rs=0.1 rsh=1000

xcell3 43 41 46 cell_2 params:area=8 j0=1e-11 j02=0
+ jsc=0.0343 rs=0.1 rsh=1000
xcell4 45 43 48 cell_2 params:area=8 j0=1e-11 j02=0
+ jsc=0.0343 rs=0.1 rsh=1000
xcell5 45 43 50 cell_2 params:area=8 j0=1e-11 j02=0
+ jsc=0.0343 rs=0.1 rsh=1000
xcell6 45 43 52 cell_2 params:area=8 j0=1e-11 j02=0
+ jsc=0.0343 rs=0.1 rsh=1000
xcell7 47 45 54 cell_2 params:area=8 j0=1e-11 j02=0
+ jsc=0.0343 rs=0.1 rsh=1000
```

```
xcell8 47 45 56 cell_2 params:area=8 j0=1e-11 j02=0
+ jsc=0.0343 rs=0.1 rsh=1000
xcell9 47 45 58 cell_2 params:area=8 j0=1e-11 j02=0
+ jsc=0.0343 rs=0.1 rsh=1000
xcell10 49 47 60 cell_2 params:area=8 j0=1e-11 j02=0
+ jsc=0.0343 rs=0.1 rsh=1000
xcell11 49 47 62 cell_2 params:area=8 j0=1e-11 j02=0
+ jsc=0.0343 rs=0.1 rsh=1000
xcell12 49 47 64 cell_2 params:area=8 j0=1e-11 j02=0
+ jsc=0.0343 rs=0.1 rsh=1000
xcell13 51 49 66 cell_2 params:area=8 j0=1e-11 j02=0
+ jsc=0.0343 rs=0.1 rsh=1000
xcell14 51 49 68 cell_2 params:area=8 j0=1e-11 j02=0
+ jsc=0.0343 rs=0.1 rsh=1000
xcell15 51 49 70 cell_2 params:area=8 j0=1e-11 j02=0
+ jsc=0.0343 rs=0.1 rsh=1000
xcell16 0 51 72 cell_2 params:area=8 j0=1e-11 j02=0
+ jsc=0.0343 rs=0.1 rsh=1000
xcell17 0 51 74 cell_2 params:area=8 j0=1e-11 j02=0
+ jsc=0.0343 rs=0.1 rsh=1000
xcell18 0 51 76 cell_2 params:area=8 j0=1e-11 j02=0
+ jsc=0.0343 rs=0.1 rsh=1000


vbias 41 0 dc 0
virrad1 42 43 dc 0
virrad2 44 43 dc 0
virrad3 46 43 dc 0
virrad4 48 45 dc 1000
virrad5 50 45 dc 0
virrad6 52 45 dc 0
virrad7 54 47 dc 1000
virrad8 56 47 dc 1000
virrad9 58 47 dc 0
virrad10 60 49 dc 1000
virrad11 62 49 dc 1000
virrad12 64 49 dc 1000
virrad13 66 49 dc 1000
virrad14 68 49 dc 1000
virrad15 70 49 dc 1000
virrad16 72 49 dc 1000
virrad17 74 49 dc 0
virrad18 76 49 dc 0


*bypass diodes
d1 43 41 diode
d2 45 43 diode
d3 47 45 diode
d4 49 47 diode
d5 51 49 diode
d6 0 51 diode
.model diode d(is=1e-6 , n=1)


.dc vbias -1 4 0.01
.probe
.plot dc i(vbias)
.end
```

```
*********************************************
*                                           *
*              GENERATOR_BEH.LIB            *
*                                           *
*********************************************

*    BEHAVIOURAL MODEL OF A PV GENERATOR
*    INPUT PARAMETERS: ,AM1.5 JSCMR,AM1.5 VOCMR,AM1.5 PMAXMR
*    AM1.5 VMMR,AM1.5 IMMR, CURRENT TEMP COEFF.,VOLTAGE TEMP.COEFF,
*    NOCT, REFERENCE TEMPERATURE


*    NODES
*    (400)    REFERENCE
*    (401) INTERNAL NODE
*    (402) INPUT, IRRADIANCE
*    (403)    INPUT, AMBIENT TEMPERATURE
*    (404) OUTPUT
*    (405)    OUTPUT, (VOLTAGE) VALUE=SHORT CIRCUIT CURRENT(A) AT
*    IRRADIANCE AND TEMPERATURE
*    (406)    OUTPUT, OPEN CIRCUIT VOLTAGE AT IRRADIANCE ANDTEMPERATURE
*    (407) OUTPUT, (VOLTAGE) VALUE=CELL OPERATING TEMPERATURE(°C)
*    (408) OUTPUT, MPP CURRENT
*    (409) OUTPUT, MPP VOLTAGE

.subckt generator_beh 400 402 403 404 405 406 407 408 409 params:
+ iscmr=1, coef_iscm=1, vocmr=1, coef_vocm=1,pmaxmr=1,
+ noct=1,immr=1 , vmmr=1, tr=1, ns=1, nsg=1 npg=1

ev402 410 400 value={if (v(402)>0.1, v(402),0.1)}
girrad 400 401 value={v(410)/1000*(npg*iscmr+npg*coef_iscm*(v(407)-25))}

eiscm 405 400 value={v(410)/1000*(iscmr+coef_iscm*(v(407)-25))}
evocm 406 400 value={if (v(405)>1e-11, (vocmr+coef_vocm*(v(407)-25)+8.66e-5*
+ (v(407)+273)*log(v(405)/(iscmr))),0)}

etcell 407 400 value={v(403)+(noct-20)/800*v(410)}

gidiode 401 400 value={npg*v(405)/(exp(v(406)/(ns*8.66e-5*(v(407)+273)))-1)*
+ (exp(v(401)/ (ns*nsg*8.66e-5*(v(407)+273)))-1)}
rsg 401 404 {nsg/npg*((vocmr/(iscmr)-pmaxmr/(iscmr**2*(vocmr/(ns*0.0258)-log
+ ((vocmr/(ns*0.0258))+0.72)/(1+vocmr/(ns*0.0258))))))}
.func frsg() {nsg/npg*((vocmr/(iscmr)-pmaxmr/(iscmr**2*(vocmr/(ns*0.0258)-log
+ ((vocmr/(ns*0.0258))+0.72))/(1+vocmr/(ns*0.0258)))))}

gimg 400 408 value={npg*(immr*v(410)/1000+coef_iscm*(v(403)-25))}
r img 408 400 1
evmg 409 400 value={if (v(410)>0.1, nsg*(ns*8.66e-5*(v(407)+273)*log(1+(v(405)-
v(408)/npg)/v(405)*(exp(v(406)/
+ (ns*8.66e-5*(v(407)+273)))-1))-v(408)*frsg/nsg),0)}
.ends generator_beh

*********************************************
*                                           *
*              GENERATOR_BEH.CIR            *
*                                           *
*********************************************

*    BEHAVIOURAL MODEL OF A PV GENERATOR
*    INPUT PARAMETERS: ,AM1.5 JSCMR,AM1.5 VOCMR,AM1.5 PMAXMR
*    AM1.5 VMMR,AM1.5 IMMR, CURRENT TEMP COEFF.,VOLTAGE TEMP.COEFF,
*    NOCT, REFERENCE TEMPERATURE
```

```
*     NODES
*     (0)        REFERENCE
*     (41) INTERNAL NODE
*     (42) INPUT, IRRADIANCE
*     (43)       INPUT, AMBIENT TEMPERATURE
*     (44) OUTPUT
*     (45)       OUTPUT, (VOLTAGE) VALUE=SHORT CIRCUIT CURRENT(A) AT
*     IRRADIANCE AND TEMPERATURE
*     (46)       OUTPUT, OPEN CIRCUIT VOLTAGE AT IRRADIANCE ANDTEMPERATURE
*     (47) OUTPUT, (VOLTAGE) VALUE=CELL OPERATING TEMPERATURE(°C)
*     (48) OUTPUT, MPP CURRENT
*     (49) OUTPUT, MPP VOLTAGE


.include generator_beh.lib
xgenerator 0 42 43 44 45 46 47 48 49 generator_beh params:
+ iscmr=0.37, coef_iscm=0.13e-3, vocmr=21, coef_vocm=-0.1,pmaxmr=5,
+ noct=47,immr=0.32 , vmmr=15.6, tr=25, ns=33, nsg=2 npg=5
.param irrad=1
vbias 44 0 dc 0
virrad 42 0 dc {irrad}
vtemp 43 0 dc 0
.dc vbias 0 50 1
.step param irrad list 10 100 400 1000
.probe
.end
```

Annex 5

PSpice Files Used in Chapter 5

Table A5.1 PSpice files used in Chapter 5

File name	Description	Section	Includes	Figures
Water_pump_t transient.cir	PSpice simulation of a PV array-series DC motor-centrifugal pump system	5.2.4	Pump.lib	5.3, 5.4, 5.5
Bat.cir	Lead–acid battery model	5.3.2		5.11
Example 5.3.cir	Evolution of the battery voltage and the state of charge considering a battery driven by a sinusoidal current source of amplitude 20 A and a 1 kHz frequency	Example 5.3	Bat.cir	5.12, 5.13, 5.14
7TSE.cir	Modified subcircuit netlist for this commercial battery model 7TSE 70 by ATERSA	5.3.3 Example 5.4		
Example 5.5.cir	Standalone PV system simulation Load and PV generator: current sources	5.3.4 Example 5.5	Bat.cir exportmed3V.stl exportmedI.stl	5.15, 5.16, 5.17, 5.18, 5.19
Example 5.6.cir	Standalone PV system simulation Load: resistive load	5.3.4 Example 5.6	Module_1.lib Bat.cir Irrad2d.stl	5.6, 5.20, 5.21, 5.22, 5.23, 5.24
Batstd.cir	Simplified PSpice battery model	5.3.5		5.25

```
*********************************************
*                                           *
*              Example 5.6 Netlist          *
*                                           *
*********************************************

* NODES
* (0) Reference
* (1) Irradiance profile input
* (3) PV modules output & Battery ( positive) & Load
* (7) Battery SOC ( %)
.inc module_1.lib
xmodule_1      0    3    1    module_1    params:    iscmr=5,
+ vocmr=22.3,ns=36,np=1,nd=1,pmaxmr=85, ta=20,tr=1
xbat1 3 0 7 bat params: ns=6, SOCm=576, k=.7, D=1e-3, SOC1=0.85
.inc bat.cir
R1 3 0 4
.inc irrad2d.stl
vmesur 1 0 stimulus Vgg
.tran 1s 120000s 0.1s
.probe
.end
```

Annex 6

PSpice Files Used in Chapter 6

Table A6.1   PSpice files used in Chapter 6

File name	Description	Section	Includes	Figures
Example 6.1.cir	Standalone PV system with battery shunt regulation	6.3.1 Example 6.1	module_1.lib bat.cir opamp.lib irrad.stl	6.3, 6.4, 6.5, 6.6, 6.7, 6.8
Example 6.2.cir	Standalone PV system with series charge regulation	6.3.2 Example 6.2 Annex 6	module_1.lib batstd.cir opamp.lib irrad.stl	6.12, 6.13, 6.14
dcdcf.cir	DC/DC converter model	6.4.3		6.19, 6.20, 6.21
ppm.cir	Standalone including a DC/DC converter (PV generator + DC/DC+ load)	6.4.3 Example 6.3 Annex 6	generator_beh.lib dcdcf.cir irrad.stl	6.22, 6.23, 6.24, 6.25, 6.26, 6.27
Inverter1.cir	Inverter topological PSpice model	6.5.1 Example 6.4 Annex 6		6.33, 6.34, 6.35, 6.36, 6.37
Inverter 2.cir	Inverter behavioural PSpice model	6.5.2		6.38, 6.39, 6.40,
Example 6.5.cir	Inverter simulation	6.5.2 Example 6.5	Generator_beh.lib Irradprueba2.stl Inverter2.cir	6.41, 6.42, 6.43, 6.44, 6.45, 6.46
Example 6.6.cir	Inverter model for direct battery connection	6.5.3 Example 6.6		
Moduleppt.cir	Simplified PV generator model	6.5.3 Example 6.7		
Example 6.7.cir	Standalone PV system with AC output, including PV generator, DC/DC converter, battery, inverter and load	6.5.3 Example 6.7	Moduleppt.cir April.stl Dcdcf.cir Batstdif.cir Inverter3.cir	6.47, 6.48, 6.49, 6.50, 6.51, 6.52

```
*********************************************
*                                           *
*              Example 6.2.cir               *
*                                           *
*********************************************

* module, charge regulator and battery connection
* NODES
* (0) Refernece
* (1) Irradiance profile input
* (3) PV modules output
* (4) Battery ( positive terminal )
* (7) Battery SOC (%)
* (8) (12) A.O. inputs
* (13) (14) A.O. polarization
* (10) A.O. output

xmodule      0      3      1     module_1     params:     iscmr=5,     tr=25,
+vocmr=22.3,ns=36,np=1,nd=1,pmaxmr=85,ta=25
.inc module_1.lib
.inc irrad.stl
vmesur 1 0 stimulus Virrad
xbat1 4 0 7 batstd params: ns=6, SOCm=1000, k=.8, D=1e-5, SOC1=.45
.inc batstd.cir
Rbat 4 0 800
R1 4 8 10000
R2 8 0 10000
x741 12 8 13 14 10 ad741
.inc opamp.lib
Vcc 13 0 dc 15
Vee 14 0 dc -15
vref 16 0 dc 6.8
R6 12 16 4100
R7 12 100 220000
Wch 3 4 vcurrent sw1mod
.model sw1mod iswitch (ioff=-10e-5, ion=10e-6, Roff=1.0e+8, Ron=0.01)
vcurrent 10 100 dc 0
.tran 1s 140000s
.probe
.end


*********************************************
*                                           *
*              ppm.cir                       *
*                                           *
*********************************************

* NODES
* (0) Reference
* (2) Temperature profile input
* (1) Irradiance profile input
* (3) PV modules output & dc-dc converter input
* (40) dc-dc converter output & Load connection

.inc generator_beh.lib
xgen 0 1 2 3 405 406 407 20 30 generator_beh params:
+ iscmr=5.2, coef_iscm=0.13e-3, vocmr=21.2, coef_vocm=-0.1,pmaxmr=85,
+ noct=47,immr=4.9, vmmr=17.3, tr=25, ns=36, nsg=1, npg=1
vtemp 2 0 dc 25
.inc irrad.stl
vmesur 1 0 stimulus Virrad
.inc dcdcf.cir
```

```
xconv 0 3 20 30 40 dcdcf params: n =1, vo1=12
Rload 40 0 5
.tran 1s 20000s
.probe
.end


*******************************************
*                                         *
*            inverter1.cir                *
*                                         *
*******************************************


vindc   2    0    24v
q1      2    9       6  q2n3055
d1      6    2  diode
r1      9    7  4
vb1     7    6  pulse 0 14  0 1ps 1ps 9.9ms 20ms

q2      2    1    3    q2n3055
d2      3    2  diode
r2      12   1  4

vb2 12  3 pulse 0  14  10ms 1ps 1ps 9.9ms 20ms

q3   6     11   0   q2n3055
r3   11    13  4
d3   0     6   diode
vb3  13    0   pulse 0 14 10ms 1ps 1ps 9.9ms 20ms

q4   3     4    0  q2n3055
d4   0     3   diode
r4   8     4   4
vb4  8     0    pulse 0 14  0 1ps 1ps 9.9ms 20ms

r    6   5   4
c    10  5   199uf
l    3   10  50.92mh
.model diode d( tt=1e-6)
.model q2n3055 npn
.tran  1ms  1
.four  50hz  12 i(r)  v(6,3)
.options abstol=0.5ma reltol=0.01 vntol=0.001v
.probe
.end


*******************************************
*                                         *
*            Batstdif.cir                 *
*                                         *
*******************************************


*    NODES
*    (1)       SOC
*    (2)       Battery V+
*    (3)       Battery V-

.subckt batstdif 3 2 1 PARAMS: ns=1, SOCm=1, k=1, D=1, SOC1=1
evch   4 2 Value={(2+(0.16*v(1)))*ns}
evdch  5 2 value={(1.926+(0.248*v(1)))*ns}
rserie 8 88 {rs}
```

```
.fun rs() {(0.7+(0.1/(abs(SOC1-0.2)))*ns*162/SOCm}
ebat 88 2  value={IF (i(vcurrent)>0, v(4), v(5))}
vcurrent 3 8 dc 0
eqt 13 2 value={SOC1+1e6*(sdt(v(9))/SOCm)}
eqt4 1 2 value={limit (v(13), 0 , 1)}
evcalculsoc 9 2 value={(k*v(10)*i(vcurrent)/3600)-(D*SOm*v(13)/3600)}
ecoch 10 2 value={IF (i(vcurrent)>0, v(4), v(5))}
.ends batstdif
```

Annex 7

PSpice Files Used in Chapter 7

Table A7.1 PSpice files used in Chapter 7

File name	Description	Section	Includes	Figures
irrad_jan_16.lib	Hourly radiation data for one day	7.2 Ex 7.1		
irrad_jan_16.cir	Plots the hourly radiation values	7.2	irrad_jan_16.lib	7.2
psh.cir	Illustrates the PSH concept		irrad_jan_16.lib	
			temp_jan_16.lib	
			module_beh.lib	
mismatch.cir	Daily energy balance	7.4 Ex 7.4	irrad_jan_16.lib	7.5
			temp_jan_16.lib	
			module_beh.lib	
nightload.cir	Daily energy balance for a night-load scenario	7.4.2	irrad_jan_16.lib	7.6
			temp_jan_16.lib	
			module_beh.lib	
dayload.cir	Daily energy balance for a day-load scenario	7.4.3	irrad_jan_16.lib	7.7
			temp_jan_16.lib	
			module_beh.lib	
monthly_radiation.lib	Monthly radiation values	7.5		
Seasonal.cir	Energy balance			
madrid.stl	Random daily radiation time series	7.7 Ex 7.6		7.10
				7.11
madrid.cir	Plots the radiation series for Madrid	7.7 Ex. 7.6		7.10
				7.11
generation.lib	Generation pulses for LLP calculation	Ex. 7.7		7.13
aux_gen.lib	Generation of auxilliary energy for LLP calculation	Ex.7.7		7.13
cons.lib	Consumption pulses for LLP calculation	Ex.7.7		7.13
llp.cir	Calculates the LLP	Ex. 7.7		7.17
				7.18
				7.19

Table A7.1 *(continued)*

File name	Description	Section	Includes	Figures
(*)vbatapril.stl	Measured battery voltage values	Ex 7.8		
(*) imodapril.stl	Measured PV current output	Ex. 7.8		
(*)load.stl	Measured load profile	Ex 7.8		7.20
(*) irrad.stl	Measured irradiance profile	Ex 7.8		7.21
Stand_alone.cir	Comparison measurement– simulation	7.9	Irrad.stl	7.22
			Iload.stl	7.23
			Imodapril.stl	7.24
			Vbatapril.stl	
			Batstd.cir	
Madrid_hour.stl	Hourly random radiation values for one year	7.10		
Madrid_temp.stl	Ambient temperature for one year			
Generator_beh_2.lib				
Stand_alone_Madrid.cir			Madrid_hour.stl	7.26
			Madrid_temp.stl	7.27
			Generator_beh_2.lib	7.28
			Batstdif.cir	
Pump_quasi_steady-lib	Water pump model	7.11		
Water_pump.cir	Long-term simulation of a water pumping system	7.11 Ex 7.9	Pump_quasi_ steady-lib	7.29

* Available on the web.

```
**********************************************
*                                            *
*            TEMP_JAN_16.LIB                 *
*                                            *
**********************************************

.subckt temp_jan_16 13 10

vtemp 13 10 pwl  0u 8.72,1u 7.99,2u 7.32,3u 6.73,4u 6.24,5u 5.86
+6u 5.62,7u 5.50,8u 5.72,9u 6.76,10u 8.46,11u 10.43,12u  12.25
+13u 13.53,14u 14,15u 13.92,16u  13.72,17u 13.38,18u 12.92,19u 12.35
+20u 11.70,21u  10.99,22u 10.23,23u 9.46,24u 8.70

.ends temp_jan_16



**********************************************
*                                            *
*            NIGHTLOAD.CIR                   *
*                                            *
**********************************************


*
*  NODES
*  (0)    REFERENCE
*  (72) INPUT, IRRADIANCE
*  (73)   INPUT, AMBIENT TEMPERATURE
*  (74) OUTPUT
*  (75)   OUTPUT, (VOLTAGE) VALUE=SHORT CIRCUIT CURRENT(A) AT
*  IRRADIANCE AND TEMPERATURE
```

```
* (76)  OUTPUT, OPEN CIRCUIT VOLTAGE AT IRRADIANCE ANDTEMPERATURE
* (77) OUTPUT, (VOLTAGE) VALUE=CELL OPERATING TEMPERATURE(°C)
* (78) OUTPUT, MPP CURRENT
* (79) OUTPUT, MPP VOLTAGE
* (80) LOAD
.include irrad_jan_16.lib
.include temp_jan_16.lib
.include module_beh.lib
xtemp 73 0 temp_jan_16
xirrad 72 0 irrad_jan_16

xmodule 0 72 73 74 75 76 77 78 79 module_beh params: iscmr=5,coef_iscm=9.94e-6,
+vocmr=22.3, coef_vocm=-0.0828, pmaxmr=85,noct=47,immr=4.726, vmmr=17.89, tr=25,
+ns=36, np=1

vbat 74 0 dc 12
rload 80 0 3.263

sload 74 80 81 0 switch1
.model switch1 vswitch roff=1e8 ron=0.01 voff=0 von=5
vcontrols1 81 0 0 pulse (0,5,19u,0,0,4u,24u)

.tran 0.01u 24u 0 0.01u
.option stepgmin
.probe
.end



*********************************************
*                                           *
*           DAYLOAD.CIR                      *
*                                           *
*********************************************


*
* NODES
* (0)    REFERENCE
* (72) INPUT, IRRADIANCE
* (73)    INPUT, AMBIENT TEMPERATURE
* (74) OUTPUT
* (75)    OUTPUT, (VOLTAGE) VALUE=SHORT CIRCUIT CURRENT(A) AT
* IRRADIANCE AND TEMPERATURE
* (76)    OUTPUT, OPEN CIRCUIT VOLTAGE AT IRRADIANCE ANDTEMPERATURE
* (77) OUTPUT, (VOLTAGE) VALUE=CELL OPERATING TEMPERATURE(°C)
* (78) OUTPUT, MPP CURRENT
* (79) OUTPUT, MPP VOLTAGE
* (80) LOAD

.include irrad_jan_16.lib
.include temp_jan_16.lib
.include module_beh.lib
xtemp 73 0 temp_jan_16
xirrad 72 0 irrad_jan_16

xmodule 0 72 73 74 75 76 77 78 79 module_beh params: iscmr=5,coef_iscm=9.94e-6,
+vocmr=22.3, coef_vocm=-0.0828, pmaxmr=85,noct=47,immr=4.726, vmmr=17.89,
+tr=25, ns=36, np=1

vbat 74 0 dc 12
rload 80 0 6.528

sload 74 80 81 0 switch1
```

```
.model switch1 vswitch roff=1e8 ron=0.01 voff=0 von=5
vcontrols1 81 0 0 pulse (0,5,8u,0,0,8u,24u)

.tran 0.01u 24u 0 0.01u
.option stepgmin
.probe
.end


*******************************************
*                                         *
*           SEAONAL.CIR                    *
*                                         *
*******************************************


*NODES
*    (0) REFERENCE
*    (72) INPUT, MONTHLY RADIATION  (KWH/M2_MONTH)
*    (74) OUTPUT, MONTHLY GENERATION (KWH/M2_MONTH)
*    (75) OUTPUT, MONTHLY CONSUMPTION (KWH/M2_MONTH)
*    (76) OUTPUT, TIME INTEGRAL OF THE GENERATION-CONSUMPTION
.include monthly_radiation.lib
.include monthly_gen_cons.lib
xradiation 72 0 monthly_radiation
xgen_cons 0 74 75 72 monthly_gen_cons params:eday=3000, nsg=2 , npg={NPG}, pmaxmr=85
.param NPG=4

eout 76 0 value={sdt(1e6*(v(74)-v(75)))}
.step param NPG list 3 4 5 6
.tran 0.1u 13u

.probe
.end


*******************************************
*                                         *
*           GENERATION.LIB                 *
*                                         *
*******************************************


*      PROVIDES PULSE ( WIDE 0.2 UNITS OF TIME .HEIGHT 5* DAILY ENERGY)
*      TIME SERIES OF DAILY PV GENERATION
*      USED IN THE LLP.CIR FILE TO CALCULATE LOSS OF LOAD PROBABILITY

*      NODES
*      (700) REFERENCE
*      (730) OUTPUT, PV GENERATION PULSES
*      (731) INPUT, DAILY RADIATION VALUES
*      (740) OUTPUT, TIME INTEGRAL OF GENERATION (kWh)

.subckt generation 700 730 731 740 params: pmaxr=1, nsg=1, npg=1

sw1 731 720 711 700 switch
.model switch vswitch roff=1e8 ron=0.0001 voff=0 von=5
vctrl1 711 700 0 pulse (0,5,0,0,0,0.01u,1u)
csh 720 700 10u
sw2 720 700 721 700 switch
vctrl2 721 700 0 pulse (0,5,0.2u,0,0,0.01u,1u)
ggen 700 730 value={5*v(720)/1000*pmaxr/1000*nsg*npg}
egen_out 740 700 value={sdt(1e6*5*v(720)/1000*pmaxr/1000*nsg*npg)}
.ends generation
```

```
*********************************************
*                                           *
*           CONS.LIB                        *
*                                           *
*********************************************

*    PROVIDES PULSE ( WIDE 0.2 UNITS OF TIME .HEIGHT 5* DAILY ENERGY)
*    TIME SERIES OF DAILY LOAD
*    USED IN THE LLP.CIR FILE TO CALCULATE LOSS OF LOAD PROBABILITY

*    NODES
*    (700) REFERENCE
*    (730) OUTPUT, PV GENERATION PULSES
*    (731) INPUT, DAILY RADIATION VALUES
*    (770) OUTPUT, TIME INTEGRAL OF LOAD (kWh)
*    PARAMETERS
*                           EDAY= DAILY LOAD ( kwh_day)

.subckt CONS 700 730 770 params: eday=1

vcons 741 700 dc {5*eday}
sw1 741 743 742 700 switch
.model switch vswitch roff=1e8 ron=0.0001 voff=0 von=5
vctrl3 742 700 0 pulse (0,5,0.7u,0,0,0.2u,1u)
r740 743 700 1k
gcons 730 700 value={v(743)}
econs_out 770 700 value={sdt(1e6*v(743))}
.ends cons


*********************************************
*                                           *
*           AUX_GEN.LIB                     *
*                                           *
*********************************************

*    PROVIDES PULSE ( WIDE VARIABLE .HEIGHT 100* DAILY ENERGY)
*    OF AUXILIARY GENERATOR DELIVERY ENERGY WHEN THE STORAGE IS LESS THAN
*    THE DAILY LOAD TO FILL-UP THE STORAGE
*    USED IN THE LLP.CIR FILE TO CALCULATE LOSS OF LOAD PROBABILITY

*    NODES
*    (700) REFERENCE
*    (730) OUTPUT, AUXILIARY GENERATION PULSES (CURRENT)
*    (731) INPUT, BATTERY ENERGY
*    (760) OUTPUT, TIME INTEGRAL OF AUXILIARY ENERGY (kWh)
*    PARAMETERS
*                           EDAY= DAILY LOAD ( kwh_day)
*                           EBAT= TOTAL SIZE OF STORAGE ( kWh)

.subckt aux_gen 700 730 731 760 params: eday=1 ebat=1

vaux 741 700 dc {100*eday}
sw1 741 743 742 700 switch
r743 743 700 1k
.model switch vswitch roff=1e10 ron=0.0001 voff=0 von=2
vctrl3 742 700 0 pulse (0,5,0.35u,0,0,0.1u,1u)

gaux 700 730 value={if (v(751)<-0.0010 & v(731)<ebat-0.2 & v(742)>0, v(743),0)}
eaux2 750 700 value={if ( (v(731)-eday)<0, v(731)-eday, 0)}
sw4 750 751 752 700 switch
vctrl4 752 700 0 pulse ( 0,5,0.15u,0,0,0.05u,1u)
caux2 751 700 10u
sw5 751 700 753 700 switch
```

```
vctrl5 753 700 0 pulse( 0,5,0.6u,0,0,0.1u,1u)

eaux3 761 700 value={if (v(751)<-0.0010 & v(731)<ebat-0.2 & v(742)>0, v(743),0)}
eaux_out 760 700 value={sdt(v(761)*1e6)}
.ends aux_gen




*********************************************
*                                           *
*            LIP.CIR                         *
*                                           *
*********************************************


*    CALCULATES THE VALUE OF THE LOSS OF LOAD PROBABILITY
*    FROM THE DATA OF TOTAL ENERGY SIZE OF BATTERY, DAILY ENERGY
*    LOAD AND STOCHASTICALLY GENERATED DAILY RADIATION VALUES FOR ONE
*     YEAR


*    WARNING, INTERNAL TIME UNITS ARE MICROSECONDS
*    REAL TIME UNIT ARE DAYS
*    WARNING, TIME STEP SHOULD BE 0.01us FOR ACCURATE RESULTS
*    LEADING TO LONG SIMULATION TIME ( tipycally 12 minutes in a standard PC)
*    PARAMETERS EDAY= DAILY ENERGY LOAD (kWh_day)
*    EBAT= TOTAL ENERGY STORAGE SIZE OF BATTERY (kWh)


*    NODES
*    0    REFERENCE
*    71 OUTPUT, BATTERY ENERGY AT TIME T ( kWh)
*    73    INPUT, DAILY RADIATION VALUES (kWh/m2_day)
*    74 OUTPUT, TOTAL ENERGY PROVIDED BY AUXILIARY GENERATOR
*                             IN THE YEAR (kWh_year)
*    75 OUTPUT, TOTAL ENERGY CONSUMED IN THE YEAR (kWh_year)
*    76 OUTPUT, TOTAL ENERGY GENERATED BY THE PV GENERATOR


.include generation.lib
.include cons.lib
.include aux_gen.lib
.include radiation_1year.lib


xrad 0 73 radiation_1year
.param ebat=20
xgen 0 71 73 76 generation params: pmaxr=85,nsg=2, npg=5
xcons 0 71 75 cons params: eday=2

xaux_gen 0 71 71 74 aux_gen params:eday=2 ebat=20
cbat 71 0 1u
diode 71 72 diode
.model diode d
vmax 72 0 dc {ebat}
.tran 0.01u 365u 0.1u 0.001u
.option stepgmin reltol=0.001 vntol=1u abstol =1m gmin=1e-10
.IC v(71)=10
.probe
.end
```

```
*********************************************
*                                           *
*            GENERATOR_BEH_2.LIB            *
*                                           *
*********************************************


*    LEVEL 2 INCLUDING SMOOTHING DATA FILTER
*    BEHAVIOURAL MODEL OF A PV GENERATOR
*    INPUT PARAMETERS: ,AM1.5 JSCMR,AM1.5 VOCMR,AM1.5 PMAXMR
*     AM1.5 VMMR,AM1.5 IMMR, CURRENT TEMP COEFF.,VOLTAGE TEMP.COEFF,
*    NOCT, REFERENCE TEMPERATURE


*    NODES
*     (400)     REFERENCE
*    (401) INTERNAL NODE
*    (402) INPUT, IRRADIANCE
*    (403)    INPUT, AMBIENT TEMPERATURE
*    (404) OUTPUT
*    (405)     OUTPUT, (VOLTAGE) VALUE=SHORT CIRCUIT CURRENT(A) AT
*    IRRADIANCE AND TEMPERATURE
*    (406)    OUTPUT, OPEN CIRCUIT VOLTAGE AT IRRADIANCE ANDTEMPERATURE
*    (407) OUTPUT, (VOLTAGE) VALUE=CELL OPERATING TEMPERATURE(°C)
*    (408) OUTPUT, MPP CURRENT
*    (409) OUTPUT, MPP VOLTAGE

.subckt generator_beh_2 400 402 403 404 405 406 407 408 409 params:
+ iscmr=1, coef_iscm=1, vocmr=1, coef_vocm=1,pmaxmr=1,
+ noct=1,immr=1 , vmmr=1, tr=1, ns=1, nsg=1 npg=1

ev402 411 400 value={v(402)+1}
rfilt 411 410 0.01
cfilt 410 400 1000p

girrad 400 401 value={v(410)/1000*(npg*iscmr+npg*coef_iscm*(v(407)-25))}

eiscm 405 400 value={v(410)/1000*(iscmr+coef_iscm*(v(407)-25))}
evocm 406 400 value={if (v(405)>1e-11 & v(410)>10, (vocmr+coef_vocm*
+(v(407)-25)+ns*8.66e-5*+ (v(407)+273)*log(v(405)/(iscmr))),0)}

etcell 407 400 value={v(403)+(noct-20)/800*v(410)}

gidiode 401 400 value={if (v(405)>1e-6 & v(406)>0.1,npg*v(405)/
+ (exp(v(406)/(ns*8.66e-5*(v(407)+273)))-1)*
+ (exp(v(401)/ (ns*nsg*8.66e-5*(v(407)+273)))-1),0)}



rsg 401 404 {nsg/npg*((vocmr/(iscmr)-pmaxmr/(iscmr**2*(vocmr/(ns*0.0258)-log
+ ((vocmr/(ns*0.0258))+0.72))/(1+vocmr/(ns*0.0258)))))}
.func frsg() {nsg/npg*((vocmr/(iscmr)-pmaxmr/(iscmr**2*(vocmr/(ns*0.0258)-log
+ ((vocmr/(ns*0.0258))+0.72))/(1+vocmr/(ns*0.0258)))))}


gimg 400 408 value={npg*(immr*v(410)/1000+coef_iscm*(v(403)-25))}
rimg 408 400 1
evmg 409 400 value={if (v(410)>1, nsg*(ns*8.66e-5*(v(407)+273)*log(1+(v(405)-
+v(408)/npg)/v(405)*(exp(v(406)/
+ (ns*8.66e-5*(v(407)+273)))-1))-v(408)*frsg/nsg),0)}
.ends generator_beh_2
```

```
*********************************************
*                                           *
*    STAND_ALONE_MADRID.CIR                  *
*                                           *
*********************************************

*    BEHAVIOURAL MODEL OF A PV GENERATOR
*    INPUT PARAMETERS:,AM1.5 JSCMR,AM1.5 VOCMR,AM1.5 PMAXMR
*    AM1.5 VMMR,AM1.5 IMMR, CURRENT TEMP COEFF.,VOLTAGE TEMP.COEFF,
*    NOCT, REFERENCE TEMPERATURE

*    NODES
*     (0)    REFERENCE

*    (42) INPUT, IRRADIANCE
*    (43)    INPUT, AMBIENT TEMPERATURE
*    (44) OUTPUT
*    (45)    OUTPUT, (VOLTAGE) VALUE=SHORT CIRCUIT CURRENT(A) AT
*    IRRADIANCE AND TEMPERATURE
*    (46)    OUTPUT, OPEN CIRCUIT VOLTAGE AT IRRADIANCE ANDTEMPERATURE
*    (47) OUTPUT, (VOLTAGE) VALUE=CELL OPERATING TEMPERATURE(°C)
*    (48) OUTPUT, MPP CURRENT
*    (49) OUTPUT, MPP VOLTAGE

.include madrid_hour.stl
.include madrid_temp.stl
.include generator_beh_2.lib
.include batstdif.cir
xgenerator 0 42 43 44 45 46 47 48 49 generator_beh_2 params:
+ iscmr=5, coef_iscm=9.94e-6, vocmr=23, coef_vocm=-0.0828,pmaxmr=85,
+ noct=47,immr=4.726 , vmmr=17.89, tr=25, ns=36, nsg=2 npg=6

xbat 145 0 50 batstdif params: ns=12 SOCm=33975 k=0.8 D=1e-5, SOC1=0.9
d1 44 145 diode
.model diode d (is=1e-9)
iload 145 0 dc 5.2

econtrol_soc 7000 0 value={if (v(50)>0.95, 15, -1)}
r2 7000 7001 100
c2 7000 7001 1n
r1 145 7002 4
q1 7002 7001 0 q40240
.model q40240 npn (is=1e-9)

virrad 42 0 stimulus vmadrid_hour
vtemp 43 0 stimulus vtemp_madrid
.tran 1u 8000u 0u 1u
.options abstol=1 reltol=0.01 vntol=1 gmin=1e-6 itl4=350
.options stepgmin
.probe
.end

*********************************************
*                                           *
*            PUMP_QUASI_STEADY.LIB           *
*                                           *
*********************************************

*    NODES
*    (500) REFERENCE
*    (501) INPUT (VOLTAGE)
*    (570) OUTPUT ( FLOW) LITER/SECOND
***********
```

```
*    PARAMETERS RA:ARMATURE RESISTOR(OHM),LA: ARMATURE INDUCTANCE(H)
*    KM:MOTOR CONSTANT,RF: FIELD RESISTOR, LF:FIELD INDUCTANCE
*     A AND B CENTRIFUGAL PUMP TORQUE CONSTANTS
*    A1,B1 AND C1 PUMP CHARACTERISITIC CURVE CONSTANTS
*     H:HEAD(METER)
*pump_quasi_steady.lib
.subckt pump_quasi_steady 500 501 570 PARAMS: RA=1,KM=1,A=1,B=1
+    F=1, RF=1, A1=1, B1=1, C1=1, H=1

ra 501 503 {RA}

econ 503 504 value={{KM}*v(508)*v(507)}
rf 504 506 {RF}

vs 506 0 dc 0
gte 0 507 value={{KM}*v(508)*v(508)}
gtl 507 0 value = {A+B*v(507)*v(507)}
rdamping 507 0 {1/{F}}

d2 0 507 diode
.model diode d
gif 0 508 value={v(504)/{RF}}
rif 508 0 1

.IC v(507)=0

*** rpm= omega*(60/2/pi)
erpm 540 0 value={v(507)*60/6.28}
eflow 550 0 value={(-{B1}*v(540)-sqrt(({B1}^2)*(v(540)^2)-4*{C1}*(A1*(v(540)^2)
+-{H})))/(2*{C1})}
eraiz 560 0 value ={({B1}^2)*(v(540)^2)-4*{C1}*(A1*(v(540)^2)-{H})}
eflow2 570 0 value={if (v(560)>0, v(550),0)};checks the sign under the square root
.ends pump_quasi_steady

*********************************************
*                                           *
*            WATER_PUMP.CIR                 *
*                                           *
*********************************************

*    NODES
*    (44) OUTPUT OF GENERATOR AND INPUT TO PUMP
*    (50) OUTPUT, WATER FLOW ( LITRE/SECOND)
*    (60) OUTPUT, TIME INTEGRAL OF WATER FLOW ( VOLUME, LITRE)
*    (42) INPUT, HOURLY RADIATION VALUES ( KWH/M2_DAY)
*    (43) INPUT, AMBIENT TEMPERATURE (°C)

.include madrid_hour.stl
.include madrid_temp.stl
.include generator_beh_2.lib
.include pump_quasi_steady.lib
xgenerator 0 42 43 44 45 46 47 48 49 generator_beh_2 params:
+ iscmr=5, coef_iscm=9.94e-6, vocmr=23, coef_vocm=-0.0828,pmaxmr=85,
+ noct=47,immr=4.726, vmmr=17.89, tr=25, ns=36, nsg=10 npg=4

xpump 0 44 50 pump_quasi_steady params: RA=0.15,KM=0.37,A=0.0925,B=7.26e-6
+    F=0.00083, RF=0.15, A1=1.35e-6, B1=0.0015, C1=-3.32, H=11.5
virrad 42 0 stimulus vmadrid_hour
vtemp 43 0 stimulus vtemp_madrid
e_volume 60 0 value={sdt(v(50)*1e6*3600)}
.tran 1u 8000u 0u 1u
.options abstol=1 reltol=0.01 vntol=1 gmin=1e-6 itl4=350
.options stepgmin
.probe
.end
```

Annex 8
PSpice Files Used in Chapter 8

Table A8.1 PSpice files used in Chapter 8

File name	Description	Section	Includes	Figures
Inverter7.cir	Inverter model for grid-connected PV systems	8.4		8.2, 8.3
Example 8.1.cir	Grid-connected PV system simulation	8.4 Example 8.1	generator_beh.lib inverter7.cir grid.cir aprilmicro.stl	8.4, 8.5, 8.6, 8.7, 8.8, 8.9, 8.10
Acgenerator_be.lib	AC modules model	8.5	Generator_beh.lib	8.11
ACgen.cir	Example of AC modules simulation	8.5 Example 8.2	Acgenerator_be.lib grid.cir irrad2E.stl	8.11, 8.12, 8.13
Inverter9.cir	Inverter PSpice model considering DC output current	8.6		8.2,
GCPV1.cir	Inverter behavioural PSpice model Grid-connected PV system simulation using inverter9.cir	8.6	Generator_beh.lib Inverter9.cir Madrid1.stl MadridT1.stl	8.14, 8.15, 8.16, 8.17, 8.18
Example 8.3.cir	Medium-size grid-connected PV system Design example	8.6 Example 8.3	Generator_beh.lib Irradmadrid.stl TempM.stl Inverter9.cir	8.19, 8.20, 8.21
Example 8.4.cir	Small-size grid-connected PV system Design example	8.6 Example 8.4	Generator_beh.lib Irradmadrid.stl TempM.stl Inverter9.cir	8.23, 8.24, 8.25, 8.26, 8.27

```
**********************************************
*                                            *
*              EXAMPLE 8.1.CIR                *
*                                            *
**********************************************

*     NODES
*     (0) Reference
*     (1) Irradiance profile input
*     (3) PV generator output
*     (4) Inverter Input
*     (6) Inverter output
*     (20) PV generator current at the maximum power point
*     (30) PV generator voltage at the maximum power point

** PV generator
.include generator_beh.lib
xgenerator 0 1 43 3 45 46 47 20 30 generator_beh params:
+ iscmr=5.2, coef_iscm=0.13e-3, vocmr=21.2, coef_vocm=-0.1,pmaxmr=85,
+ noct=47,immr=4.9 , vmmr=17.3, tr=25, ns=36, nsg=2 npg=3

** Irradiance and temperature profiles
.inc aprilmicro.stl
vmesur 1 0 stimulus Virrad
vtemp 43 0 dc 12

** Blocking diode
d2 3 4 diode
.model diode d(n=1)

** inverter
.inc inverter7.cir
xinvert 4 0 6 20 30 inv params: nf=0.9

** grid
.inc grid.cir
xgr1 6 0 grid


.tran 1ms 0.14s 0 1u
.probe
.end
```

Annex 9

PSpice Files Used in Chapter 9

Table A9.1 PSpice files used in Chapter 9

File name	Description	Section	Includes	Figures
Cie.stl	CIE photopic response of the human eye	9.5		9.1
Fluorescent_rel.stl	Fluorescent light spectral density normalized	9.5		9.3
Normalization.cir	Computes the normalization constant to 1 lux	9.5	Fluorescent_rel.stl Cie.stl	9.4
Jsc_art.lib	Library file with equations to compute the short circuit current	9.5		
Jsc_silicon_art.cir	Calculates the short circuit current under artificial light	9.5	silicon_abs.lib fluorescent_rel.stl wavelength.lib jsc_art.lib	9.5
Illuminace_am15g.cir	Calculates the illuminance equivalent value of an AM1.5G spectrum	9.7	Am15g.lib Cie.stl	
Irradiance_art.cir	Calculates the irradiance of a fluorescent light	Ex. 9.3	Fluorescent_rel.stl	
Irradiance_eff.lib	Calculates the effective irradiance of a mixture of artificial and natural light	9.8		
Montecarlo.cir	Calculates the random Monte Carlo I (V) characteristics of a solar PV module	9.8	Irradiance_eff.lib Module_beh.lib	9.6 9.7 9.8
Pocket_calculator.cir	Simulates the circuit of a solar pocket calculator	9.9	Moduel_beh.lib	9.10, 9.11
Random_gen.cir	Serves to calculate random numbers	9.11.1		
Irrad_h_31.lib	Generates random daily radiation values	9.11.1		9.16
Hour_rad_31_bartoli.lib	Hourly radiation values for one month	9.11.2		9.17

Table A9.1   *(continued)*

File name	Description	Section	Includes	Figures
Flash_light.cir	Simulates a flash light circuit	9.11.2		
			Hour_rad_31_bartoli.lib	9.18
			Generator_beh.lib	9.19
			Batstdif.cir	
Street_light_bartoli.cir	Simulates a street light	9.12	Hour_rad_31_bartoli.lib	9.20
			Generator_beh.lib	
			Temp_profile.lib	
			Batstdif.cir	
Temp_profile.lib	Generates the values of temperature at a given time	Annex 12		

```
*******************************************
*                                         *
*           CIE.STL                       *
*                                         *
*******************************************

*    CIE PHOTOPIC EYE RESPONSIVITY
*    NORMALIZED TO A MAXIMUM OF 1 AT 555NM AS A FUNCTION OF
     THE WAVELENGTH IN MICRON

.stimulus vcie pwl
+0.380u 0.000039
+ 0.39u 0.00012
+ 0.4u 0.000396
+ 0.41u 0.00121
+ 0.42u 0.004
+ 0.43u 0.0116
+ 0.44u 0.023
+ 0.45u 0.038
+ 0.46u 0.06
+ 0.47u 0.09098
+ 0.48u 0.13902
+ 0.49u 0.20802
+ 0.5u 0.323
+ 0.507u 0.44431
+ 0.51u 0.503
+ 0.52u 0.71
+ 0.53u 0.862
+ 0.54u 0.954
+ 0.55u 0.99495
+ 0.555u 1
+ 0.56u 0.995
+ 0.57u 0.952
+ 0.58u 0.870
+ 0.59u 0.757
+ 0.6u 0.631
+ 0.61u 0.503
+ 0.62u 0.381
+ 0.63u 0.265
+ 0.64u 0.175
+ 0.65u 0.107
+ 0.66u 0.061
+ 0.67u 0.032
```

```
+ 0.68u 0.0107
+ 0.69u 0.00821
+ 0.7u 0.004102
+ 0.71u 0.002091
+ 0.72u 0.001047
+ 0.73u 0.00052
+ 0.74u 0.000249
+ 0.75u 0.00012
+ 0.76u 0.00006
+ 0.77u 0.00003


*********************************************
*                                           *
*           FLUORESCENT_REL.STL             *
*                                           *
*********************************************


*    RELATIVE SPECTRAL IRRADIANCE OF A FLUORESCENT LAMP
*    WAVELENGTH IN MICRON AND VALUE IN RELATIVE NORMALIZED
*    TO A MAXIMUM VALUE OF 1

.stimulus vfluorescent_rel pwl

+    0u          0.00000
+    3.4e-1u     0.00001
+    3.5e-1u     6e-3
+    3.81E-01u   7.22E-02
+    3.85e-01u   9.11e-02
+    3.90e-01u   1.01e-01
+    3.95e-01u   1.29e-01
+    4.01e-01u   1.51e-01
+    4.011e-01u  4.43e-01
+    4.10e-01u   4.43e-01
+    4.101e-01u  1.92e-01
+    4.15e-01u   2.23e-01
+    4.19e-01u   2.48e-01
+    4.22e-01u   2.70e-01
+    4.27e-01u   2.92e-01
+    4.32e-01u   3.11e-01
+    4.322e-01u  10e-01
+    4.42e-01u   10e-01
+    4.43e-01u   3.77e-01
+    4.46e-01u   3.96e-01
+    4.49e-01u   4.11e-01
+    4.57e-01u   0.427
+    4.64e-01u   0.443
+    4.71e-01u   0.455
+    4.78e-01u   0.46
+    4.84e-01u   0.44
+    4.91e-01u   0.44
+    4.99e-01u   0.427
+    5.03e-01u   0.411
+    5.07e-01u   0.396
+    5.13e-01u   0.380
+    5.23e-01u   0.380
+    5.32e-01u   0.389
+    5.41e-01u   0.405
+    5.411e-01u  0.772
+    5.52e-01u   0.772
+    5.521e-01u  0.465
+    5.57e-01u   0.49
+    5.63e-01u   0.518
```

```
+    5.69e-01u   0.543
+    5.75e-01u   0.562
+    5.76e-01u   0.584
+    5.84e-01u   0.584
+    5.841e-01u  0.546
+    5.92e-01u   0.515
+    5.99e-01u   0.465
+    6.05e-01u   0.411
+    6.12e-01u   0.361
+    6.18e-01u   0.311
+    6.28e-01u   0.261
+    6.34e-01u   0.22
+    6.39e-01u   0.195
+    6.45e-01u   0.163
+    6.52e-01u   0.135
+    6.61e-01u   0.107
+    6.68e-01u   0.0816
+    6.75e-01u   0.0628
+    6.82e-01u   0.0408
+    6.91e-01u   0.022
+    6.99e-01u   0.00628
```

```
**********************************************
*                                            *
*              IRRADIANCE_ART.CIR            *
*                                            *
**********************************************
```

```
*   CALCULATES THE IRRADIANCE OF AN ARTICIAL LIGHT SPECTRUM
*   OF GIVEN ILLUMINACE
*   NODES
*   (90) fluorescent light spectrum
*   (92) irradiance

.include fluorescent_rel.stl
.param gv=1
.param k=0.0292

vfluor 90 0 stimulus vfluorescent_rel

eIRRAD 92 0 value={sdt(v(90)*k*gv*1e6)};irradiance
.tran 0.01u 0.770u 0.38u 0.01u
.probe
.end
```

```
**********************************************
*                                            *
*              JSC_ART.LIB                   *
*                                            *
**********************************************
```

```
*   IS THE SAME SUBCIRCUIT AS JSC.LIB
*   SUBCIRCUIT CALCULATING THE ANALYTICAL SOLAR CELL MODEL*
    FOR SPECTRAL SHORT CIRCUIT CURRENT DENSITY, TOTAL SHORT*
    CIRCUIT CURRENT DENSITY AND
*   QUANTUM EFFICIENCY AND SPECTRAL RESPONSE

*NODES
*(200) REFERENCE
*(201) INPUT,   WAVELENGTH
*(202) INPUT, ABSORPTION COEFFICIENT
```

```
*(203) INPUT, SPECTRAL IRRADIANCE
*(204) INPUT, REFLECTION COEFFICIENT
*(205) OUTPUT, EMITTER SPECTRAL SHORT CIRCUIT CURRENT DENSITY
*(206) OUTPUT, BASE SPECTRAL SHORT CIRCUIT CURRENT DENSITY
*(207) OUTPUT, TOTAL SHORT CIRCUIT CURRENT DENSITY
*(208) OUTPUT, TOTAL INTERNAL QUANTUM EFFICIENCY
*(209) OUTPUT, TOTAL INTERNAL SPECTRAL RESPONSE

*DE_NORMALIZATION FACTOR TO ILLUMINANCE=NLUX IS 0.0292*NLUX

******* subcircuit
.param q=1.6e-19
.subckt jsc_art 200 201 202 203 204 205 206 207 208 209
PARAMS:we=1,lp=1,dp=1,se=1,
+ wb=1, ln=1, dn=1, sb=1

**emitter component

egeom0 230 200 value={1.6e-19*v(202)*v(203)*(0.0292*nlux)*v(201)*(1e16/19.8)*
+ lp*(1-V(204))/(v(202)*lp+1)}

egeom1 231 200 value={cosh(we/lp)+se*(lp/dp)*sinh(we/lp)}
egeom2 232 200 value={se*(lp/dp)*cosh(we/lp)+sinh(we/lp)}
egeom3 233 200 value={(se*(lp/dp)+v(202)*lp-exp(-v(202)*we)*v(232))}


ejsce 205 200 value={v(230)/(v(202)*lp-1)*(-v(202)*lp*
+ exp(-v(202)*we)+v(233)/v(231))};short circuit

eqee 234 200 value={v(205)*19.8/(q*v(203)*(0.0292*nlux)*
+ (1-V(204))*v(201)*1e16)};quantum efficiency

**base component

egeom30 240 200 value={cosh(wb/ln)+sb*ln/dn*sinh(wb/ln)}
egeom31 241 200 value={sb*ln/dn*(cosh(wb/ln)-exp(-v(202)*wb))}
egeom32 242 200 value={sinh(wb/ln)+v(202)*ln*exp(-v(202)*wb)}
egeom33 243 200 value={q*v(202)*v(203)*v(201)*1e16/19.8*0.0292*nlux*
+ (1-v(204))*ln*exp(-we*v(202))/((ln*v(202))^2-1)}
ejscb 206 200 value={v(243)*(v(202)*ln-(v(241)+v(242))/v(240))};short circuit
eqeb 244 200 value={v(206)*19.8/(q*v(203)*(0.0292*nlux)*(1-v
(204))*v(201)*1e16)};quantum efficiency

******* total short circuit current
ejsc 207 200 value={1e6*sdt(v(205)+v(206))}
eqe   208 200 value={(v(234)+v(244))*100}; total quantum efficiency in %
******* quantum efficiency
esr 209 200 value={(v(234)+v(244))*0.808*v(201)}

.ends jsc_art


*******************************************
*                                         *
*          JSC_SILICON_ART.CIR            *
*                                         *
*******************************************

*
*   CALCULATES THE SHORT CIRCUIT CURRENT DENSITY, QUANTUM EFFICIENCY
*   AND SPECTRAL RESPONSE UNDER A FLUORESCENT LIGHT
**   NODES
*   (0) REFERENCE
```

```
*    (21) INPUT,    WAVELENGTH
*    (22) INPUT, ABSORPTION COEFFICIENT
*    (23) INPUT, FLUORESCENT LIGHT SPECTRAL IRRADIANCE
*    (24) INPUT, REFLECTION COEFFICIENT
*    (25) OUTPUT, EMITTER SPECTRAL SHORT CIRCUIT CURRENT
**   DENSITY
*    (26) OUTPUT, BASE SPECTRAL SHORT CIRCUIT CURRENT DENSITY
*    (27) OUTPUT, TOTAL SHORT CIRCUIT CURRENT DENSITY
*    (28) OUTPUT, TOTAL INTERNAL QUANTUM EFFICIENCY
*    (29) OUTPUT, TOTAL INTERNAL SPECTRAL RESPONSE


*jsc_silicon_art.cir

.param nlux=257000; illuminance
.include silicon_abs.lib
.include fluorescent_rel.stl
.include wavelength.lib
.include jsc_art.lib

****** circuit

xwavelength 21 0 wavelength
*caux1 21 0 1p
*caux2 25 0 0.1p
xabs 22 0 silicon_abs
vfluor 23 0 stimulus vfluorescent_rel

xjsc_art 0 21 22 23 24 25 26 27 28 29 jsc_art params: we=0.3e-4 lp=0.43e-4 dp=3.4
+se=20000
+ wb=300e-4 ln=162e-4 dn=36.63 sb=1000

vr 24 0 dc 0.1; value of the reflection coefficient

.options itl4=350 reltol=0.01
.tran 0.01u 1.2u 0.3u 0.01u
.option stepgmin
.probe

.end


*********************************************
*                                           *
*        POCKET_CALCULATOR.CIR              *
*                                           *
*********************************************

*    SIMULATES THE CIRCUIT OF A POCKET CALCULATOR
*    UNDER A MIXTURE OF IRRADIANCE AND ILLUMINANCE VALUES

.include module_beh.lib
.param  g=10
.param  gv=500


xmodule 0 92 93 94 95 96 97 98 99 module_beh params: iscmr=8.5e-3 coef_iscm=0 vocmr=2.8
+ coef_vocm=0 pmaxmr=14.2e-3 tr=25 noct=47 ns=4 immr=7.06e-3 vmmr=2.01

virradeff 92 0 pulse (0.1 {g+3.8e-3*gv} 50u 2u 2u 150u 300u); sweep of irradiance
values
vtemp 93 0 dc 25
c1 94 0 1n
```

```
vbat 94 100 dc 1.5
d1 0 100 diode
.model diode d (is=1e-9)
iload 94 0 dc 3.5u
.ic v(94)=1.5
.tran 0u 1000u 0.001u 0.1u
.probe
.end


*********************************************
*                                           *
*    HOUR_RAD_31_BARTOLI.LIB                 *
*                                           *
*********************************************


*    NODES
*    (1000) HOUR ANGLE
*    (14)    NUMBER OF DAY
*    (12) RANDOM NUMBER NOT NORMALIZED
*    (11) HORIZONTAL DAILY RADIATION DAY n (Wh/m2_day)
*    (74) HOURLY RADIATION INCLINED SURFACE (Wh/m2_hour)
*    (10) REFERENCE
*    (7)    SUNSET HOUR ANGLE FOR HORIZONTAL SURFACE (RAD)
*    (6)    SUN DECLINATION (RAD)
*    (106) IN-PLANE SUNRISE ANGLE (RAD)
*    (107)  IN-PLANE SUNSET ANGLE (RAD)
*    (50)    EXTRA-ATMOSPHERIC RADIATION HO
*    PARAMETERS N= FIRST DAY OF MONTH CONSIDERED, N=1 JANUARY
*

.subckt hour_rad_31_bartoli 74 10 params: n=1

***** number of day after day n
v_n 14 10 pwl 0u 0, 0.01u {n} ,23.99u {n},24.01u {n+1}, 47.99u {n+1},
+ 48.01u {n+2},71.99u {n+2},72.01u {n+3},95.99u {n+3},96.01u {n+4},
+ 119.99u {n+4},120.01u {n+5},143.99u {n+5},144.01u {n+6},167.99u {n+6},
+ 168.01u {n+7},191.99u {n+7}, 192.01u {n+8},215.99u {n+8},216.01u {n+9},
+ 239.99u {n+9},240.01u {n+10},263.99u {n+10},264.01u {n+11},287.99u {n+11},
+ 288.01u {n+12}, 311.99u {n+12},312.01u {n+13},335.99u {n+13},336.01u {n+14},
+ 359.99u {n+14}, 360.01u {n+15},383.99u {n+15},384.01u {n+16},407.99u {n+16},
+ 408.01u {n+17},431.99u {n+17},432.01u {n+18},455.99u {n+18},456.01u {n+19},
+ 479.99u {n+19},480.01u {n+20},503.99u {n+20},504.01u {n+21},527.99u {n+21},
+ 528.01u {n+22},551.99u {n+22},552.01u {n+23},575.99u {n+23}, 576.01u {n+24},
+ 599.99u {n+24}, 600.01u {n+25}, 623.99u {n+25},624.01u {n+26},647.99u {n+26},
+ 648.01u {n+27},671.99u {n+27},672.01u {n+28},695.99u {n+28},696.01u {n+29},
+ 719.99u {n+29},720.01u {n+30},743.99u {n+30}

***** sinthetic radiation series , average 1734W/m2_day
v_random 12 10 pwl 0u 0,0.1u 917, 23.99u 917, 24.01u 1756,47.99u 1756
+ 48.01u 2213,71.99u 2213,72.01u 1857,95.99u 1857,96.01u 1927,
+ 119.99u 1927,120.01u 2167,143.99u 2167,144.01u 2529,167.99u 2529,
+ 168.01u 1709,191.99u 1709, 192.01u 1974,215.99u 1974,216.01u 2772,
+ 239.99u 2772,240.01u 1401,263.99u 1401,264.01u 1556,287.99u 1556,
+ 288.01u 1875, 311.99u 1875,312.01u 343,335.99u 343,336.01u 1911,
+ 359.99u 1911, 360.01u 402,383.99u 402,384.01u 694,407.99u 694,
+ 408.01u 1368,431.99u 1368,432.01u 2327,455.99u 2327,456.01u 3395,
+ 479.99u 3395,480.01u 2785,503.99u 2785,504.01u 2650,527.99u 2650,
+ 528.01u 1429,551.99u 1429,552.01u 2439,575.99u 2439, 576.01u 1255,
+ 599.99u 1255, 600.01u 953, 623.99u 953,624.01u 888,647.99u 888,
+ 648.01u 1073,671.99u 1073,672.01u 1355,695.99u 1355,696.01u 1468,
+ 719.99u 1468,720.01u 2006, 743.99u 2006
```

```
************ average 2000 W/m2day
*v_random 12 10 pwl 0u 0,0.1u 2145, 23.99u 2145, 24.01u 1198,47.99u 1198
*+ 48.01u 2072,71.99u 2072,72.01u 2577,95.99u 2577,96.01u 2158,
*+ 119.99u 2158,120.01u 2237,143.99u 2237,144.01u 2515,167.99u 2515,
*+ 168.01u 2936,191.99u 2936, 192.01u 1984,215.99u 1984,216.01u 2292,
*+ 239.99u 2292,240.01u 3219,263.99u 3219,264.01u 1627,287.99u 1627,
*+ 288.01u 1807, 311.99u 1807,312.01u 2177,335.99u 2177,336.01u 398,
*+ 359.99u 398, 360.01u 2219,383.99u 2219,384.01u 466,407.99u 466,
*+ 408.01u 805,431.99u 805,432.01u 1588,455.99u 1588,456.01u 2702,
*+ 479.99u 2702,480.01u 3942,503.99u 3942,504.01u 3233,527.99u 3233,
*+ 528.01u 3076,551.99u 3076,552.01u 1659,575.99u 1659, 576.01u 2832,
*+ 599.99u 2832, 600.01u 1457, 623.99u 1457,624.01u 1106,647.99u 1106,
*+ 648.01u 1031,671.99u 1031,672.01u 1246,695.99u 1246,696.01u 1573,
*+ 719.99u 1573,720.01u 1704,743.99u 1704

*******dayly horizontal plane radiation and average value
e_hh 11 10 value={v(12)};dayly radiation horizontal plane

*calculation of the irradiance profiles in 31 days

.param pi=3.141592
.param ro=0.5;albedo factor
.func b() {(pi/180)*30};inclination of the collecting surface,(rad)
.func g() {(pi/180)*(0)};azimuth of the collecting surface,g=0 is facing south
.func l() {(pi/180)*41.2};latitude
e_sc 5 10 value={1353*(1+0.0033*cos(pi/180*360/365*v(14)))}

e_a 108 10 value={0.409+0.516*sin(v(7)-1.047)}
e_bb 109 10 value={0.6609-0.467*sin(v(7)-1.047)}
e_d 6 10 value={(pi/180)*23.45*sin((pi/180)*360/365*(v(14)+284))};declination

e_ws 7 10 value={acos(-tan(l)*tan(v(6)))};sunset hour angle at horizontal surface

.func a1() {cos(g)*sin(b)*sin(l)+cos(b)*cos(l)}
.func a2() {sin(b)*sin(g)}
e_a3 8 10 value={-tan(v(6))*(cos(b)*sin(l)-cos(g)*sin(b)*cos(l))}

**** ratio angle of collection over zenith distance
ecoll 20 10 value={(1/(cos(v(1000))-cos(v(7))))*(cos(v(1000))*(cos(g)*sin(b)*
+tan(l)+cos(b))
+ +sin(v(1000))/cos(l)*sin(b)*sin(g)+tan(v(6))*(cos(b)*tan(l)-cos(g)*sin(b)))}

********* in-plane sunset and sunrise angles
e_wps 101 10 value={if(g>0,acos((a1*v(8)-a2*sqrt(a1**2-
+v(8)**2+a2**2))/(a1**2+a2**2)),
+ acos((a1*v(8)-a2*sqrt(a1**2-v(8)**2+a2**2))/(a1**2+a2**2)))};in-plane sunset
angle
e_wss 100 10 value={if(g>0,-acos((a1*v(8)+a2*sqrt(a1**2-v(8)**2+a2**2))/
+(a1**2+a2**2)),
+ -acos((a1*v(8)+a2*sqrt(a1**2+v(8)**2+a2**2))/(a1**2+a2**2)))};in-plane
sunrise angle,takes negative values
****************************************************************
e_wsss 106 10 value={if (v(100)>-v(7), v(100),-v(7))};real in-plane sunrise angle
e_wpss 107 10 value={if( v(101)>v(7),v(7),v(101))};real in-plane sunset angle

e_ho 50 10
value={24/pi*v(5)*(cos(v(6))*cos(l)*sin(v(7))+v(7)*sin(v(6))*
sin(l))};extraatmospheric radiation Ho

***** correlations

e_rh 51 10 value={pi/24*(v(108)+v(109)*cos(v(1000)))*(cos(v(1000))-
+cos(v(7)))/(sin(v(7))-v(7)*cos(v(7)))}
```

```
e_rd 52 10 value={pi/24*(cos(v(1000))-cos(v(7)))/(sin(v(7))-v(7)*cos(v(7)))}

*********** radiation at the collecting surface

e_hcoll 60 10 value={1e6*sdt(v(74)/31)}; average value in kwh_day

*********** radiation components

e_hd 70 10 value={v(11)*(1-1.13/v(50)*v(11))};diffuse radiation horizontal
e_id 71 10 value={v(52)*v(70)}; diffuse
e_ih 72 10 value={v(51)*v(11)}; global horizontal
e_ib 73 10 value={(v(72)-v(71))/(sin(l)*sin(v(6))+cos(l)*cos(v(6))*
+cos(v(1000)))}

********* in-plane radiation

e_icolla 741 10 value={(v(72)-v(71))*v(20)+v(71)/2*(1+cos(b))+v(72)/2*ro*
+(1-cos(b))}
e_icoll 74 10 value={if(v(741)>1 & v(1000)<v(107) & v(1000)>v(106),
+v(741),1)};limit to >10 values

***** time angle in radians, sweeps from -pi to +pi radians in 24 internal units of time

***** time angle

v_w 1000 10 pulse (-3.14 3.14 0 23.99u 0.01u 0.01u 24u);sweeps time angle in radians

.ends hour_rad_31_bartoli




*******************************************
*                                         *
*            FLASH_LIGHT.CIR              *
*                                         *
*******************************************

xirradiance 920 0 hour_rad_31_bartoli params: n=1
.inc hour_rad_31_bartoli.lib


r1 920 92 100
c1 92 0 1000p


xmodule 0 92 93 94 95 96 97 98 99 module_beh params: iscmr=0.126 coef_iscm=20.16e-6
+vocmr=10.4
+ coef_vocm=-0.0368 pmaxmr=0.9828 tr=25 noct=47 ns=16 immr=0.1224 vmmr=8.32
.inc module_beh.lib

vtemp 93 0 dc 5
d1 94 31 diode
.model diode d (is=2e-10)

xbat1 31 0 7 batstdif params: ns=3, SOCm=36, k=.8, D=1e-5, SOC1=0.9
.inc batstdif.cir
if 31 0 dc 17.5e-3
.options abstol=10p reltol=0.01
.tran 1u 744u 0u 0.1u
.probe
.end
```

```
**********************************************
*                                            *
*    STREET_LIGHT_BARTOLI.CIR                *
*                                            *
**********************************************

*    SIMULATES A STREET LIGHT SYSTEM COMPOSED OF BATTERY PV
*     GENERATOR
*    AND A LOAD OF 8 HOURS AT 3.08 AMPS AT 12 VOLTS
*    NODES
*    (1000) HOURLY RADIATION VALUES FOR 31 DAYS OF JANUARY
*    (93) TEMPERATURE PROFILE (°c)
*    (94) OUTPUT, pv GENERATOR OUTPUT NODE
*    (31) BATTERY OUTPUT NODE
*    (7) BATTERY STATE OF CHARGE (SOC)

xirradiance 1000 0 hour_rad_31_bartoli params: n=1
.inc hour_rad_31_bartoli.lib
.inc generator_beh.lib
.include temp_profile.lib
.inc batstdif.cir
r1 1000 92 100
c1 92 0 1000p

xtemp 93 0 temp_profile params:nday=17 tmax=15 tmin=5.5
xgenerator 0 92 93 94 95 96 97 98 09 generator_beh params:
+ iscmr=10.14, coef_iscm=20.6e-6, vocmr=21.6, coef_vocm=-0.0828,pmaxmr=165,
+ noct=47,immr=9.48 , vmmr=17.4, tr=25, ns=36, nsg=1 npg=1

d1 94 31 diode
.model diode d
xbat1 31 0 7 batstdif params: ns=6, SOCm=2400, k=.8, D=1e-5, SOC1=0.3
if 31 0 pulse (0 3.08 20u 0.1u 0.1u 8u 24u)
.options abstol =10p reltol=0.01
.tran 1u 744u 0u 0.1u
.probe
.end

**********************************************
*                                            *
*           TEMP_PROFILE.LIB                 *
*                                            *
**********************************************

*    CALCULATES THE AMBIENT TEMPERATURE PROFILE FOR A GIVEN
*     DAY
*    PARAMETERS NDAY=NUMBER OF DAY AFTER 1ST OF JANUARY
*              TMAX= MAXIMUM TEMPERATURE
*              TMIN= MINIMUM TEMPERATURE
*       LATITUDE 41.2° NORTH (BARCELONA)

*    NODES
*    (20) OUTPUT, TEMPERATURE (°C )
*    (10)    REFERENCE
.subckt temp_profile 20 10 params: nday=1 tmax=1 tmin=1

.param pi=3.141592


v_w 1000 10 pulse (-3.141592 3.141592 0 24u 0u 0u 24u);time angle in radians


e_temp 20 10 value={if (v(1000)<-v(7), tmax-(tmax-tmin)/2*(1+cos(pi/(pi/6+v(7)-
+2*pi)*
```

```
+ (v(1000)+v(7)))),if( v(1000)<pi/6,
+ tmin+(tmax-tmin)/2*(1+cos(pi/(-v(7)-pi/6)*(v(1000)-pi/6))),
+ tmax-(tmax-tmin)/2*(1+cos(pi/(2*pi-v(7)-pi/6)*v(1000)-
+ (pi+pi*pi/6/(2*pi-v(7)-pi/6))))))}
.func l() {(pi/180)*41.2};latitude

e_d 6 10 value={(pi/180)*23.45*sin((pi/180)*360/365*(nday+284))};declination
e_ws 7 10 value={acos(-tan(l)*tan(v(6)))};sunset hour angle at horizontal surface

.ends temp_profile
```

Annex 10

Summary of Solar Cell Basic Theory

Solar cells are semiconductor devices in which the photovoltaic effect takes place, converting the photon energy into electron-hole pairs, which after travelling, are collected by an electric field located at a homojunction or heterojunction.

The transport equations governing the carrier concentrations are a set of five differential equations, namely two current equations for minority and majority carriers, two continuity equations also for minority and majority carriers and Poisson's law. These coupled equations do not have an analytical solution, however, in very simple cases, in particular if:

- steady-state conditions are satisfied;

- the electric field is only different from zero at the junction (space charge region, SCR);

- there is low injection meaning that the minority carrier concentrations are much smaller than the majority carrier concentrations;

- doping concentrations are constant, i.e. N_A acceptor concentration for the base and N_D donor concentration for the emitter;

then the differential equations can be written for minority carriers in each of the two regions of the solar cell bulk and the total current can be calculated by addition of the minority carrier currents at the boundaries of the SCR.

If we assume a geometry as the one shown in Figure A10.1, we can write the simplified equations for the emitter and the base layers.

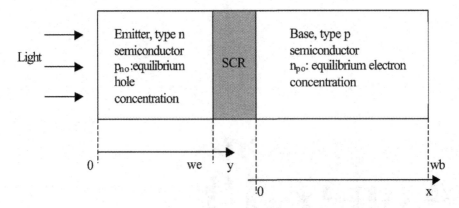

Figure A10.1   Schematic diagram of the cross-section of a solar cell

Emitter layer

$$\frac{1}{q}\frac{\mathrm{d}J_p}{\mathrm{d}y} + \frac{p_n - p_{no}}{\tau_p} - G(y) = 0 \tag{A10.1}$$

$$J_p = -qD_p\frac{\mathrm{d}p}{\mathrm{d}y} \tag{A10.2}$$

Substituting equation (A10.1) in equation (A10.2) it follows,

$$D_p\frac{\mathrm{d}^2p}{\mathrm{d}y^2} - \frac{p_n - p_{no}}{\tau_p} = -\alpha\phi_0 e^{-\alpha y} \tag{A10.3}$$

where the generation term $G(y)$ has been substituted by:

$$G(y) = \alpha\phi_0 e^{-\alpha y} \tag{A10.4}$$

where ϕ_0 is the photon flux at the semiconductor surface, and α is the absorption coefficient of the material. Both magnitudes depend on the wavelength. Equation (A10.3) is the diffusion equation for the holes, which are minority carriers in the emitter layer, and can be solved if two boundary conditions are known. The boundary conditions at a semiconductor p–n junction are known as:

$$(p_n - p_{no})_{w_e} = p_{no}(e^{V/V_T} - 1) \tag{A10.5}$$

where V is the applied voltage, assuming that there are no losses from the contacts to the junction boundaries, and V_T is the thermal potential $V_T = kT/q$, with k the Boltzmann constant.

The second boundary condition is applied at the semiconductor surface as:

$$J_p(y = 0) = -qS_e(p_n - p_{no}) \tag{A10.6}$$

where S_e is the emitter surface recombination velocity having units of velocity (m/s). In order to calculate the short circuit current the boundary condition (A10.5) further simplifies by setting $V = 0$, and hence at $y = w_e$,

$$p_n - p_{no} = 0 \tag{A10.7}$$

Equation (A1.3) has an analytical solution of the kind:

$$p_n'(y) = p_n - p_{no} = ASh\left(\frac{y}{L_p}\right) + BCh\left(\frac{y}{L_p}\right) + Ce^{-\alpha y} \tag{A10.8}$$

which includes natural and forced terms. The three constants are easily determined by applying the two boundary conditions and substituting the forced response into the differential equation, resulting in the following expressions:

$$A = BS_e\frac{L_p}{D_p} - \frac{\alpha\phi\tau_p\frac{L_p}{D_p}(S_e + \alpha D_p)}{L_p^2\alpha^2 - 1} \tag{A10.9}$$

$$B = \frac{\alpha\phi\tau_p}{L_p^2\alpha^2 - 1}\frac{e^{-\alpha w_e} + \frac{L_p}{D_p}Sh\left(\frac{w_e}{L_p}\right)(S_e + \alpha D_p)}{S_e\frac{L_p}{D_p}Sh\left(\frac{w_e}{L_p}\right) + Ch\left(\frac{w_e}{L_p}\right)} \tag{A10.10}$$

$$C = -\frac{\alpha\phi\tau_p}{L_p^2\alpha^2 - 1} \tag{A10.11}$$

With the values of these three constants the hole concentration is fully known. We need the value of the hole current at $y = w_e$ that is calculated from

$$J_p(y = w_e) = -qD_p\left(\frac{dp}{dy}\right)_{y=w_e} \tag{A10.12}$$

which after derivation of equation (A10.8) gives equation (2.2) in Chapter 2.

Base layer

The minority carriers in the base layer are the electrons and then the semiconductor equations are, taking into account the geometry definitions in Figure A10.1:

$$\frac{1}{q}\frac{dJ_p}{dy} + \frac{p_n - p_{no}}{\tau_p} - G(y) = 0 \tag{A10.13}$$

$$J_p = -qD_p\frac{dp}{dy} \tag{A10.14}$$

Substituting equation (A10.14) in equation (A10.13) it follows,

$$D_n \frac{d^2n}{dx^2} - \frac{n_p - n_{po}}{\tau_n} = -\alpha_0 \phi_0' e^{-\alpha x} \qquad (A10.15)$$

where ϕ_0' is the photon flux at the boundary of the base layer. Equation (A10.15) is the diffusion equation for the electrons and can be solved if two boundary conditions are known. The boundary condition at the p–n junction boundary of the base layer is:

$$(n_p - n_{po})_0 = n_{po}(e^{V/V_T} - 1) \qquad (A10.16)$$

The second boundary condition is applied at the other base layer boundary as:

$$J_n(x = w_b) = -qS_b(n_p - n_{po}) \qquad (A10.17)$$

where S_b is the base surface recombination velocity also having units of velocity (m/s). In order to calculate the short circuit current the boundary condition (A10.16) further simplifies by setting $V = 0$, and hence at $x = 0$,

$$n_p - n_{po} = 0 \qquad (A10.18)$$

Equation (A10.15) has an analytical solution of the kind:

$$n_p'(x) = n_p - n_{po} = A' Sh\left(\frac{x}{L_n}\right) + B' Ch\left(\frac{x}{L_n}\right) + C' e^{-\alpha x} \qquad (A10.19)$$

where the constants are derived from the boundary conditions,

$$A' = \frac{\alpha \phi_0' \tau_n}{L_n^2 \alpha^2 - 1} \frac{e^{-\alpha w_b}(\alpha D_n - S_b) + \frac{D_n}{L_n} Sh\left(\frac{w_b}{L_n}\right) + S_b Ch\left(\frac{w_b}{L_n}\right)}{S_b Sh\left(\frac{w_b}{L_n}\right) + \frac{D_n}{L_n} Ch\left(\frac{w_b}{L_n}\right)} \qquad (A10.20)$$

$$B' = \frac{\alpha \phi_0' \tau_n}{L_n^2 \alpha^2 - 1} \qquad (A10.21)$$

$$C' = -B' \qquad (A10.22)$$

From the values of these constants equation (2.3) in Chapter 2 can be derived.

Dark Characteristics

The dark characteristics of a solar cell are easily derived solving the same differential equations (A10.3) and (A10.15) making $\phi_0 = 0$ and $\phi_0' = 0$, and $V \neq 0$.

Then the solution for emitter and base are equations (2.13) and (2.14) in Chapter 2.

Parameter Values

The dark and illuminated solar cell equations require the values of a number of semi-conductor parameters, namely:

- Emitter layer
  - hole mobility, μ_p
  - hole lifetime, τ_p
  - equilibrium hole concentration, n_{po}

- Base layer
  - electron mobility, μ_n
  - electron lifetime, τ_n
  - equilibrium hole concentration, p_{no}

The values of the minority carrier mobilities and lifetimes depend on the doping concentration and several models have been proposed and the ones summarized in Table A10.1 are widely used (N is the doping concentration).

Finally the equilibrium carrier concentration is given:

For a type N semiconductor

$$p_{no} = \frac{n_i^2}{N_{Deff}}$$

Table A10.1 Parameter models

	Type p	Type n
Mobility(cm²/v.s)	$\mu_n = 232 + \dfrac{1180}{\left(1 + \frac{N}{8 10^{18}}\right)^{0.9}}$	$\mu_p = 130 + \dfrac{370}{\left(1 + \frac{N}{8 10^{17}}\right)^{1.25}}$
	Reference [A10.1]	Reference [A10.3]
Bandgap narrowing (eV)	$\Delta Eg = 910^{-3}(F + \sqrt{0.5 + F^2})$	$\Delta Eg = 18.710^{-3} \ln\left(\dfrac{N}{710^{17}}\right)$
	$F = \ln\left(\dfrac{N}{110^{17}}\right)_{N>110^{17}}$	$N > 710^{17}$
	Reference [A10.2]	Reference [A10.3]
Lifetime(s)	$\frac{1}{\tau} = 3.4510^{-12}N + 0.9510^{-31}N^2$	$\frac{1}{\tau} = 7 \times 10^{-13}N + 1.810^{-31}N^2$
	Reference [1.10.1]	Reference [A10.3]

where

$$N_{Deff} = N_D e^{-\frac{\Delta E_g}{kT}}$$

For a P-type layer

$$n_{po} = \frac{n_i^2}{N_{Aeff}}$$

where

$$N_{Aeff} = N_A e^{-\frac{\Delta E_g}{kT}}$$

n_i is the intrinsic carrier concentration, N_A is the acceptor doping concentration and N_D is the donor doping concentration.

References

[A10.1] Swirhun, S.E., Kwark, Y.H., Swanson R.M., 'Measurement of elecron mobility and bandgap narrowing in heavily doped p-type silicon', *IEDM Technical Digest*, p. 24, 1986.

[A10.2] Slotboom, J.W., de Graaf H.C., 'Measurement of bandgap narrowing in silicon bipolar transistors', *Solid State Electronics*, **19**, p. 857, 1976.

[A10.3] del Alamo, J., Swirhun, S.E., Swanson, R. M., 'Simultaneous measurement of hole lifetime, hole mobility and bandgap narrowing in heavily doped n-type silicon', *IEDM Technical Digest*, p. 24, 1986.

Annex 11

Estimation of the Radiation in an Arbitrarily Oriented Surface

The calculation of the irradiance and irradiation in an arbitrarily oriented surface is something required in the large majority of PV system designs mainly due to two reasons: (a) the data generally known is the global irradiation received at a horizontal surface; and (b) few applications require a horizontal collecting surface.

The estimation of the radiation received from the sun at inclined surfaces is a complicated matter and involves many parameters and computational procedures, along with extensive data from measurements and observations at a given area of the earth. International bodies have promoted studies where experts report the procedures and results, such as for example in reference [A11.1] and many works have been produced [A11.2], [A11.3].

The purpose of this book is limited to the basics of modelling photovoltaic systems, and a detailed description of the solar radiation and inclined surface calculations is out of the scope of this book. However, a brief summary of concepts is given here to provide a PSpice code that implements a simplified procedure of calculation.

Coordinates of the Sun

Relative to the observer the position of the sun is identified by two different sets of celestial coordinates, one set is the azimuth (a) and elevation (h) in horizontal spherical coordinates, and the declination (δ) and the hour angle (ω) in equatorial spherical coordinates.

The declination of the sun depends on the day of the year and sweeps between $-23.45°$ (winter solstice) and $+23.45°$ (summer solstice). The declination is zero at the two equinoxes (spring and autumn). There are several analytical approximations, which provide the value of the declination for a given day, and we have used the following:

$$\delta = 23.45 \sin\left[\frac{360}{365}(N + 284)\right] \tag{A11.1}$$

in degrees. N is the day number of the year starting Ist January [A11.4].

The origin of the hour angle is considered at the south orientation in northern latitudes and at the north in southern latitudes. The relation between the hour angle and the official time on our watches depends on what is called the equation of time and on the distance of the site to the origin of the time zone. Moreover in some countries a convention is adopted to introduce an additional time shift and the clock has to be advanced or delayed twice a year.

Several analytical approximations exists for the time equation (ET), also named AnalemA11. One of them is the following [A11.4]:

$$ET = 9.87 \sin 2B - 7.53 \cos B - 1.5 \sin B \tag{A11.2}$$

ET is given in minutes and:

$$B = \frac{360}{364}(N - 81) \tag{A11.3}$$

The relationship between the hour angle and the official time on our watches is the following:

$$OT = 12 + TA + \frac{180}{\pi}\frac{\omega}{15} - \frac{ET}{60} + \frac{LL - LH}{15} \tag{A11.4}$$

where OT is the official time in hours, TA is the seasonal time advance in hours, and ET is the value of the time equation in minutes. LL is the longitude of the site and LH is the longitude of the origin of the time in that time zone.

The horizontal coordinates of the sun are given by:

$$\sinh = \sin\lambda\sin\delta + \cos\lambda\cos\delta\cos\omega \tag{A11.5}$$

$$\sin a = \frac{\cos\delta\sin\omega}{\cosh} \tag{A11.6}$$

where λ is the latitude of the site. The angle known as the zenith distance is the angle complementary to the elevation of the sun, h:

$$\cos\vartheta_z = \cos\left(\frac{\pi}{2} - h\right) = \sinh \tag{A11.7}$$

Sunset Angle

The sunset angle is obtained by making $h = 0$ in equation (A11.5). It follows that:

$$\cos\omega_s = -tg\,\lambda\,tg\,\delta \tag{A11.8}$$

Irradiance Components

The value of the global irradiance at the collector surface I_{coll} has three components: (a) the direct irradiance, which is the result of the irradiance received from a point source representing the sun; (b) the diffuse irradiance resulting from the energy received from a distributed source, which is the sky; and (c) the albedo component coming from the reflecting surfaces near the site.

These three magnitudes can be calculated from three primary magnitudes:

I_b Beam irradiance in the direction of the sun.
I_h Global irradiance at a horizontal surface.
I_d Diffuse irradiance at a horizontal surface.

The use of these magnitudes makes the calculation easier due to the availability of data. The total global irradiance in an arbitrarily oriented surface is given by [A11.5]:

$$I_{coll} = I_b \cos \vartheta_{coll} + \frac{I_d}{2}(1 + \cos \beta) + \frac{I_h}{2}\rho(1 - \cos \beta) \qquad (A11.9)$$

where ϑ_{coll} is the angle between the normal to the collecting surface and the direction of the sun at this moment and ρ is the value of the albedo reflection coefficient. The three primary magnitudes are further related by making $\beta = 0$ in equation (A11.9):

$$I_h = I_b \cos \vartheta_z + I_d \qquad (A11.10)$$

Instantaneous values of the irradiance components are usually unknown unless detailed monitoring has been set in place. In order to estimate the radiation availability or to select an orientation for the azimuth and inclination angle of a given photovoltaic array, it is imperious to make use of more data which are, generally, radiation values, that is time integrals of irradiance averaged for a number of years. Therefore handbooks or tables usually provide average values of the irradiation received in an average day of each month of the year. From these values estimates can be made about the irradiation and irradiance components, as described in the sections below.

Extra-atmospheric Irradiation

A magnitude which is very useful in the calculation of irradiation values is the extra-atmospheric irradiation H_o which is the irradiation received at a surface, placed horizontally, in the absence of atmosphere in one day. After integration over all angles (from $-\omega_s$ to $+\omega_s$) it results:

$$H_0 = \frac{24}{\pi} SC(\omega_s \sin \lambda \sin \delta + \cos \lambda \cos \delta \sin \omega_s) \qquad (A11.11)$$

where SC is the solar constant.

The value of H_o averaged over a month is known as $\bar{H}_0$.

Radiation Correlations

It is common practice to relate the components of the irradiance with the values of the irradiation components, that is the values of the monthly average of daily global irradiation at a horizontal surface H_h, and the value of the extra-atmospheric irradiation H_o, which can be calculated from equation (A11.11). There are several correlations proposed in the literature for which several definitions are required:

- $\bar{I}_h$: (kWh/m^2_hour) monthly average value of the global hourly irradiation at a horizontal surface. This value can be approximated to the value of the global horizontal irradiance at the middle of the time interval.

- $\bar{H}_h$: (kWh/m^2_day) monthly average value of the global daily irradiation at a horizontal surface.

There is a correlation between these two magnitudes which depends on the hour angle and on the day of the year as follows:

$$\frac{\bar{I}_h}{\bar{H}_h} = r_h = \frac{\pi}{24}(a + b\cos\omega)\frac{\cos\omega - \cos\omega_s}{\sin\omega_s - \omega_s\cos\omega_s} \tag{A11.12}$$

where

$$a = 0.409 + 0.516\sin(\omega_s - 1.047) \tag{A11.13}$$

and

$$b = 0.6609 - 0.467\sin(\omega_s - 1.047) \tag{A11.14}$$

Moreover, the diffuse components of the hourly and daily average radiation values are defined by:

- $\bar{I}_d$: (kWh/m^2_hour) monthly average value of the diffused hourly irradiation at a horizontal surface. This value can be approximated to the value of the diffused horizontal irradiance at the middle of the time interval.

- $\bar{H}_d$: (kWh/m^2_day) monthly average value of the daily diffused irradiation at a horizontal surface.

$$\frac{\bar{I}_d}{\bar{H}_d} = r_d = \frac{\pi}{24}\frac{\cos\omega - \cos\omega_s}{\sin\omega_s - \omega_s\cos\omega_s} \tag{A11.15}$$

Finally the diffused and global monthly averages of the daily irradiation are correlated, among other models, by that of Page [A11.6]:

$$\frac{\bar{H}_d}{\bar{H}_h} = 1 - 1.13\bar{k}_h \tag{A11.16}$$

where

$$\bar{k}_h = \frac{\bar{H}_h}{\bar{H}_o}$$

is known as the clearness index of the atmosphere.

Irradiance at the Collecting Surface (I_{coll})

Taking into account equations (A11.1) to (A11.16) the irradiance at the tilted surface arbitrarily oriented can be calculated from the knowledge of the values of the magnitudes defined above and of the ratio of the cosine of the angles ϑ_{coll} and, ϑ_z which is given by:

$$\frac{\cos \vartheta_{coll}}{\cos \vartheta_z} = \frac{1_s}{\cos \omega - \cos \omega_s} \left[\cos \omega (\cos g \sin \beta \cdot tg \, \lambda + \cos \beta) + \frac{\sin \omega}{\cos \lambda} \sin \beta \sin g \right.$$
$$\left. + tg \, \delta (\cos \beta \cdot tg \lambda - \cos g \sin \beta) \right] \tag{A11.17}$$

where g is the azimuth angle of the collecting surface considering the origin facing south and positive to the west.

The irradiance at the collecting surface is given by:

$$I_{coll} = I_b \cos \vartheta_{coll} + \frac{I_d}{2}(1 + \cos \beta) + \frac{I_h}{2}\rho(1 - \cos \beta) \tag{A11.18}$$

where ρ is the albedo reflection coefficient (estimated from the site location). Typical values ranging from 0.15 to 0.22 are common for ground sites.

From equation (A11.10):

$$I_b = \frac{I_h - I_d}{\cos \vartheta_z} \tag{A11.19}$$

thereby,

$$I_{coll} = (I_h - I_d)\frac{\cos \vartheta_{coll}}{\cos \vartheta_z} + \frac{I_d}{2}(1 + \cos \beta) + \frac{I_h}{2}\rho(1 - \cos \beta) \tag{A11.20}$$

which finally gives the value of the irradiance for a given hour angle, day of the year and latitude, at an arbitrarily oriented surface.

In-plane Sunset and Sunrise Angles

The result shown in equation (A11.20) is valid within the length of that particular day, and to be more precise, within the so-called 'apparent' day length at the collecting surface. The apparent length is defined as the angle between apparent sunrise and apparent in-plane sunset which are defined when the angle ϑ_{coll} is 90°. If we now apply this condition to equation (A11.17).

$$0 = \left[\cos \omega (\cos g \sin \beta \cdot tg\lambda + \cos \beta) + \frac{\sin \omega}{\cos \lambda} \sin \beta \sin g + tg\delta (\cos \beta \cdot tg\lambda - \cos g \sin \beta) \right]$$

(A11.21)

letting

$$A = \cos \omega$$ (A11.22)

$$a_1 = \cos g \sin \beta \sin \lambda + \cos \beta \cos \lambda$$ (A11.23)

$$a_2 = \sin \beta \sin g$$ (A11.24)

$$a_3 = -tg\, \delta (\cos \beta \sin \lambda + \cos g \sin \beta \cos \lambda)$$ (A11.25)

then, solving for A,

$$A = \frac{a_1 a_3 \pm a_2 \sqrt{a_1^2 - a_3^2 + a_2^2}}{a_1^2 + a_2^2}$$

(A11.26)

and finally

$$\omega_{ps} = \cos^{-1} \frac{a_1 a_3 - a_2 \sqrt{a_1^2 - a_3^2 + a_2^2}}{a_1^2 + a_2^2}$$

(A11.27)

and

$$\omega_{ss} = -\cos^{-1} \frac{a_1 a_3 + a_2 \sqrt{a_1^2 - a_3^2 + a_2^2}}{a_1^2 + a_2^2}$$

(A11.28)

where ω_{ps} is the sunset 'apparent' angle and ω_{ss} is the sunrise 'apparent' angle.

Global Irradiation at an Arbitrarily Oriented Surface

The irradiation is the time integral of the irradiance, typically over one day. To calculate its value, the irradiance I_{coll} in equation (A11.20) has to be integrated. Now the limits of the integral will be the smaller values of the sunset and sunrise angles considering that both are apparent at the collector surface, ω_{ps} and ω_{ss} and at the horizontal surface $\pm\omega_s$. In order to implement that, two new sunrise and sunset angles are generated:

$$\omega_{sss} = MIN(\omega_{ss}, -\omega_s) \tag{A11.29}$$

$$\omega_{pss} = MIN(\omega_{ps}, \omega_s) \tag{A11.30}$$

Integrating equation (A11.20):

$$\bar{H}_{coll} = \bar{H}_h \int_{\omega_{sss}}^{\omega_{pss}} r_h \frac{\cos \vartheta_{coll}}{\cos \vartheta_z} d\omega - \bar{H}_d \int_{\omega_{sss}}^{\omega_{pss}} r_d \frac{\cos \vartheta_{coll}}{\cos \vartheta_z} d\omega + \bar{H}_d \int_{\omega_{sss}}^{\omega_{pss}} \frac{r_d}{2}(1 + \cos \beta)d\omega$$

$$+ \bar{H}_h \int_{\omega_{sss}}^{\omega_{pss}} \frac{r_h}{2}\rho(1 - \cos \beta)d\omega$$

This gives the value for the average day of the month if we take the monthly average values of the sunset and sunrise angles. The correlations between daily and hourly irradiation values have been used. The PSpice netlist defines the integrals by the names f1, f2, f3 and f4.

Radiation Profiles for the Average Day of a Month

In general, calculations are performed over 12 average days, one for each month. It is also common to identify one particular day of each month as the closest to the average day. These are the 10th June, 11th December, 15th February, April, May and November, 16th March, August, September and October and 17th of January and July.

In order to estimate the value of the global irradiance at a particular time of the average day of the month, we know from the hour-day correlations that,

$$\bar{I}_h = r_h \bar{H}_h$$

as $\bar{I}_h$ is given by

$$\bar{I}_h = \frac{1}{T} \int_0^T I_h(t) dt$$

with $T = 1$ hour, the value of $I_h(T/2)$ at the middle of the time interval, if the instantaneous value of the irradiance within the hour is assumed to depend linearly on the hour angle, can be approximated by:

$$\bar{I}_h = \frac{1}{T} \int_0^T I_h(t) dt \approx \frac{1}{T} T I_h(T/2) = I_h(T/2)$$

meaning that the numerical value of the hourly irradiation (in Wh/m^2_hour) equals the value of the irradiance at the middle of the time interval (in W/m^2):

$$I_h(T/2) \approx \bar{I}_h = r_h \bar{H}_h$$

The same reasoning applies to the diffuse component and then

$$I_d(T/2) \approx \bar{I}_d = r_d \bar{H}_d$$

These values are now used in equation (A11.20) to compute the value of the irradiance at the collecting plane I_{coll} and the procedure generalized to any time value.

Two examples of the calculations are shown in Figures A11.1 and A11.2 where two different orientations of the collecting surface are considered for the same day and location. As can be seen, a surface facing south has a symmetrical irradiance profile, whereas a surface facing a given angle from the south gives asymmetrical irradiance profiles. Of course, the integral values, that is the radiation, are also different, the surface facing south collecting more radiation for the same tilt angle.

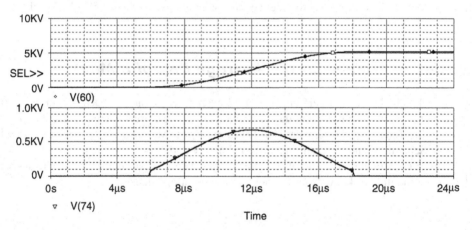

Figure A11.1   Irradiance (bottom) in W/m^2 and irradiation in Wh (top) for 21st June at a latitude of 41.2° N taking as data the average horizontal surface irradiation in the month of June (5638 Wh/day). Warning: the x-axis shows the internal time (in internal units of time in μs), which correspond to a real time of hours, covering a whole day (24 hours)

PSpice Performance Analysis and Goal Functions

PSpice has the capability to run what is called 'performance analysis', which is based on a parametric analysis type and the evaluation of a goal function.

A parametric analysis is an analysis either DC or transient, which is carried out for a series of values of a given parameter. After this analysis PSpice allows the user to evaluate a 'goal function', which can be defined or selected from a list of built-in functions. We are using this

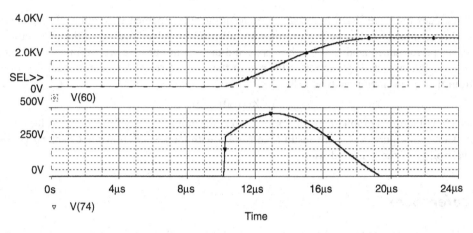

Figure A11.2 Iradiance (bottom) and irradiation (top) for 21st June at a location 41.2° N at a vertical surface facing 30° W

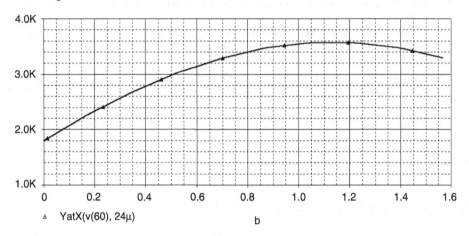

Figure A11.3 Plot of the global irradiation collected at a surface facing south ($g = 0$) for several tilt angles (b). Y-axis is given in kWh/m^2_day and the x-axis is the angle b in radians. The results are for a location at 41.2° N latitude and January data of the horizontal global irradiation $H_h = 1800\,\text{Wh/m}^2$_day

capabilities to evaluate the value of the irradiation collected at a surface oriented at a fixed azimuth and variable inclination in such a way that PSpice will return a function showing the effect of the angle b on the irradiation collected at the site.

First we carry the parametric analysis making the angle β (b in the PSpice code) sweep from 0 degrees to 90 degrees, adding a line to the PSpice code in file irradiation_1day.cir, as follows:

```
.step param b list 0, 0.1744, 0.3488, 0.5232,0.6976, 0.872, 1.046, 1.2208, 1.3952,1.569
```

The values for b are listed in radians. The resulting file is named irradiation_abacus.cir.

Once the analysis has been performed for all listed β values, we set in the axis settings menu 'performance analysis' for the x-axis and then as in our case we are interested in the value of the collected irradiation at the end of the day, we select 'evaluation goal function' from the 'trace' menu, and we select:

$$\text{YatX}(\text{v}(60),24\text{u})$$

which will return a graph of the values of v(60) at the end of the day for all listed values of β. As shown in Figure A11.3 this result is consistent with the results shown in Chapter 1 on the availability of the solar radiation.

References

[A11.1] Commission of the European Communities 'Prediction of solar radiation on inclined surfaces', *Solar Energy R&D in the European Community*, Series F, volume 3, D. Riedel, 1986.

[A11.2] Klein, S., 'Calculation of monthly average insolation on tilted surfaces' *Solar Energy*, **19**, 325–9, 1977.

[A11.3] Meinel, B. and Meinel, M.P., *Applied Solar Energy*, Addison Wesley, 1977.

[A11.4] Duffie, J.A. and Beckman, W.A., *Solar Engineering of Thermal Processes*, Wiley, New York, 1991.

[A11.5] Hernandez, F., Ramos, F., Tinaut, D., Rodríguez, M., Diaz-Salgado, C., Macías, M. Blanco, M.E., *Metodología y cálculo de radiación para colectores concentradores' CSIC, Madrid, Spain 1987.

[A11.6] Page, J.K., *Proceedings of the UN Conference on New Energy Sources*, **4**, 378, 1964.

```
************************************
*                                  *
*         RADIATION_1DAY.CIR       *
*                                  *
************************************


*   CALCULATES THE RADIATION COMPONENTS AT AN ARBITRARILY ORIENTED
*   SURFACE ALONG ONE DAY

*   NODES
*   (10) HOUR ANGLE, RADIANS
*   (20) ANGLE RATIO IN EQUATION A11.17
*   (50) EXTRAATMOSPHERIC RADIATION
*   (51) CORRELATION RATIO RH
*   (52) CORRELATION RATIO RD
*   (60) GLOBAL RADIATION AT INCLINED SURFACE (WH/M2_DAY)
*   (70) DIFFUSE RADIATION AT INCLINED SURFACE (WH/M2_DAY)
*   (71) DIFFUSE COMPONENT OF IRRADIANCE AT INCLINED SURFACE, ID (W/M2)
*   (72) GLOBAL IRRADIANCE AT HORIZONTAL SURFACE, IH (W/M2)
*   (74) GLOBAL IRRADIANCE INCLINED SURFACE, ICOLL (W/M2)
*   (106) SUNSET ANGLE, INCLINED SURFACE, RADIANS
*   (107) SUNRISE ANGLE AT INCLINED SURFACE, RADIANS
*   (108)   DECLINATION, RADIANS

**** parameter values and data
.param pi=3.14159265
.param hh=5638;global radiation in horizontal surface
.param n=172;day of the year starting january first
```

```
.param ro=0.5;albedo factor
.func b() {(pi/180)*90};inclination of the collecting surface,(rad)
.func g() {(pi/180)*(30)};azimuth of the collecting surface,g=0 is facing south
.func l() {(pi/180)*41.2};latitude
.func sc() {1353*(1+0.0033*cos(pi/180*360/365*n))}
.func a() {0.409+0.516*sin(ws-1.047)}
.func bb() {0.6609-0.467*sin(ws-1.047)}
.func d() {(pi/180)*23.45*sin((pi/180)*360/365*(n+284))};declination
.func ws() {acos(-tan(l)*tan(d))};sunset hour angle at horizontal surface
.func a1() {cos(g)*sin(b)*sin(l)+cos(b)*cos(l)}
.func a2() {sin(b)*sin(g)}
.func a3() {-tan(d)*(cos(b)*sin(l)-cos(g)*sin(b)*cos(l))}

**** ratio angle of collection over zenith distance
ecoll 20 0 value={(1/(cos(v(10))-cos(ws)))*(cos(v(10))*(cos(g)*sin(b)*tan(l)
+cos(b))
++sin(v(10))/cos(l)*sin(b)*sin(g)+tan(d)*(cos(b)*tan(l)-cos(g)*sin(b)))}

********* in-plane sunset and sunrise angles
.func wps() {if(g>0,acos((a1*a3-a2*sqrt(a1**2-a3**2+a2**2))/(a1**2+a2**2)),
+ acos((a1*a3-a2*sqrt(a1**2-a3**2+a2**2))/(a1**2+a2**2)))};in-plane sunset
*angle
.func wss() {if(g>0,-acos((a1*a3+a2*sqrt(a1**2-a3**2+a2**2))/(a1**2+a2**2)),
+ -acos((a1*a3+a2*sqrt(a1**2+a3**2+a2**2))/(a1**2+a2**2)))};in-plane sunrise
*angle,takes negative values
****************************************************************************
.func wsss() {if (wss>-ws, wss,-ws)};real in-plane sunrise angle
.func wpss() {if( wps>ws,ws,wps)};real in-plane sunset angle

e_ho 50 0 value={24/pi*sc*(cos(d)*cos(l)*sin(ws)+ws*sin(d)*sin(l))};
*extraatmospheric radiation Ho

***** Irradiance -irradiation correlations

e_rh 51 0 value={pi/24*(a+bb*cos(v(10)))*(cos(v(10))-cos(ws))/(sin(ws)-
+ws*cos(ws))}
e_rd 52 0 value={pi/24*(cos(v(10))-cos(ws))/(sin(ws)-ws*cos(ws))}

****** Irradiation integrals F1, F2, F3 and F4

e_f11 531 0 value={if( v(10)>wsss & v(10)<wpss, v(51)*v(20), 0)};integrand F1
e_f1 53 0 value={sdt(v(531))}; integral F1

e_f21 541 0 value={if(v(10)>wsss & v(10)<wpss, v(52)*v(20), 0)};integrand F2
e_f2 54 0 value={sdt(v(541))};integral F2

e_f31 551 0 value={if(v(10)>wsss & v(10)<wpss, v(52)*(1+cos(b))/2, 0)};integranf F3
e_f3 55 0 value={sdt(v(551))};integral F3

e_f41 561 0 value={if(v(10)>wsss & v(10)<wpss, v(51)*ro/2*(1-cos(b)), 0)};inte-
+grand F4
e_f4 56 0 value={sdt(v(561))};integralF4


*********** Irradiation at the collecting surface

e_hcoll 60 0 value={1e6*hh*((v(53)+v(56))-(v(54)-v(55))*(1-1.13 *
+(hh/v(50))))};global radiation



*********** irradiance components
```

```
e_hd 70 0 value={hh*(1-1.13/v(50)*hh)};diffuse radiation
e_id 71 0 value={v(52)*v(70)}; irradiance diffuse
e_ih 72 0 value={v(51)*hh};irradiance global horizontal
e_ib 73 0 value={(v(72)-v(71))/(sin(l)*sin(d)+cos(l)*cos(d)*cos(v(10)))}
```

********* in-plane irradiance

```
e_icolla 741 0 value={(v(72)-v(71))*v(20)+v(71)/2*(1+cos(b))+v(72)/2*ro*
+(1-cos(b))}
e_icoll 74 0 value={if(v(741)>0 &v(10)<wpss & v(10)>wsss, v(741),0)};limit to posi-
*tive values
```

***** time angle in radians, sweeps from -pi to +pi radians in 24 internal units of time

```
v_w 10 0 pulse (-3.14 3.14 0 23.99u 0.01u 0.01u 24u);time angle in radians

.probe
.tran 0u 24u 0u 0.1u
```

***** check and print magnitudes

```
e_wss 100 0 value={wss}
e_wps 101 0 value={wps}
e_ws 102 0 value={ws}
e_a1 103 0 value={a1}
e_a2 104 0 value={a2}
e_a3 105 0 value={a3}
e_wsss 106 0 value={wsss}
e_wpss 107 0 value={wpss}
e_d 108 0 value={d}
. print tran v(100) v(101)v(102)v(103) v(104) v(105) v(106) v(107) v(741)
+v(74) v(108)
.end
```

```
***********************************
*                                 *
*      RADIATION_ABACUS.CIR       *
*                                 *
***********************************
```

**** CALCULATION OF THE VALUES OF HCOLL AS A FUNCTION OF THE TILT ANGLE b FOR A
*** CONSTANT AZIMUTH AND A GIVEN DAY
*   AFTER RUNNING THE SIMULATION, SELECT IN THE AXIS SETTINGS MENU
*   IN X-AXIS, PERFORMANCE ANALYSIS, AND THEN
*   FROM TRACE MANU, SELECT EVALUATION OF GOAL FUNCTION
*   CHOOSE YatX(v(60),24u)
*** parameter values and data
```
.param pi=3.14159265
.param hh=1800;global radiation in horizontal surface
.param n=355;day of the year starting january first
.param ro=0.5;albedo factor
.param b=0
.func g() {(pi/180)*(0)};azimuth of the collecting surface,g=0 is facing south
.func l() {(pi/180)*41.2};latitude
.func sc() {1353*(1+0.0033*cos(pi/180*360/365*n))}
.func a() {0.409+0.516*sin(ws-1.047)}
.func bb() {0.6609-0.467*sin(ws-1.047)}
.func d() {(pi/180)*23.45*sin((pi/180)*360/365*(n+284))};declination
.func ws() {acos(-tan(l)*tan(d))};sunset hour angle at horizontal surface
e_a1 103 0 value={cos(g)*sin(b)*sin(l)+cos(b)*cos(l)}
e_a2 104 0 value={sin(b)*sin(g)}
```

```
e_a3 105 0 value={-tan(d)*(cos(b)*sin(l)-cos(g)*sin(b)*cos(l))}

**** ratio angle of collection over zenith distance
ecoll 20 0 value={(1/(cos(v(10))-cos(ws)))*(cos(v(10))*(cos(g)*sin(b)*tan(l)+-
+cos(b))
++sin(v(10))/cos(l)*sin(b)*sin(g)+tan(d)*(cos(b)*tan(l)-cos(g)*sin(b)))}

********* in-plane sunset and sunrise angles
e_wps 101 0 value={acos((v(103)*v(105)-v(104)*sqrt(v(103)**2-v(105)**2
+v(104)**2))/(v(103)**2+v(104)**2))};in-plane sunset angle
e_wss 100 0 value={-acos((v(103)*v(105)+v(104)*sqrt(v(103)**2-v(105)**2
+v(104)**2))/(v(103)**2+v(104)**2))};in-plane sunrise angle,takes negative
*values
******************************************************************
e_wsss 106 0 value={if (v(100)>-ws, v(100),-ws)};real in-plane sunrise angle
e_wpss 107 0 value={if( v(101)>ws,ws,v(101))};real in-plane sunset angle

e_ho 50 0 value={24/pi*sc*(cos(d)*cos(l)*sin(ws)+ws*sin(d)*sin(l))};extraatmo-
*spheric radiation Ho

***** Irradiance -irradiation correlations

e_rh 51 0 value={pi/24*(a+bb*cos(v(10)))*(cos(v(10))-cos(ws))/(sin(ws)-
+ws*cos(ws))}
e_rd 52 0 value={pi/24*(cos(v(10))-cos(ws))/(sin(ws)-ws*cos(ws))}

****** Irradiation integrals F1, F2, F3 and F4

e_f11 531 0 value={if( v(10)>v(106) & v(10)<v(107), v(51)*v(20), 0)};integrand F1
e_f1 53 0 value={sdt(v(531))}; integral F1

e_f21 541 0 value={if(v(10)>v(106) & v(10)<v(107), v(52)*v(20), 0)};integrand F2
e_f2 54 0 value={sdt(v(541))};integral F2

e_f31 551 0 value={if(v(10)>v(106) & v(10)<v(107), v(52)*(1+cos(b))/2, 0)};
*integranf F3
e_f3 55 0 value={sdt(v(551))};integral F3

e_f41 561 0 value={if(v(10)>v(106) & v(10)<v(107), v(51)*ro/2*(1-cos(b)), 0)};
*integrand F4
e_f4 56 0 value={sdt(v(561))};integralF4


*********** Irradiation at the collecting surface

e_hcoll 60 0 value={1e6*hh*((v(53)+v(56))-(v(54)-v(55))*(1-1.13 *(hh/
+v(50))))};global radiation


************ irradiance components

e_hd 70 0 value={hh*(1-1.13/v(50)*hh)};diffuse radiation
e_id 71 0 value={v(52)*v(70)}; irradiance diffuse
e_ih 72 0 value={v(51)*hh};irradiance global horizontal
e_ib 73 0 value={(v(72)-v(71))/(sin(l)*sin(d)+cos(l)*cos(d)*cos(v(10)))}

********* in-plane irradiance

e_icolla 741 0 value={(v(72)-v(71))*v(20)+v(71)/2*(1+cos(b))+v(72)/2*ro*
+(1-cos(b))}
e_icoll 74 0 value={if(v(741)>0 & v(10)<v(107) & v(10)>v(106), v(741),0)};limit to
*positive values
```

```
***** time angle in radians, sweeps from -pi to +pi radians in 24 internal units of time

v_w 10 0 pulse (-3.14 3.14 0 23.99u 0.01u 0.01u 24u);time angle in radians

.probe
.tran 0u 24u 0u 0.1u
.step param b list 0, 0.1744, 0.3488, 0.5232,0.6976, 0.872, 1.046, 1.2208,
+1.3952,1.569


.end
```

Index